AF576041

Advances in Ecohydrology for Water Resources Optimization in Arid and Semi-arid Areas

Advances in Ecohydrology for Water Resources Optimization in Arid and Semi-arid Areas

Editors

Mirko Castellini
Simone Di Prima
Ryan Stewart
Marcella Biddoccu
Mehdi Rahmati
Vincenzo Alagna

MDPI • Basel • Beijing • Wuhan • Barcelona • Belgrade • Manchester • Tokyo • Cluj • Tianjin

Editors
Mirko Castellini
Council for Agricultural Research and Economics—Research Center for Agriculture and Environment (CREA-AA)
Italy

Simone Di Prima
University of Sassari
Italy

Ryan Stewart
Virginia Polytechnic Institute and State University
USA

Marcella Biddoccu
Institute of Sciences and Technologies for Sustainable Energy and Mobility of National Research Council of Italy (CNR-STEMS)
Italy

Mehdi Rahmati
University of Maragheh
Iran
Forschungszentrum Jülich GmbH
Germany

Vincenzo Alagna
University of Palermo
Italy

Editorial Office
MDPI
St. Alban-Anlage 66
4052 Basel, Switzerland

This is a reprint of articles from the Special Issue published online in the open access journal *Water* (ISSN 2073-4441) (available at: https://www.mdpi.com/journal/water/special_issues/Ecohydrology).

For citation purposes, cite each article independently as indicated on the article page online and as indicated below:

LastName, A.A.; LastName, B.B.; LastName, C.C. Article Title. *Journal Name* **Year**, *Volume Number*, Page Range.

ISBN 978-3-0365-4747-3 (Hbk)
ISBN 978-3-0365-4748-0 (PDF)

Cover image courtesy of Mirko Castellini

Contents

About the Editors

Mirko Castellini

Mirko Castellini (Ph.D., 2005) is a researcher at Council for Agricultural Research and Economics, Agriculture and Environment Research Center (CREA-AA) in Bari. His main scientific interests focus on soil physical and hydraulic properties. Specific research topics are: (i) soil physical quality, (ii) soil management for sustainable agriculture, (iii) land use change impact on soil properties, (iv) use of soil conditioners (e.g., amendments, composts) to improve the soil water retention, (v) water fluxes in saturated and unsaturated soil conditions, (vi) temporal and spatial variability of physical and hydraulic properties of the soil, and (vii) the main factors affecting soil physical degradation processes (soil surface crusting, soil compaction, etc.).

Simone Di Prima

Simone Di Prima (Ph.D., 2016) is a researcher at the Department of Agricultural Sciences of the University of Sassari (Italy). He got a postdoctoral research position at the Laboratoire d'Ecologie des Hydrosystèmes Naturels et Anthropisés (LEHNA, ENTPE, Université Lyon 1, France) focused on stormwater management and the use of non-intrusive geophysical techniques to catch nano-tracers and the pathways of water in urban soils. His main scientific interests focus on soil hydrology and water resources management with specific regard to laboratory and field determination of soil hydraulic properties, infiltration processes, simulation of water flow in the vadose zone. He authored more than 60 scientific papers on international peer reviewed journals. He served as reviewer for several international scientific journals.

Ryan Stewart

Ryan Stewart is an Associate Professor in the School of Plant and Environmental Sciences at Virginia Tech. He currently leads the Critical Zone Lab at Virginia Tech, which focuses on understanding the movement of water, gases, nutrients, and contaminants in various landscapes, with a primary focus on agroecosystems. His teaching program covers topics related to pollution science, hydrology, and soil properties and functioning. Prior to arriving at Virginia Tech in 2013, he obtained his B.S. degree (2002) in Mechanical Engineering from Cal Poly, San Luis Obispo, and his M.S. (2010) and Ph.D. (2013) degrees in Water Resources Engineering from Oregon State University. In his spare time, Ryan enjoys family outings, photography, gardening, and the occasional biking adventure.

Marcella Biddoccu

Marcella Biddoccu is Researcher at Institute of Sciences and Technologies for Sustainable Energy and Mobility of National Research Council of Italy (CNR-STEMS). Her interests spans across the environmental impacts of human activities, especially related to hydrological processes, from the field to catchment scale. She obtained the M.Sc.Eng in Environmental Engineering at Politecnico di Torino and then her PhD at Università degli Studi di Torino (PhD programme "Agricultural, Forest and Food Sciences"). Her research interests are related to: assessment of the environmental impacts and technical and management solutions for sustainability of agricultural activities, especially related to effects of vineyard's management on hydrological and soil degradation processes (soil erosion and soil compaction). She is member of the EUSO Erosion Working Group (JRC-EU) and GSERmap Working Group (FAO-GSP).

Mehdi Rahmati

Mehdi Rahmati is an associate professor at University of Maragheh, Iran and a Senior Researcher at Forschungszentrum Jülich GmbH, Institute of Bio- and Geosciences: Agrosphere (IBG-3), Jülich, Germany. His Research Background and Interests include soil physics and hydrology, spatio-temporal modeling and digital soil mapping as well as data mining and data Handling. During last years, he was teaching several courses for both undergraduate and postgraduate students. He has also participated in supervision of several M.Sc. and PhD students' thesis. He has published more than 70 peer-reviewed papers in scientific journals that more than 50 of those papers are published in JCR and high-ranked journals.

Vincenzo Alagna

Vincenzo Alagna, PhD. Since 2021, he has been a researcher of Agricultural Hydraulics and Watershed Management at the Department of Agriculture, Food and Forest Sciences (SAAF) of the University of Palermo. In 2011, he obtained the MSc in Agricultural Sciences and Technologies at the University of Palermo and in 2017 he was awarded with the PhD in Agricultural, Forest and Environmental Sciences. In 2017, his PhD thesis received the "Giuseppe Benini" award for the best thesis in Agricultural Hydraulics and Watershed Management. The award was assigned by the Italian Society of Agricultural Engineering during the XI National Congress "Biosystems Engineering Addressing the Human Challenges of the 21st Century". His research interests are mainly related to soil hydrology and agricultural water management, with particular focus on the implementation of strategies aimed at optimizing water resources and to increase water use efficiency.

Preface to "Advances in Ecohydrology for Water Resources Optimization in Arid and Semi-arid Areas"

Current climate change requires a rethinking and updating of all relations between hydrology and the environment, mainly where changes have the strongest impacts: arid and semi-arid regions. Often, these environments are characterized, all over the world, by an uneven distribution of water resources, posing limitations in development and socio-political issues. Our knowledge of ecohydrology is forcibly incomplete due to the complex nature of ecosystems and interactions among their compartments, which are constantly changing under different stresses: from deforestation to urbanization, from water pollution to water scarcity, from air pollution to climate change, and from soil erosion to soil pollution.

Furthermore, the hydrology of arid and semiarid regions is a combination of characters: water supply/losses irregularly distributed in time and space, heterogeneous topography and landscape, and high anthropogenic pressure. Hence, since water distribution follows the prevailing conditions, the development of relationships among ecosystems and hydrology is needed at different layers of knowledge to overcome all barriers for optimizing water use and water use efficiency at suitable time and space scales.

In the face of these difficulties, it is evident that there is a necessity to increase, improve and advance our knowledge about the interactions between ecology and hydrological regimes of arid and semiarid areas, and, in particular, of:

- Soil water retention and availability;
- Water supply and evapotranspiration spatial and temporal variability;
- Surface and groundwater transfer processes;
- Water flows between soil and vegetate surfaces at catchment scale.

Water is evidently central to the framework of ecosystems, both for the theoretical comprehension of processes and for practical applications in solving environmental issues in the twenty-first century. The works here presented develop basic principles and case studies around the world that show the importance of understanding ecohydrological processes and better assessing ecosystem services.

In summary, the Special Issue of *Water* Journal "Advances in Ecohydrology for Water Resources Optimization in Arid and Semi-arid Areas" aims to suitably face current ecohydrological research in these fields, drawing upon the main current lines of research for the foreseeable future. The different layers of knowledge were assured by the strong interdisciplinarity of the afforded analysis, carried out in different sites of the world united by arid and semiarid climates.

These papers may stimulate further discussion on understanding about how ecological processes interact with ecosystem functions, so improved models may be developed to serve future users, land managers and society.

Gianfranco Rana, PhD
Council for Agricultural Research and Economics

Editorial

Advances in Ecohydrology for Water Resources Optimization in Arid and Semi-Arid Areas

Mirko Castellini [1,*], Simone Di Prima [2], Ryan Stewart [3], Marcella Biddoccu [4], Mehdi Rahmati [5,6] and Vincenzo Alagna [7]

[1] Council for Agricultural Research and Economics–Research Center for Agriculture and Environment (CREA–AA), Via C. Ulpiani, 570125 Bari, Italy
[2] Agricultural Department, University of Sassari, Viale Italia, 39, 07100 Sassari, Italy; sdiprima@uniss.it
[3] School of Plant and Environmental Sciences, Virginia Polytechnic Institute & State University, Blacksburg, VA 24061, USA; ryan.stewart@vt.edu
[4] Institute for Agricultural and Earthmoving Machines (IMAMOTER), National Research Council of Italy, Strada delle Cacce, 73, 10135 Torino, Italy; marcella.biddoccu@cnr.it
[5] Department of Soil Science and Engineering, Faculty of Agriculture, University of Maragheh, Maragheh 83111-55181, Iran; mehdirmti@gmail.com
[6] Forschungszentrum Jülich GmbH, Institute of Bio- and Geosciences: Agrosphere (IBG-3), 52428 Jülich, Germany
[7] Department of Agricultural, Food and Forest Sciences, University of Palermo, Viale delle Scienze, Ed. 4 Ingr. E., 90128 Palermo, Italy; vincenzo.alagna01@unipa.it
* Correspondence: mirko.castellini@crea.gov.it

Abstract: Conserving water resources is a current challenge that will become increasingly urgent in future due to climate change. The arid and semi-arid areas of the globe are expected to be particularly affected by changes in water availability. Consequently, advances in ecohydrology sciences, i.e., the interplay between ecological and hydrological processes, are necessary to enhance the understanding of the critical zone, optimize water resources' usage in arid and semi-arid areas, and mitigate climate change. This Special Issue (SI) collected 10 original contributions on sustainable land management and the optimization of water resources in fragile environments that are at elevated risk due to climate change. In this context, the topics mainly concern transpiration, evapotranspiration, groundwater recharge, deep percolation, and related issues. The collection of manuscripts presented in this SI represents knowledge of ecohydrology. It is expected that ecohydrology will have increasing applications in the future. Therefore, it is realistic to assume that efforts to increase environmental sustainability and socio-economic development, with water as a central theme, will have a greater chance of success.

Keywords: groundwater; transpiration; evapotranspiration; deep infiltration; gully erosion; runoff; forest restoration; compost; silicon; soil water retention

Citation: Castellini, M.; Di Prima, S.; Stewart, R.; Biddoccu, M.; Rahmati, M.; Alagna, V. Advances in Ecohydrology for Water Resources Optimization in Arid and Semi-Arid Areas. *Water* **2022**, *14*, 1830. https://doi.org/10.3390/w14121830

Received: 27 May 2022
Accepted: 30 May 2022
Published: 7 June 2022

Publisher's Note: MDPI stays neutral with regard to jurisdictional claims in published maps and institutional affiliations.

1. Introduction

In recent decades, most surveys related to soil physics and hydrology have shifted from the laboratory to the field, and have grown from a limited vision, considering only one aspect, i.e., only the physical characteristics of hydrology, to a comprehensive vision that interfaces with the domains of related disciplines, such as meteorology and climatology, engineering, pedology, ecology and geochemistry [1].

Ecohydrology is "... an interdisciplinary scientific field studying the interactions between water and ecological systems. It is considered a sub-discipline of hydrology, with an ecological focus. These interactions may take place within water bodies, such as rivers and lakes, or on land, in forests, deserts, and other terrestrial ecosystems. Areas of research in ecohydrology include transpiration and plant water use, adaption of organisms to their water environment, influence of vegetation and benthic plants on stream flow and function, and feedbacks between ecological processes, the soil carbon sponge and the hydrological cycle" [2].

Conserving water resources is current challenge that will become increasingly urgent in future, due to climate change [3]. The arid and semi-arid areas of the globe are expected to be particularly affected by changes in freshwater availability [4,5]. Consequently, advances in ecohydrology sciences are necessary to enhance understanding of the critical zone hydrology, to optimize water resource usage in arid and semi-arid areas and mitigate climate change [6,7].

The main goal of this Special Issue (SI) was to present advanced research and applications in ecohydrology, with a focus on water resources' optimization in arid and semi-arid areas. Overall, the contributions gathered in this SI account for the ecohydrology of (i) natural areas, including deserts [8,9], wetlands [10], grasslands–rangelands [11] and forests [12], (ii) cultivated areas, including orchards [13] or arable lands [14], or (iii) without specific focus on the environment–vegetation interaction [15–17]. The most common contributions were studies on groundwater and transpiration–evapotranspiration.

Based on recommendations by the scientific editors of MDPI journals, a paper [15] was included among the Editor's choice articles, namely, investigations deemed of particular interest for readers, or important in their field.

A synthesis of the main results archived by the collected investigations is reported in the following section.

2. Overview of This Special Issue

This Special Issue collected 10 original contributions focused on ecohydrology. Almost all concerned field investigations took place in arid and semi-arid areas [8–11,13–16], although some of them were from more humid environments [12]; only one investigated the optimization of water resource in a controlled laboratory environment [17]. Furthermore, half of them were developed in China [8–11,14], while the rest were from Kansas, United States [15], Korea [16], Brazil [12] and Mediterranean Europe (Italy and Portugal) [13,17].

For the reader's convenience, the publications are grouped by general theme: (1) transpiration/evapotranspiration, (2) groundwater/deep percolation, and (3) miscellaneous.

Topic 1 comprises four papers. Darouich et al. [13] estimated the evapotranspiration fluxes and water use in two Mediterranean vineyards, one in northern Italy with a sloping, rainfed crop, and one in southern Portugal in a flat area with irrigation. The SIMDualKc model was successfully applied for both study cases to simulate water use, crop evapotranspiration and the soil–water balance.

Tang et al. [14] investigated the spatiotemporal variability in evapotranspiration and the effects of water and heat on water-use efficiency (WUE), in a natural area of the Eurasian hinterland and on the northwestern border of China (Xinjiang). The results showed that spatial variations in WUE in different hydrological processes were significant, and the WUE trend in most regions showed an upward trend. They also found that, in arid areas, temperature has a positive effect on WUE, while precipitation has a lagging effect on WUE.

Xu and Yu [9] conducted their investigation on the environmental control of the transpiration of a desert ecosystem within an oasis–desert ecotone, where the sap flow from three dominant shrub species and concurrent environmental variables were determined during two growing seasons. Their work provided insights into environmental controls on the water flux of arid and semi-arid regions at the ecosystem scale and implications for diurnal hydrology modeling, specifically diurnal transpiration and water stress modeling.

Jia et al. [10] reported the research results for an environment at risk of desertification in Inner Mongolia, northwestern China. They investigated: (i) the evapotranspiration of groundwater (ETG) using estimates from multiple observation wells and soil moisture fluctuations at different sites, (ii) the temporal and spatial variations in ETG, and (iii) the sensitivity of ETG to water table depth for different types of vegetation (i.e., Phreatophytes).

Topic 2 comprises three papers. Ternes [15] examined how private water wells influenced the conservation of a high Plains aquifer in Midwestern states such as Kansas, and how this groundwater resource was tapped beyond its natural replenishment rates. Comparing the watering technologies of well-owners to those of non-well-owners across the state, and analyzing well ownership, facilitated an assessment of the relationships that exist between infrastructural conditions and water-conservation.

Park et al. [16] quantified the different silicon fractions in volcanic ash soils on Jeju Island (south Korea) that may affect groundwater silicon content, and compared them with those in forest soils in mainland Korea. The idea comes from the fact that silicon may be dissolved and leached through the soil profile into the groundwater. A high concentration of silicon, along with calcium and potassium in drinking water, improves water taste and benefits human health. Their results indicate that silicon is more soluble in the Andisols of high-precipitation regions and that Andisols on Jeju Island potentially affect groundwater silicon concentration.

The paper by Cheng et al. [8] aimed to explore the redistribution process of precipitation moisture in shallow soil in an arid sandy region (Ulanbuh desert, northern China), as fragile ecosystems in arid sandy regions are extremely sensitive to water deficits. The authors identified the characteristics of water dynamics and transformation in the arid area of the Ulanbuh Desert. Their results can serve as a guideline for the quantitative assessment of water resources in arid sandy regions.

Topic 3 comprises three papers that outlined some other important research issues. Li et al. [11] dealt with the general issue of the sustainable development of arid rangeland in Central Asia, and how gully erosion, flooding, and droughts jointly restrict the sustainable development of arid rangeland. This paper shows that the applied rainwater harvesting (RWH) system in a gully was a flexible practice that alleviated complex environmental problems. The authors proposed some suitable low-cost RWH techniques to restore degraded grassland and promote community development. Their study could provide suggestions for ecological restoration and pasture management in arid regions of Central Asia.

There has been an increase in secondary tropical forests in recent years, due to forest restoration in degraded areas. However, the success of forest restoration processes depends on the presence of a healthy soil environment in which seedlings and trees can successfully develop. In this research context, the study by Pereira al. [12] aimed to answer this question: is passive restoration capable of improving soil-saturated hydraulic conductivity (K_s), as well as the physical attributes of the soil superficial horizon? Consequently, they investigated the behavior of K_s and some soil physical properties in forests of different ages in Rio Claro (São Paulo State, Southeast Brazil), which were undergoing passive restoration by natural regeneration, a degraded forest fragment, a pasture, and a sugarcane field. The authors confirmed what has been suggested in the literature, namely that the recovery of soil hydro-physical functioning is a slow process and varies according to the type of soil attribute, previous land-use and its degradation legacy.

Finally, Bondì et al. [17] investigated the reliability of indicators of soil physical quality, including pore-size distribution parameters, to assess the effectiveness of compost amendment on hysteretic water retention curves in a sandy loam soil. Hysteresis refers to a soil-attributed phenomenon wherein the soil water content at a given pressure head is higher during drying than during wetting. The authors concluded that compost addition could trigger positive effects on soil hydrological processes and agronomic services, since water infiltration was favored during wetting and water storage during drying.

3. Conclusions

The 10 original manuscripts contained in this SI have reported experimental results for sustainable land management and water resources' optimization in fragile or at-risk environments. Overall, environments investigated from the ecohydrology perspective included European vineyards, desert areas and fragile ecosystems in Asia, South American forests, and various other contexts. Most of them have not been extensively investigated to date, probably due to relatively low levels of economic interest. However, the growing awareness of the critical need to safeguard marginal or fragile environments, or restore degraded ones, for future generations, can multiply research efforts.

The collected manuscripts were summarized by grouping them into three main topics to show research advances in specific fields, such as (1) transpiration/evapotranspiration [9,10,13,14], (2) groundwater or deep percolation [8,15,16], and (3) miscellaneous [11,12,17]. The research aims and main results were also summarized.

The manuscripts collected in the SI provided results from specific environments worldwide, but all have stressed the possibility of extending the results to similar environments. More broadly, the collected results represent an enhancement of our understanding of critical zone hydrology, to optimize water resources' usage in arid and semi-arid areas, and to mitigate the impacts of climate change.

Funding: This research received no external funding.

Acknowledgments: This manuscript represents an activity within the research project: "Water4AgriFood, Miglioramento delle produzioni agroalimentari mediterranee in condizioni di carenza di risorse idriche", PNR 2015–2020", funded by MIUR, PON ARS01_00825 "Ricerca e Innovazione" 2014–2020.

Conflicts of Interest: The authors declare no conflict of interest.

References

1. Hillel, D. 1. Soil Physics and Soil Physical Characteristics. In *Introduction to Soil Physics*; Elsevier Academic Press: Amsterdam, The Netherlands, 1983.
2. Wikipedia. Ecohydrology. Available online: https://en.wikipedia.org/w/index.php?title=Ecohydrology&oldid=1077116263 (accessed on 17 May 2022).
3. Di Prima, S.; Castellini, M.; Pirastru, M.; Keesstra, S. Soil Water Conservation: Dynamics and Impact. *Water* **2018**, *10*, 952. [CrossRef]
4. Mahmood, R.; Jia, S.F. Assessment of hydro-climatic trends and causes of dramatically declining stream flow to Lake Chad, Africa, using a hydrological approach. *Sci. Total Environ.* **2019**, *675*, 122–140. [CrossRef] [PubMed]
5. Ventrella, D.; Giglio, L.; Charfeddine, M.; Lopez, R.; Castellini, M.; Sollitto, D.; Castrignanò, A.; Fornaro, F. Climate change impact on crop rotations of winter durum wheat and tomato in southern Italy: Yield analysis and soil fertility. *Ital. J. Agron.* **2012**, *7*, 100–108. [CrossRef]
6. Ventrella, D.; Charfeddine, M.; Giglio, L.; Castellini, M. Application of DSSAT models for an agronomic adaptation strategy under climate change in Southern Italy: Optimum sowing and transplanting time for winter durum wheat and tomato. *Ital. J. Agron.* **2012**, *7*, 109–115. [CrossRef]
7. Berger, J.; Palta, J.; Vades, V. An integrated framework for crop adaptation to dry environments: Responses to transient and terminal drought: A Review Article. *Plant Sci.* **2016**, *253*, 58–67. [CrossRef]
8. Cheng, Y.; Yang, W.; Zhan, H.; Jiang, Q.; Shi, M.; Wang, Y. On the Origin of Deep Soil Water Infiltration in the Arid Sandy Region of China. *Water* **2020**, *12*, 2409. [CrossRef]
9. Xu, S.; Yu, Z. Environmental Control on Transpiration: A Case Study of a Desert Ecosystem in Northwest China. *Water* **2020**, *12*, 1211. [CrossRef]
10. Jia, W.; Yin, L.; Zhang, M.; Yu, K.; Wang, L.; Hu, F. Estimation of Groundwater Evapotranspiration of Different Dominant Phreatophytes in the Mu Us Sandy Region. *Water* **2021**, *13*, 440. [CrossRef]
11. Li, Z.; Zhang, W.; Aikebaier, Y.; Dong, T.; Huang, G.; Qu, T.; Zhang, H. Sustainable Development of Arid Rangelands and Managing Rainwater in Gullies, Central Asia. *Water* **2020**, *12*, 2533. [CrossRef]
12. Pereira, N.A.; Di Prima, S.; Bovi, R.C.; da Silva, L.F.S.; de Godoy, G.; Naves, R.P.; Cooper, M. Does the Process of Passive Forest Restoration Affect the Hydrophysical Attributes of the Soil Superficial Horizon? *Water* **2020**, *12*, 1689. [CrossRef]
13. Darouich, H.; Ramos, T.B.; Pereira, L.S.; Rabino, D.; Bagagiolo, G.; Capello, G.; Simionesei, L.; Cavallo, E.; Biddoccu, M. Water Use and Soil Water Balance of Mediterranean Vineyards under Rainfed and Drip Irrigation Management: Evapotranspiration Partition and Soil Management Modelling for Resource Conservation. *Water* **2022**, *14*, 554. [CrossRef]
14. Tang, Y.-Y.; Chen, J.-P.; Zhang, F.; Yuan, S.-S. Spatiotemporal Analysis of Evapotranspiration and Effects of Water and Heat on Water Use Efficiency. *Water* **2021**, *13*, 3019. [CrossRef]
15. Ternes, B. Technological Spaces in the Semi-Arid High Plains: Examining Well Ownership and Investment in Water-Saving Appliances. *Water* **2021**, *13*, 365. [CrossRef]
16. Park, W.-P.; Hyun, H.-N.; Koo, B.-J. Silicon Fractionation of Soluble Silicon in Volcanic Ash Soils That May Affect Groundwater Silicon Content on Jeju Island, Korea. *Water* **2020**, *12*, 2686. [CrossRef]
17. Bondì, C.; Castellini, M.; Iovino, M. Compost Amendment Impact on Soil Physical Quality Estimated from Hysteretic Water Retention Curve. *Water* **2022**, *14*, 1002. [CrossRef]

Article

Compost Amendment Impact on Soil Physical Quality Estimated from Hysteretic Water Retention Curve

Cristina Bondì [1,*], Mirko Castellini [2] and Massimo Iovino [1]

[1] Department of Agricultural, Food and Forest Sciences, University of Palermo, 90128 Palermo, Italy; massimo.iovino@unipa.it

[2] Council for Agricultural Research and Economics–Research Center for Agriculture and Environment (CREA–AA), Via C. Ulpiani 5, 70125 Bari, Italy; mirko.castellini@crea.gov.it

* Correspondence: cristina.bondi@unipa.it

Abstract: Capacity-based indicators of soil physical quality (SPQ) and pore distribution parameters were proposed to assess the effects of compost amendment but their determination was limited to desorption water retention experiments. This study also considered the pore size distribution obtained from adsorption experiments to establish the effectiveness of compost amendment in modifying the physical and hydrological attributes of a sandy loam soil. Repacked soil samples with different compost to soil ratios, *r*, were subjected to a wetting–drying cycle, and the water retention data were fit to the van Genuchten model to obtain the pore volume distribution functions. The soil bulk density was minimally affected by the wetting–drying cycle but a significant negative correlation with *r* was obtained. The sorption process involved larger and more heterogeneous pores than the desorption one thus resulting in an estimation of the air capacity SPQ indicators (P_{mac} and *AC*) that were higher for the wetting–water retention curve (WWRC) than the drying one (DWRC). The opposite result was found for the water storage SPQ indicators (*PAWC* and *RFC*). In general, SPQ indicators and pore distribution parameters were generally outside the optimal range but estimates from the DWRC were closer to the reference values. The water entry potential increased and the air entry potential decreased with an increase in the compost rate. Significant correlations were found between the SPQ indicators estimated from the DWRC and *r* but the same result was not obtained for the WWRC. It was concluded that compost addition could trigger positive effects on soil hydrological processes and agronomic service as both water infiltration during wetting and water storage during drying are favored. However, the effectiveness of the sorption process for evaluating the physical quality of soils needs further investigation.

Keywords: soil structure; pore volume distribution function; bulk density; macroporosity; air capacity; plant available water capacity; relative field capacity; S-index

Citation: Bondì, C.; Castellini, M.; Iovino, M. Compost Amendment Impact on Soil Physical Quality Estimated from Hysteretic Water Retention Curve. *Water* **2022**, *14*, 1002. https://doi.org/10.3390/w14071002

Academic Editor: Jan Wesseling

Received: 23 February 2022
Accepted: 20 March 2022
Published: 22 March 2022

1. Introduction

Application of compost is one efficient way to increase soil organic matter level and indirectly improve soil structure and hydrological functions [1,2]. A large number of studies have documented positive effects of compost application on different soil physical and hydrological attributes: total porosity [3], bulk density [4], soil resistance to penetration [5], pore size distribution [6], aggregation and aggregate stability [1,7,8], water retention capacity [9,10], and saturated and unsaturated hydraulic conductivity [11,12]. As a consequence of improved water retention capacity, compost addition increases the plant available water capacity (PAWC) of soils [13]. Comparative analysis of twenty-five studies showed that compost incorporation had positive effects on degraded urban soils in terms of reduced compaction, enhanced infiltration and hydraulic conductivity, and increased water content and PAWC [14].

The above-mentioned positive effects depend on the compost application rate but also on the feedstock type, compost maturity, and compost quality. Application rate usually ranges from 30 to 150 Mg ha^{-1} [3,6,10] but values up to 750 Mg ha^{-1} have been reported [12]. A variety of organic wastes have been proposed as soil amendment materials, also with the aim to find sustainable disposal for urban and agricultural byproducts. Compost types include sewage sludge and municipal waste [15,16], maize and sewage sludge [9], pruning waste [17], farm crop residues [18], yard waste [3,19], orange juice processing wastes [20], and mixtures of these materials. As an example, Głąb et al. [21] reported the effects of co-composted maize, sewage sludge, and biochar mixtures addition on the hydrological and physical quality of a loamy sandy soil. Compared with the control soil, the physical properties of the amended soil were significantly improved with beneficial effects that were dependent on the rate of compost application and the type of feedstock. Likewise, Rivier et al. [22] investigated the effect of compost and vermicompost on soil structure, water retention, aggregate stability, and plant water use efficiency, compared with that of mineral fertilizers and food-waste digestate.

Binding primary particles into stable aggregates, organic matter influences attributes of soil pores such as size, distribution, shape, and connectivity [18]. The soil–water retention curve (SWRC), relating volumetric water content, θ, to pressure head, *h*, provides an indirect method to estimate soil porosity and, therefore, to assess the effectiveness of compost for improving the conditioning of soil. A complete description of the pore volume distribution function can be obtained by fitting appropriate semi-empirical functions to $\theta(h)$ data [23,24]. Reynolds et al. [25] classified pore size distribution on the basis of selected "location" and "shape" parameters and proposed optimal ranges for each parameter. Another approach relies on estimation of capacity-based indicators of soil physical quality (SPQ) accounting for soil ability to store air and water [23,26]. In any case, comparison of measured pore distribution parameters and SPQ indicators with optimal ranges deduced from literature allows a straightforward assessment of compost application effects [27,28]. To date, such an approach has proved useful for studying different agro-environments and evaluating their sustainability [29–34].

The SWRC is not unique because of hysteresis [35]. Briefly, hysteresis is a phenomenon depending on several factors, including non-uniformity in pore cross-sections, variation of dynamic contact angles in the advancing or receding water–air interface menisci, and entrapped air effects, air volume changes, and aging phenomena [36]. Due to hysteresis, the soil water content at a given pressure head is higher during the drying than the wetting process. Nevertheless, estimation of pore distribution parameters and capacitive SPQ indicators has been generally conducted from $\theta(h)$ data obtained from desorption experiments under the simplified assumption that hysteresis can be neglected in field conditions because its influence is often masked by heterogeneities and spatial variability [37]. Another reason for using desorption SWRC is that capacity-based SPQ indicators are generally associated with the soil's ability to store water and transfer it from wet to dry periods, i.e., across a relatively long temporal scale. Adsorption SWRC, i.e., the water retention curve determined for a wetting process, can give additional information on the soil's ability to store water over the shorter time scale related to the infiltration process. However, to the best of our knowledge, wetting SWRC was never used to determine SPQ nor the effects of compost addition on water retention hysteresis evaluated.

The study investigates the reliability of capacity-based indicators of SPQ and pore size distribution parameters for assessing the effectiveness of compost amendment on a sandy loam soil. With the aim to move forward from the traditional approach based on the analysis of the desorption SWRC, the effects of water retention hysteresis on the estimation of soil physical quality were evaluated. The assumed hypothesis was that analysis of the pore distribution system obtained from the wetting SWRC could complete our knowledge of active porosity under different hydrological processes.

2. Materials and Methods

2.1. Sample Preparation and SWRC Measurement

Soil samples were collected in a citrus orchard at the Department of Agriculture and Forestry Sciences of the University of Palermo, Italy (UTM 33S 355511E-4218990N). The soil (Typic Rhodoxeralf) was classified as sandy loam with a relatively high gravel content (13% by weight) [38] and moderate organic carbon content at the time of sampling (Table 1). The soil samples were air-dried, gently crushed, and passed through a 2 mm sieve before mixing with compost.

Table 1. Physicochemical attributes of soil and compost.

Soil		Compost	
Clay (%)	17.6	pH	7.2
Silt (%)	29.8	EC (dS/m)	0.54
Sand (%)	52.6	C (%)	9.91
OM (%)	2.1	N (%)	0.64
pH	7.8	P (%)	0.45
EC (dS/m)	0.48	Ash (%)	82.5
CEC (cmol Kg^{-1})	25.31	C/N ratio	15:1

The amending compost consisted of 5-months-aged compost from orange juice processing wastes (75%) and garden cleaning (25%) [20]. Agro-industrial wastes were composed of about 60% peel, 30% pulp, and 10% pips, while garden cleaning contained triturated pruning residues and mown grasses. The characteristics of the used compost are reported in Table 1. Compost was preliminarily screened through a 2 mm sieve to eliminate large vegetal residues.

Air-dried compost was mixed with soil in five different proportions by weight: 10% (M10); 20% (M20); 30% (M30); 50% (M50); and 75% (M75). For comparative purposes, the two unmixed matrices, i.e., 100% soil (M0) and 100% compost (M100), were also considered. It is worth noting that the highest compost to soil ratios, r, are far above the maximum usually applied in the field [3,39]. However, the choice to also use high compost application rates was considered reasonable given the ash content of the compost is above and the total carbon content is below the average values for usual food waste and/or dairy animal manure composts [3,18].

Replicate samples of each mixture were obtained by compacting into 5 cm diameter by 5 cm height cylinders a dry mass of the two constituents given by:

$$M_s = \frac{V\, BD_c BD_s}{BD_c + rBD_s} \qquad M_c = rM_s \tag{1}$$

in which M_c (g) and M_s (g) are the dry masses of compost and soil, respectively, BD_s (g cm^{-3}) and BD_c (g cm^{-3}) are the dry bulk densities of the two constituents [40], V (cm^3) is the sample volume, and r is the compost to soil ratio. The air-dried masses were corrected to account for the initial soil water content and then gently mixed for 5 min by means of a mechanical sieve to obtain a homogeneous mixture. Sample compaction was conducted in four successive steps by beating the mixture with five strokes from a height of 5 cm followed by five rotations with a pestle at each increment. The same treatment was applied to M0 and M100 to avoid any artifacts due to sample preparation. The samples were then weighted to check that the measured initial bulk density was the same as the theoretical one determined from Equation (1). This sample preparation procedure allowed to obtain highly replicable results even with a limited sample number (n = 2).

The water retention curve was determined by the tension hanging water column apparatus [41] for pressure head, h (m), values ranging from −0.01 to −1 m, and the pressure plate extractors [42] for h values ranging from −1 to −150 m. Each sample was placed on the porous plate of a glass funnel and saturated from the bottom by progressively raising the water level in a graduated burette that allowed to measure the volume of water adsorbed by or drained from the sample. Initial saturation was obtained in four equilibrium steps of 24 h each at h values of −0.2, −0.1, and −0.05 m followed by submersion. Then, the sample was drained by lowering h in several successive steps of 24 h each and finally equilibrated at h = −1 m.

The water retention curve was measured by applying a wetting/drainage cycle consisting of a sequence of 14 h values applied in ascending/descending order (h = −0.01 m, −0.02 m, −0.03 m, −0.05 m, −0.075 m, −0.10 m, −0.15 m, −0.20 m, −0.25 m, −0.30 m, −0.40 m, −0.50 m, −0.70 m, and −1 m). At each h level, the volume of water adsorbed or drained from/into the burette was recorded. The volumetric water content, θ (m^3m^{-3}), at each equilibrium stage was calculated by adding the drained or adsorbed volumes to the final θ_{-1} value determined at h = −1 m by weighting the sample after oven-drying at 105 °C for 24 h. The sample height was measured at the end of the experiment (h = −1 m) and the sample dry soil bulk density, *BD* (g cm^{-3}) was calculated from the oven-dried weight of the soil sample [43]. Additional replicate samples (n = 3), prepared by the same procedure using only soil (M0) and compost (M100), were intensively monitored during the water retention experiment to evaluate changes in bulk density due to swelling or consolidation processes. At this aim, the sample height was measured at nine fixed points of the sample surface by using a gauge with a precision of 0.5 mm and an average value was determined by the arithmetic mean. Measurements were conducted after sample preparation (H0), at initial saturation (HS), at the initial equilibrium pressure head h = −1 m (H1), at the end of the wetting process when the sample was equilibrated at h = −0.01 m (H2), and at the end of the drainage process for the final equilibrium pressure head h = −1 m (H3).

Water retention data at pressure heads of −1 m, −3.3 m, −10 m, −30 m, and −150 m were determined in pressure plate extractors on three replicated samples of 5 cm diameter by 1 cm height. For each of the seven considered mixtures, the dry masses of compost and soil were calculated from Equation (1) to obtain the same theoretical bulk density value of the 5 cm by 5 cm samples. Determination of volumetric water content at h = −1 m was included in pressure plate experiments for comparison with the θ value measured at the same potential in the tension apparatus. All the measurements were conducted under temperature-controlled conditions at 22 ± 1 °C.

2.2. SWRC Parameterization

Experimental data were fitted by the van Genuchten [24] model (VGN):

$$\theta(h) = \theta_r + (\theta_s - \theta_s)\left(1 + |\alpha h|^n\right)^{-m} \quad (2)$$

in which θ_s (m^3m^{-3}) and θ_r (m^3m^{-3}) are the saturated and residual volumetric water contents, respectively, α is a scale parameter, and n and m with $m = 1-1/n$ are shape parameters. Equation (2) was fitted to experimental data by the SWRC Fit software [44]. Separate fitting was conducted for the two replicates of a given mixture and for the wetting (WWRC) and drainage (DWRC) water retention data. In the second case (DWRC), both the tension and the pressure plate data (n = 19) were considered given the latter were obtained for a drainage process. The VGN model shape and scale parameters (α, n, θ_s, and θ_r) were estimated without any constraint to their possible range. For the WWRC, only the tension data (n = 14) were considered and the VGN model was fitted to the data with θ_r fixed at the value determined for the DWRC. The reliability of estimates was evaluated by the coefficient of correlation, *R*, the mean error, *ME*, and the root mean square error, *RMSE* [45]:

$$R = \frac{\Sigma(O_i - \overline{O})(P_i - \overline{P})}{\Sigma(O_i - \overline{O})^2 \Sigma(P_i - \overline{P})^2} \quad (3)$$

$$ME = \frac{\Sigma_{i=1}^{n}(P_i - O_i)}{N} \quad (4)$$

$$RMSE = \sqrt{\frac{\sum_{i=1}^{n}(P_i - O_i)^2}{N}} \quad (5)$$

in which P_i is the value of the volumetric water content at a given pressure head estimated from Equation (2), O_i is the corresponding measured value, and $\overline{P}$ and $\overline{O}$ are, respectively, the mean of estimated and measured θ values. Only the van Genuchten model was considered in this investigation as a preliminary visual analysis of the θ(*h*) data highlighted a typical S-shaped curve that could be adequately fitted by a unimodal water retention curve model.

2.3. Estimation of Soil Physical Quality

The fitted VGN water retention curves were used to estimate the following capacity-based indicators of soil physical quality [25,27,28,46]:

$$Macroporosity \quad Pmac = Ts - Tm \quad (6)$$

where θ_m (m^3m^{-3}) is the saturated volumetric water content of the soil matrix ($h = -0.1$ m). The P_{mac} parameter gives the volume of large (macro) pores (i.e., >300 μm equivalent pore diameter) indirectly indicating the soil's ability to quickly drain excess water:

$$Air\ capacity \quad AC = Ts - TFC \quad (7)$$

where θ_{FC} (m^3m^{-3}) is the field capacity water content ($h = -1$ m). The *AC* parameter is an indicator of soil aeration:

$$Plant - available\ water\ capacity \quad PAWC = TFC - TPWP \quad (8)$$

where θ_{PWP} (m^3m^{-3}) is the permanent wilting point corresponding to $h = -150$ m:

$$Relative\ field\ capacity \quad RFC = TFCTs \quad (9)$$

that expresses the soil's capacity to store water (and air) relative to the soil's total pore volume (as represented by θ_s):

$$S - index \quad Sindex = -n(Us - Ur)[1 + 1m] - (m+1) \quad (10)$$

where U_s (g g^{-1}) and U_r (g g^{-1}) are the gravimetric saturated and residual water content that, under the assumption of rigid soil, can be calculated from θ_s and θ_r. The S_{index} represents the magnitude of the slope of the SWRC at the inflection point when the curve is expressed as gravimetric water content versus a natural logarithm of the pressure head [23]. The theory of the S_{index} is based on the premise that the shape of the SWRC is controlled primarily by structure pores for *h* values from saturation to the inflection point and by matrix pores for lower *h* values. While the former can be modified by soil management (including amendments), the latter mainly depends on more stable soil properties, such as texture.

Optimal soil physical quality conditions require [25,31,47]: $P_{mac} \geq 0.07$ m^3m^{-3}, $AC \geq 0.14$ m^3m^{-3}, $PAWC \geq 0.20$ m^3m^{-3}, and $0.60 \leq RFC \leq 0.70$; $S \geq 0.050$.

The pore volume distribution function, $S_v(h)$, may be defined as the slope of the SWRC expressed as volumetric water content versus ln(*h*), and plotted against equivalent pore diameter, d_e (μm) [25]:

$$S_v = \frac{d\theta}{d(lnh)} = -mn(\theta_s - \theta_r)\alpha^n h^n \left[1 + (\alpha h)^n\right]^{-(m+1)} \quad (11)$$

The capillary rise equation was used to estimate the equivalent diameter $d_e = 2980/h$ (μm) with *h* expressed in cm. The pore volume distribution function was normalized by dividing S_v by the magnitude of the slope at the inflection point, S_{vi}. Note that S_{vi} is given by $S_{vi} = (BD \times S_{index})$ for a rigid soil. The normalized soil pore volume distribution, $S^*(h)$, provides a means for comparing among different porous materials being $0 \leq S^*(h) \leq 1$. Pore volume distributions can be characterized and compared using "location" and "shape" parameters, where the location parameters include the modal diameter, d_{mode} (µm), the median diameter, d_{median} (µm), and mean diameter, d_{mean} µm). The shape parameters include standard deviation, *SD* (-), skewness, *SK* (-), and kurtosis, *KU* (-). For brevity reasons, the expressions for estimating the location and shape parameters are not given here but the reader is referred to Reynolds et al. [25]. The optimal ranges for these indicators are: d_{median} = 3–7 µm; d_{mode} = 60–140 µm; d_{mean} = 0.7–2 µm; *SD* = 400–1000; *SK*: from −0.43 to −0.41; and *KU* = 1.13–1.14.

The flow chart in Figure 1 depicts the procedures and the calculations used to estimate the SPQ indicators and the pore volume distribution functions.

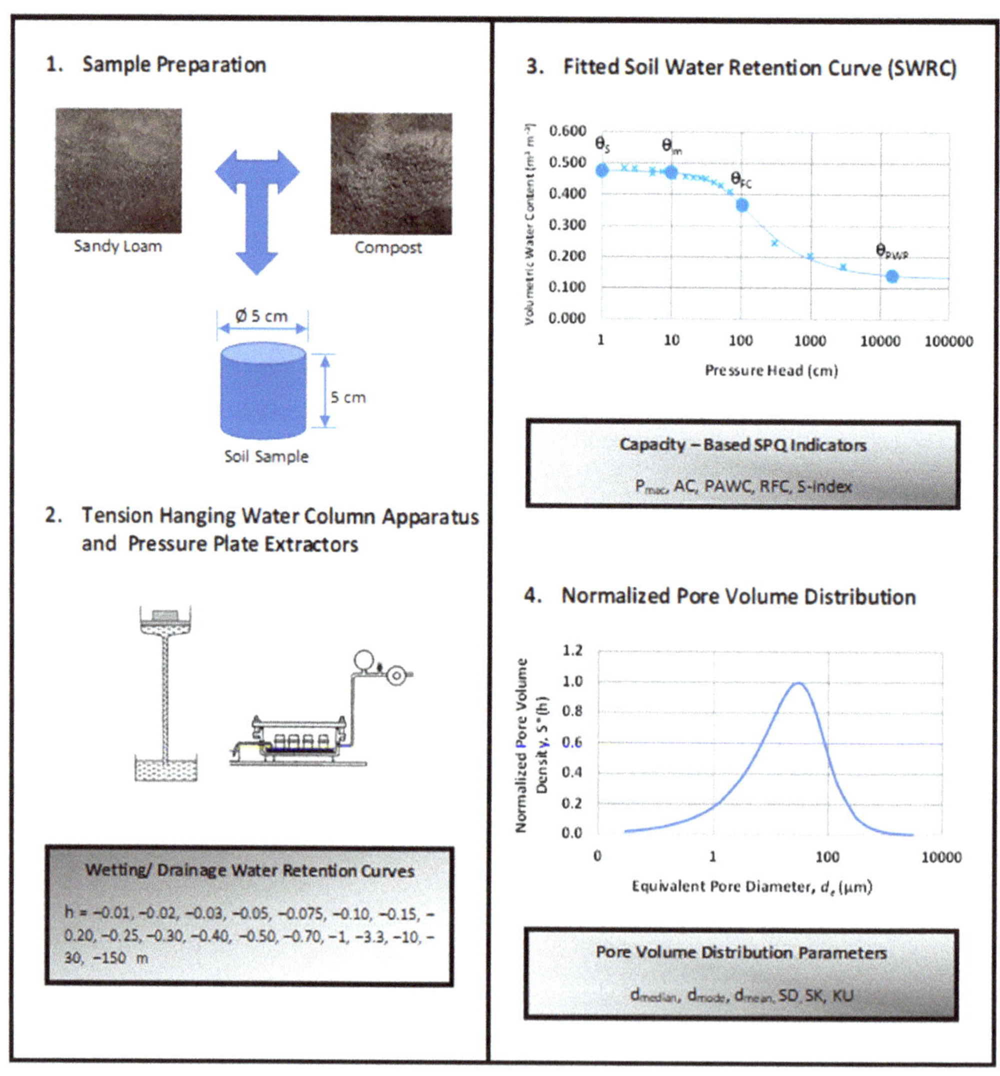

Figure 1. Flow chart of the procedural steps to estimate capacity-based indicators of SPQ and pore distribution parameters.

2.4. Data Analisys

For each considered variable and soil property, mean and associated coefficients of variation (CV) were calculated according to the assumed cumulative frequency distribution. Specifically, sample height, *BD*, P_{mac}, *AC*, *PAWC*, *RFC*, S_{index}, d_{mode}, and d_{median}, were assumed to be normally distributed whereas, according to Reynolds et al. [25], the other data (d_{mean}, *SD*, *SK*, and *KU*) were assumed to be ln distributed.

Comparison of means was conducted by the HSD Tukey test ($p = 0.05$). The influence of compost addition was investigated by assessing the significance of the regression coefficients between the considered soil variables (i.e., BD, VGN parameters, SPQ indicators, and pore distribution parameters) and the compost to soil ratio, r ($p = 0.05$).

3. Results and Discussion

3.1. Soil Bulk Density

The average height of the samples after preparation under air-dried conditions, H0, varied between 3.99 cm for M0 and 3.92 cm for M100 (Figure 2). The corresponding bulk density values were characterized by the coefficient variations (CV) of 0.29% and 2.14% that are far below the limit of 15% considered acceptable for the properties of this soil [48], thus confirming the reliability of the sampling preparation procedure. Following initial saturation (HS), the mean sample height increased to 4.21 cm for M0 (+5.6%) and 4.49 cm (+14.4%) for M100. Moreover, the CV of the M100 increased following saturation thus showing a higher susceptibility of the amended soil to undergo particle rearrangement as a consequence of the saturation process. According to an HSD Tukey test, the difference in sample heights between preparation and saturation was significant ($p = 0.05$) for both M0 and M100. After equilibration at the initial pressure head of $h = -1$ m (H1), the mean sample height significantly decreases to 4.09 cm (M0) and 4.26 cm (M100). Therefore, initial saturation followed by the first drainage cycle modified the original bulk density of the laboratory packed samples probably as a consequence of inter-particle bonds relaxation during wetting followed by settling during drainage. The M100 samples showed higher and more variable heights than M0 samples thus indicating a greater sensitivity to such modifications (Figure 2).

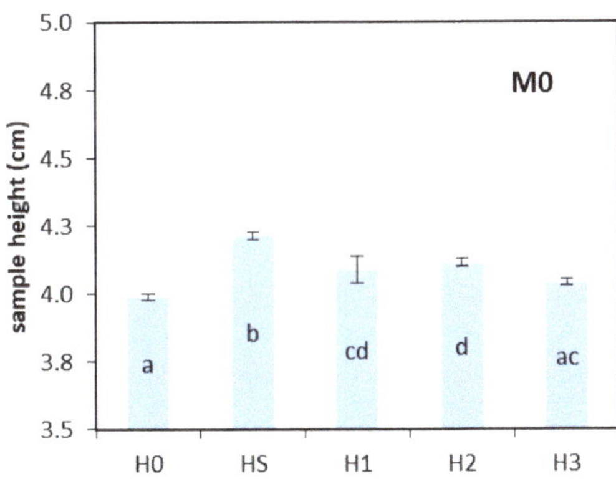

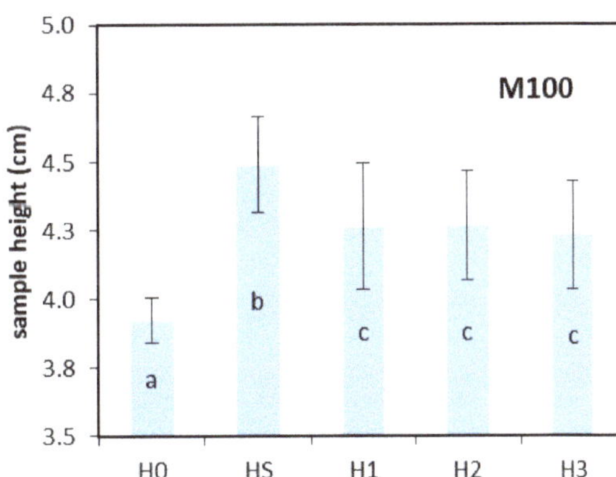

Figure 2. Sample height measured for M0 and M100 samples after preparation (H0), at initial saturation (HS), at the beginning of wetting (H1), at the end of wetting (H2), and at the end of drainage (H3).

In the subsequent wetting–drainage cycle, the samples were subjected to much fewer height modifications. For M0, height increased by a not significant 0.6% during wetting and decreased by 1.75% during drainage. The height after the wetting–drainage loop (H3 = 4.04 cm) was not statistically different from H1 measured at the same equilibrium pressure head, and it slightly increased only for the higher pressure head of the sequence

(H2). For M100, the mean height was practically the same (H1 = 4.26 cm, H2 = 4.27 cm, and H3 = 4.23 cm) being the differences well within the precision of the measurement technique (Figure 2). Therefore, provided a preliminary wetting–drainage cycle had been performed, the sample bulk density was not influenced by subsequent sorption–desorption processes and the soil samples can be considered rigid independently of the applied pressure head in the range $-1 \leq h \leq -0.01$ m. It is worth noting that the mean height of the M100 samples is on average 1.04 times the mean height of the M0 samples. The results of this preliminary investigation allow us to conclude that the laboratory repacked samples may undergo changes in particle configuration as a consequence of the applied sequence of pressure head values. These changes, however, tend to be negligible after the first wetting–drainage loop, at least in the range of *h* values that are requested for estimating the SPQ indicators. Therefore, the hypothesis of rigid porous medium was considered reasonable and, following the suggestion by Reynolds et al. [25], the *BD* measured at $h = -1$ m was assumed to assess the effects of compost amendment.

The *BD* values measured at the end of the wetting–drainage cycle (H3) ranged from 1.03 to 1.14 g cm^{-3} with a mean value of 1.08 g cm^{-3} (CV = 3.10%). Independently of the compost to soil ratio, the soil *BD* was always within the range of *BD* values considered as optimal for field crop production ($0.90 \leq BD \leq 1.2$ g cm^{-3}) [31,43]. A significant negative correlation ($p < 0.005$) was found between the sample bulk density and the compost percentage (Figure 3). Similar results were reported by Khaleel et al. [49] and Eden et al. [39] in their review articles on the long-term effects of organic waste recycling in agriculture. Mandal et al. [4] obtained a linear decreasing *BD* relationship for a silt loam soil amended with composted poultry litter up to 40% *v/v*. However, a closer examination of the plot also highlights that a threshold type behavior for *BD* vs. *r* relationship could be supposed. In particular, compost addition seems to have no effect on *BD* until a threshold value of *r* = 30% is reached. Afterward, the soil *BD* decreases at increasing *r* value. Such a different soil response was observed, among others, by Reynolds et al. [26] for the short-term effects of a single high rate addition of yard waste compost to a clay loam soil. They found significant differences only at the highest compost rate application, i.e., 300 t ha^{-1}, whereas 75 t ha^{-1} and 150 t ha^{-1} produced negligible or small improvements relative to the control. Similarly, Brown and Cotton [50] observed significant changes of a loamy soil *BD* only for application rates higher than 168 t ha^{-1}.

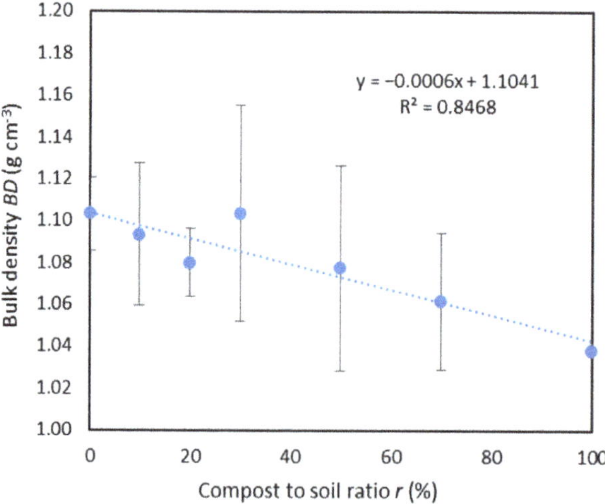

Figure 3. Relationship between soil bulk density and compost to soil ratio.

3.2. Soil Water Retention and Pore Volume Distribution

The wetting–drainage water retention curves exhibited a typical hysteretic behavior with volumetric water content at a given pressure head that was always lower for the wetting curve, θ_W, than the draining one, θ_D. As an example, Figure 4 compares the wetting (WWRC) and draining (DWRC) water retention curves obtained for M20 and M75 mixtures. The two curves were practically coincident close to saturation ($h \geq -0.05$ m). At lower h values, the differences between θ_D and θ_W first increased up to a maximum 0.156 m^3m^{-3} at h = 0.50 m and then tended to decrease but remained as high as 0.135 m^3m^{-3} at the lowest h value measured with the tension apparatus ($h = -1$ m).

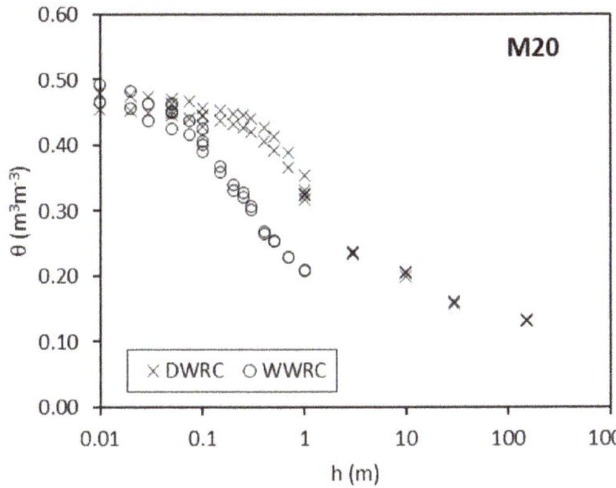

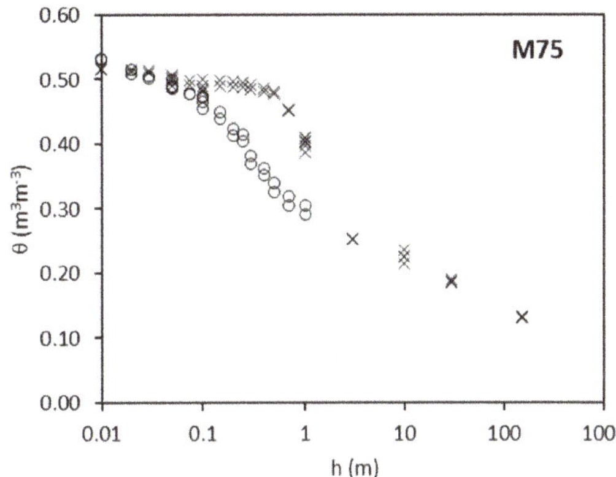

Figure 4. Measured water retention data for the wetting (WWRC) and draining (DWRC) curves of M20 and M75 mixtures.

The statistics of the estimated VGN model parameters and the indices of the fitting quality (*R*, *ME*, and *RMSE*) for both the wetting and the drainage WRC are listed in Table 2. A complete dataset of the estimated parameters is provided in the Supplementary Materials, Table S1.

Table 2. Minimum, maximum, and mean value of the estimated van Genuchten parameters for the wetting and draining water retention curve. The statistics of coefficient of correlation, *R*, mean error, *ME*, and root mean square error, *RMSE*, for the estimated water retention curves are also listed.

	α	n	θ_s	θ_r	R	ME	$RMSE$
			Wetting water retention curve WWRC				
Min	0.0613	1.406	0.457	0.110	0.9954	-6.54×10^{-4}	6.00×10^{-3}
Max	0.1203	1.894	0.536	0.161	0.9976	3.83×10^{-4}	9.00×10^{-3}
Mean	0.0871	1.639	0.500	0.138	0.9962	-3.14×10^{-6}	7.57×10^{-3}
			Draining water retention curve DWRC				
Min	0.0088	1.458	0.443	0.110	0.9931	-4.70×10^{-5}	5.90×10^{-3}
Max	0.0231	1.998	0.508	0.161	0.9985	1.13×10^{-4}	1.58×10^{-2}
Mean	0.0149	1.726	0.479	0.138	0.9959	2.62×10^{-5}	1.11×10^{-2}

The VGN model adequately fitted the water retention data as detected by the high *R* and low *ME* and *RMSE* values. The mean α value for the WWRC was 5.8 times higher than for the DWRC. This α_W/α_D value is above those usually reported ($\alpha_W/\alpha_D \approx 2$) [35]. The mean n parameter of the WWRC was 0.95 times that of the DWRC (Table 2). A significant negative correlation ($R^2 = 0.544$) was found between n_W and n_D thus indicating that, for the considered soil–compost mixture, estimation of the n parameter for one of the two branches of the water retention curve is also able to retrieve information for the other one. The estimated saturated water contents for the wetting, θ_{sW}, and drying, θ_{sD}, branches were highly correlated (R^2 = 0.917) but the slope of the regression line was significantly different from one thus indicating that the water contents close to saturation for the WWRC were higher than for the DWRC (Figure 3, Table 2). The reason for the observed discrepancies between θ_{sW} and θ_{sD} is unknown but could be associated to the observed variability of sample height during the sorption and desorption processes (Figure 2). In any case, these differences can be considered negligible for the aims of estimating the SPQ given they were always lower than 0.033 m^3m^{-3} and equal to 0.020 m^3m^{-3} on average.

The scale parameter α decreased and the shape parameter n increased in the passage from the WWRC to the DWRC (Table 2). Generally, the α and n parameters are positively correlated (e.g., [51,52]); therefore, the observed inverse relationship is a sign that active soil pore classes are different in the two processes. Comparison between the pore volume distribution functions confirms that the sorption process (WWRC) involves larger and more heterogeneous pores (Figure 5). Therefore, different information is provided by the two soil water retention curves and, consequently, by the respective estimates of the SPQ indicators.

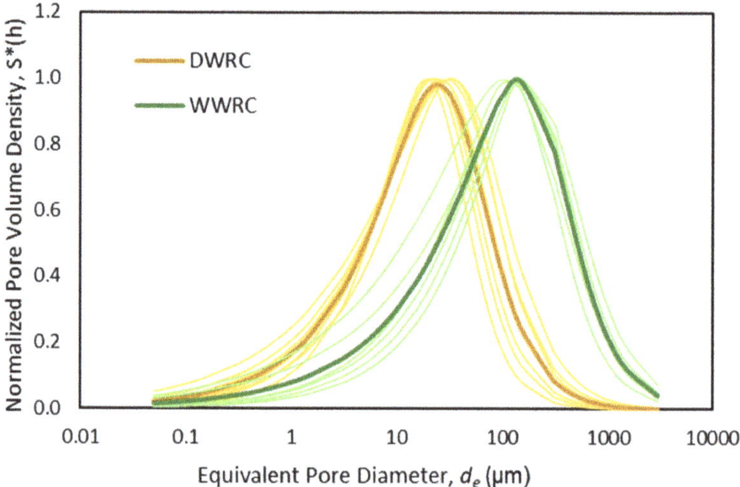

Figure 5. Normalized pore volume distribution functions for the wetting and draining processes.

3.3. Influence of Hysteresis on Soil Physical Quality

The capacity-based indicators linked to the macro- and mesoporosity (P_{mac} and AC) calculated for the WWRC were generally higher than for the DWRC (Table 3). In particular, the maximum values of P_{mac} and AC obtained from the DWRC (P_{mac} = 0.012 m^3m^{-3} and AC = 0.146 m^3m^{-3}) were lower than the minimum values of the same parameters obtained from the WWRC (P_{mac} = 0.047 m^3m^{-3} and AC = 0.224 m^3m^{-3}). Differences were also observed for indicators linked to the plant water availability (*PAWC* and *RFC*) but, in this case, the WWRC yielded lower values than the DWRC one. According to these results, use of the WWRC in spite of the DWRC yielded larger estimates of SPQ indicators related to soil aeration and lower estimation of those related to water storage. These findings were not entirely predictable and, to the best of our knowledge, they were experimentally assessed for the first time in this study. Similar estimates of the S_{index} were obtained by the two sets of water retention data (Table 3), probably because of the observed compensation between the two domains (macro- and micropores) involved in its calculation.

Table 3. Statistics of the considered SPQ indicators obtained from wetting and draining SWRC and correlation coefficients between SPQ indicators and the compost ratio *r*. The suggested optimal range for each SPQ indicator is also shown.

SPQ	Units	Wetting Water Retention Curve				Draning Water Retention Curve				Optimal Range
		Min	Max	Mean	*R*	Min	Max	Mean	*R*	
P_{mac}	m^3m^{-3}	0.047	0.087	0.072	0.1439	0.001	0.012	0.006	***−0.8508***	≥0.07
AC	m^3m^{-3}	0.224	0.306	0.267	−0.2528	0.085	0.146	0.116	***−0.9505***	≥0.14
PAWC	m^3m^{-3}	0.051	0.272	0.102	***0.6419***	0.185	0.255	0.216	***0.9580***	≥0.20
RFC	-	0.354	0.571	0.464	0.5322	0.696	0.831	0.756	***0.9790***	0.60–0.70
S-index	-	0.077	0.125	0.101	−0.4651	0.074	0.132	0.104	***0.8001***	≥0.05
d_{median}	μm	51.4	130.9	94.7	−0.3660	15.1	24.1	17.8	***−0.7754***	3–7
d_{mode}	μm	102.3	168.4	142.8	−0.1191	18.5	33.1	25.7	***−0.9523***	60–140
d_{mean}	μm	36.8	116.9	78.3	−0.4141	12.1	20.7	15.0	***−0.5329***	0.7–2
SD	-	3.7	11.7	6.1	0.5079	3.3	9.2	5.1	***−0.6303***	400–1000
SK	-	−0.29	−0.17	−0.22	***−0.5725***	−0.27	−0.16	−0.21	***0.7285***	−0.43 to −0.41
KU	-	1.14	1.16	1.15	***0.5919***	1.14	1.16	1.15	***−0.7699***	1.13–1.14

Values in bold indicate statistically significant correlation ($p = 0.05$).

According to the observed differences between the pore volume distribution curves, the location parameters (d_{median}, d_{mode}, and d_{mean}) obtained from the WWRC were larger than the DWRC. In particular, the median diameter of water-filled pores during sorption (d_{median} = 95 μm) is larger than the draining ones (d_{median} = 18 μm) in agreement with the hysteresis theory (Figure 5). Indeed, during a drainage process, the soil holds water in relatively smaller pores whereas the water entry process takes place, on average, at lower suction heads (i.e., larger pores). The shape parameters of the pore volume distribution functions were more similar but a tendency to increase *SD* and decrease *SK* for the WWRC was observed (Table 3). The mean pore volume distribution function was in both cases skewed towards small pores ($SK < 0$) whereas the two curves were classified as leptokurtic, i.e., more peaked in the center and more tailed in the extremes than the lognormal curve ($KU > 1$) [25].

According to the literature suggestions for optimal soil physical quality [23,25,31], the mean values of capacity-based SPQ indicators and pore distribution parameters were generally outside the optimal range but estimates from the DWRC were closer to the reference values (Table 3). The only exception was for P_{mac} and *AC* estimated from the WWRC that fell in the optimal range whereas the corresponding indicators estimated from the DWRC signaled an aeration deficit for soil. Location and shape parameters of pore volume distribution were always non-optimal. The observed discrepancies were not surprising given the soil samples considered in this investigation were laboratory repacked (i.e., structureless) whereas the literature guidelines for optimal SPQ were generally obtained from undisturbed soil samples. A recent study conducted on repacked soil samples of a loamy sand amended with crop residues and dairy manure compost [18] yielded location parameters, d_{median}, d_{mode}, and d_{mean}, of 21–26 μm, 28–32 μm, and 19–23 μm, respectively, and shape parameters, *SD*, *SK*, and *KU* of 3.1–3.9, −0.18 to −0.15 and 1.14, respectively. These values were outside the optimal range reported by Reynolds et al. [25] and closer to the mean values obtained in the present study from the DWRC (d_{median} = 17.8 μm, d_{mode} = 25.7 μm, d_{mean} = 15.0 μm, $SD = 5.1$, $SK = -0.21$, and $KU = 1.15$) thus confirming that the existing guidelines for optimal pore size distribution parameters need to be applied with caution to repacked soil samples. Furthermore, estimation of the SPQ indicators, as well as the definition of their optimal intervals, was almost exclusively conducted considering desorption data and our results show that different information can be obtained if sorption data are considered. Providing specific experimental information on the effect of water retention hysteresis on repacked soil samples is essential to fill the knowledge gap and provide recalibrated values for the current optimal SPQ guidelines.

3.4. Influence of Compost Amendment on Soil Water Retention and Physical Quality

The lack of reference values for the SPQ indicators obtained from the WWRC does not allow drawing of definitive conclusions about the reliability of such an approach. However, an indirect validation could be gathered from the analysis of compost amendment effects on the VGN model parameters as well as the related capacity-based SPQ indicators and pore size distribution parameters obtained by the two approaches (i.e., WWRC and DWRC).

Compost amendment influenced the shape and the scale parameters of the SWRC to different extents depending on the considered wetting or draining process (Table 4). In particular, α increased with r for the WWRC and decreased for the DWRC. However, only the latter relationship was statistically significant. Symmetrically, the n parameter decreased with r for the WWRC and increased for the DWRC. The α parameter is related to the inverse of the water or air entry potential. As expected from the hysteresis theory, the absolute value of the water entry potential (i.e., the inverse of α_W) is lower than the air entry potential (i.e., the inverse of α_D) (Table 2). However, at increasing the compost content, the water entry potential tended to decrease (Table 4) thus meaning that the compost amendment facilitates the water entry into the soil during a wetting process or, in other words, that infiltration is favored. On the contrary, the air entry potential tended to increase, that is, water loss is impeded during the soil drainage. The observed modifications have positive effects on soil hydrological and agronomic response under dryland agriculture as either infiltration is promoted during rainfall periods or storage enhanced during dry periods, in both cases, increasing the water availability for crops.

Table 4. Correlation coefficients between the van Genuchten model parameters (i.e., α, n, θ_s, and θ_r) and compost ratio r.

	α	n	θ_s	θ_r
WWRC	0.3376	**−0.5773**	**0.8188**	n.d.
DWRC	**−0.8909**	**0.7785**	**0.7905**	**0.7367**

Values in bold indicate statistically significant correlation (p = 0.05).

Compost addition tended to decrease the shape parameter n for the WWRC and to increase it for the DWRC. As the n parameter is related to the slope of the water retention curve at the inflection point, it means that due to amendment, the WWRC tended to be less S-shaped and the DWRC more S-shaped. In other words, not only did the wetting or the draining processes activate different pore systems but the compost addition seems to influence them in a different way, i.e., increasing the active pore assortment for the WWRC and decreasing it for the DWRC. Independently of the considered process (wetting or draining), θ_{sW}, θ_{sD}, and θ_{rD} were positively correlated with r indicating that the compost amendment was effective in increasing the water content of the considered sandy loam soil at, or close to, saturation as well as at the dry end of the SWRC (Table 4).

The significant correlations between the capacity-based SPQ indicators estimated from the DWRC and the compost ratio, r, strengthened the reliability of the procedure that estimates the soil physical quality from the water release experiments (Table 3). A significant reduction in P_{mac} and AC and an increase in $PAWC$, RFC, and S_{index} was observed (Figure 6) that was attributed to an increase in micropores and a decrease in macropores associated to the modification of pore volume distribution. Similar results were observed in literature with PAWC that increased [18] or was practically unaffected by compost addition [3]. On the other hand, many studies showed long-term benefits on AC that increased up to 26% in a fine-textured soil [26] and 15% in a loamy sand [3] when food waste compost was used. Our results lead to the conclusion that short-term organic matter incorporation probably increased micropore volume as a consequence of compost mineral residue (13.7% of particles less than 2 μm in diameter) but had negative effects on macroporosity.

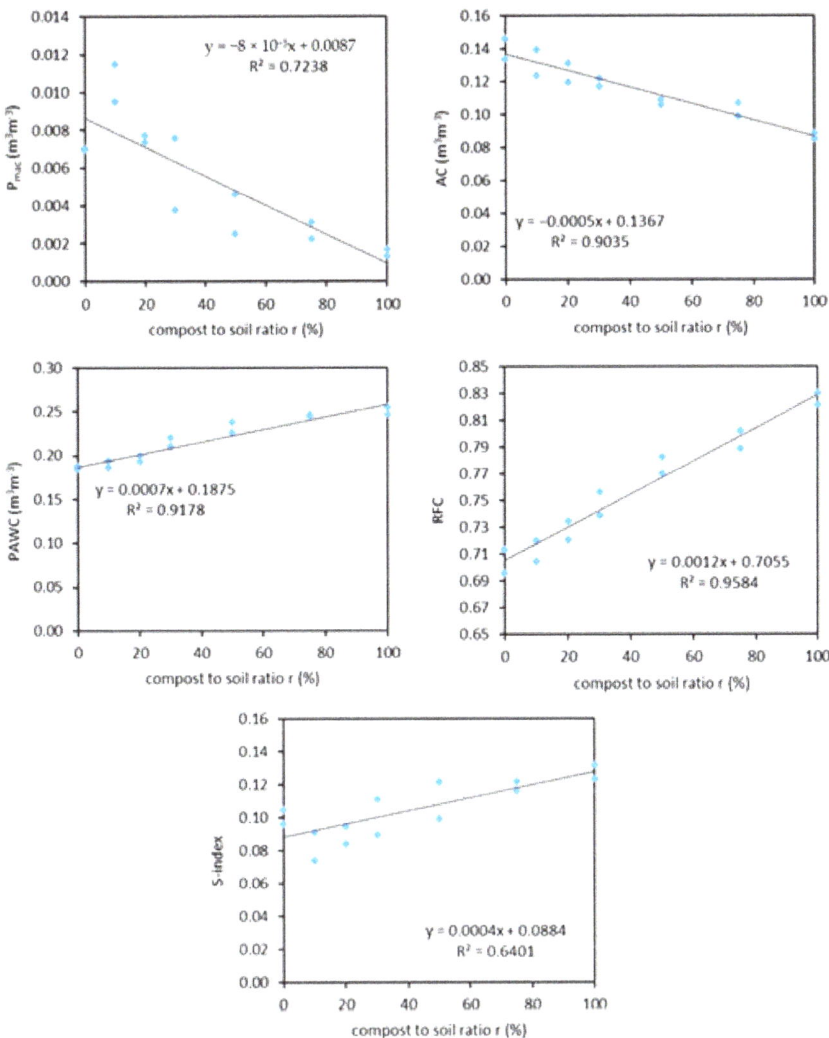

Figure 6. Regressions between the capacity-based SPQ indicators derived from the DWRC and the compost ratio.

The highly significant negative correlation observed for d_{median}, d_{mode}, and d_{mean} (Table 3) confirmed that amending was effective in modifying the pore systems. Ibrahim and Horton [18] reported a similar result that was also attributed to filling of loamy sand pore spaces by amendment material. Shape parameters were likewise significantly affected by compost amendment (Table 3). In particular, as the compost percentage increased, the pore distributions were less sorted (decreasing *SD*) and shifted toward large soil pores (increasing *SK*). Moreover, *KU* was affected by compost addition even if the range of variation was limited (Table 3). Al-Omran et al. [53] showed that *SD* and *SK* increased by 18% and 12%, respectively, whereas *KU* values were practically unaffected by compost amendment.

The results obtained from the DWRC data clearly show that compost had a positive effect on soil water storage. These results are in agreement with the extensive literature showing how the use of compost improves the physical properties of the soil. However, information is lacking for hysteretic soil water retention. Our data showed that, in most cases, correlations between SPQ indicators obtained from the WWRC and r were not significant indicating that the sorption process is probably less recommended for the aim of SPQ evaluation. The only exception was for the PAWC that exhibited a positive, even less strong (R = 0.642), correlation with r (Table 3). Determination of the wetting branch of the SWRC is more affected by experimental errors due, for example, to soil hydrophobicity [54], air entrapment [55], and slaking or swelling of soil aggregates [56]. Therefore, it is possible that one, or all these factors, hampered the sensitivity of the estimated SPQ to compost addition. On the other hand, there are signs that also support the use of the WWRC for soil physical quality evaluation. For example, the circumstance that a negative, despite not significant, trend with r was observed for AC and the location parameters. It is also worth to note, that the relationship between S_{index} and r (Table 3) followed the same trend found for the shape parameter n (Table 4). Skewness and kurtosis had a contrasting behavior with compost amendments, that is, a negative correlation was found for the WWRC when it was positive for the DWRC and vice versa.

Dynamic indicators of SPQ were not considered in this investigation, but our results suggest that the use of compost could also affect them, depending on the wetting or drainage phase considered. More in-depth analysis is therefore desirable to investigate the effect of hysteresis on dynamic indicators of SPQ, including number and volume of the hydraulically active pores [32,57]. As an example, Bagarello et al. [58] identified a clear hysteresis effect on soil water retention of a sandy loam soil that also affected soil hydraulic conductivity data [59]. Consequently, different estimates of the dynamic indicators are expected, depending whether the wetting or drying process is considered.

Another point that requires further investigation is the use of repacked soil samples that could be non-representative of field conditions. Many researchers hypothesized that incorporation of compost would influence soil structure and, consequently, soil hydraulic properties in the short term (among others, [1,22]). The addition of compost to sieved soil, to obtain a repacked laboratory sample, can be likened, to some extent, to the real soil shortly after compost incorporation followed by tillage. Further investigations are necessary to test the adopted methodology after several wetting–drainage cycles with the aim to study the combined effects of compost and the soil structural restoration.

4. Conclusions

Hysteresis of the water retention curve affected the estimation of capacity-based indicators of SPQ and pore volume distribution parameters of a sandy loam soil amended with compost obtained from orange juice processing wastes and garden cleaning. The sorption process involved larger and more heterogeneous pores thus resulting in capacity-based indicators linked to soil aeration (P_{mac} and AC) that were generally higher and plant water availability indicators ($PAWC$ and RFC) were generally lower than those determined from desorption data. The SPQ indicators estimated from the DWRC were closer to the reference literature values that were usually retrieved from the same desorption process.

The absolute value of the water entry potential decreased and the air entry potential increased at increasing the percentage of added compost. In terms of the SPQ response to compost amendment, all the selected indicators estimated from the DWRC were sensitive to compost amendment. The same result was not obtained for the SPQ indicators estimated from the WWRC. In this case, the correlations with compost percentage were generally not significant thus indicating that the sorption process is probably less recommended than the desorption one for evaluating the physical quality of soils.

Overall, the results showed that compost amendment was effective in modifying the soil pore distribution system and the related SPQ indicators. In particular, compost addition could trigger positive effects on soil hydrological processes and agronomic services as both water infiltration during wetting and water storage during drying are favored. It is necessary to extend this investigation to other soils exploring to what extent the observed modification in pore volume distribution could also affect hydrodynamic indicators of soil quality directly related to the soil's ability to transmit water down into the profile.

The perspectives opened by these studies could help to improve our understanding of the effects of organic matter (compost) incorporation on soil structure, hydrological functions, and physical quality.

Supplementary Materials: The following supporting information can be downloaded at: https://www.mdpi.com/article/10.3390/w14071002/s1, Table S1: Estimated values of the van Genuchten parameters for the wetting and draining water retention curve and corresponding fitting statistics.

Author Contributions: Conceptualization, M.C. and M.I.; methodology, M.C. and M.I.; formal analysis, M.C., C.B. and M.I.; investigation, C.B.; data curation, C.B. and M.I.; writing—original draft preparation, C.B. and M.I.; writing—review and editing, M.C. and M.I. All authors have read and agreed to the published version of the manuscript.

Funding: This work was supported by the project WATER4AGRIFOOD, contratto "Tetto Verde", Codice Progetto: CON-0375, CUP B94I20000300005.

Informed Consent Statement: Not applicable.

Data Availability Statement: The data presented in this study are available on request from the corresponding author.

Conflicts of Interest: The authors declare no conflict of interest.

References

1. Dong, L.; Zhang, W.; Xiong, Y.; Zou, J.; Huang, Q.; Xu, X.; Ren, P.; Huang, G. Impact of short-term organic amendments incorporation on soil structure and hydrology in semiarid agricultural lands. *Int. Soil Water Conserv. Res.* **2021**, *in press*. [CrossRef]
2. Reynolds, W.D.; Drury, C.F.; Yang, X.M.; Fox, C.A.; Tan, C.S.; Zhang, T.Q. Land management effects on the near-surface physical quality of a clay loam soil. *Soil Tillage Res.* **2007**, *96*, 316–330. [CrossRef]
3. Arthur, E.; Cornelis, W.M.; Vermang, J.; de Rocker, E. Amending a loamy sand with three compost types: Impact on soil quality. *Soil Use Manag.* **2011**, *27*, 116–123. [CrossRef]
4. Mandal, M.; Chandran, R.; Sencindiver, J. Amending Subsoil with Composted Poultry Litter-I: Effects on Soil Physical and Chemical Properties. *Agronomy* **2013**, *3*, 657–669. [CrossRef]
5. Negiş, H.; Şeker, C.; Gümüş, I.; Manirakiza, N.; Mücevher, O. Effects of Biochar and Compost Applications on Penetration Resistance and Physical Quality of a Sandy Clay Loam Soil. *Commun. Soil Sci. Plant Anal.* **2020**, *51*, 38–44. [CrossRef]
6. Aggelides, S.M.; Londra, P.A. Effects of compost produced from town wastes and sewage sludge on the physical properties of a loamy and a clay soil. *Bioresour. Technol.* **2000**, *71*, 253–259. [CrossRef]
7. Annabi, M.; Houot, S.; Francou, C.; Poitrenaud, M.; Bissonnais, Y.L. Soil Aggregate Stability Improvement with Urban Composts of Different Maturities. *Soil Sci. Soc. Am. J.* **2007**, *71*, 413–423. [CrossRef]
8. Ibrahim, A.; Marie, H.A.M.E.; Elfaki, J. Impact of biochar and compost on aggregate stability in loamy sand soil. *Agric. Res. J.* **2021**, *58*, 34–44. [CrossRef]
9. Głąb, T.; Żabiński, A.; Sadowska, U.; Gondek, K.; Kopeć, M.; Mierzwa-Hersztek, M.; Tabor, S.; Stanek-Tarkowska, J. Fertilization effects of compost produced from maize, sewage sludge and biochar on soil water retention and chemical properties. *Soil Tillage Res.* **2020**, *197*, 104493. [CrossRef]
10. Zemánek, P. Evaluation of compost influence on soil water retention. *Acta Univ. Agric. Silvic. Mendel. Brun. LIX* **2011**, *3*, 227–232. [CrossRef]
11. Schneider, S.; Coquet, Y.; Vachier, P.; Labat, C.; Roger-Estrade, J.; Benoit, P.; Pot, V.; Houot, S. Effect of Urban Waste Compost Application on Soil Near-Saturated Hydraulic Conductivity. *J. Environ. Qual.* **2009**, *38*, 772–781. [CrossRef] [PubMed]
12. Whelan, A.; Kechavarzi, C.; Coulon, F.; Sakrabani, R.; Lord, R. Influence of compost amendments on the hydraulic functioning of brownfield soils. *Soil Use Manag.* **2013**, *29*, 260–270. [CrossRef]
13. Şeker, C.; Manirakiza, N. Effectiveness of compost and biochar in improving water retention characteristics and aggregation of a sandy clay loam soil under wind erosion. *Carpathian J. Earth Environ. Sci.* **2020**, *15*, 5–18. [CrossRef]
14. Kranz, C.N.; McLaughlin, R.A.; Johnson, A.; Miller, G.; Heitman, J.L. The effects of compost incorporation on soil physical properties in urban soils—A concise review. *J. Environ. Manag.* **2020**, *261*, 110209. [CrossRef] [PubMed]
15. Sadeghi, S.; Ebrahimi, S.; Zakerinia, M. The Study of the Parametric Changes in Water Potential Points by Using Waste Manuciple Compost in Three Kinds of Soils. *Int. J. Basic Sci. Appl. Res.* **2014**, *3*, 254–260. Available online: http://www.isicenter.org (accessed on 22 February 2022).
16. Paradelo, R.; Basanta, R.; Barral, M.T. Water-holding capacity and plant growth in compost-based substrates modified with polyacrylamide, guar gum or bentonite. *Sci. Hortic.* **2019**, *243*, 344–349. [CrossRef]
17. Benito, M.; Masaguer, A.; Moliner, A.; de Antonio, R. Chemical and physical properties of pruning waste compost and their seasonal variability. *Bioresour. Technol.* **2006**, *97*, 2071–2076. [CrossRef]

18. Ibrahim, A.; Horton, R. Biochar and compost amendment impacts on soil water and pore size distribution of a loamy sand soil. *Soil Sci. Soc. Am. J.* **2021**, *85*, 1021–1036. [CrossRef]
19. Curtis, M.J.; Claassen, V.P. Regenerating topsoil functionality in four drastically disturbed soil types by compost incorporation. *Restor. Ecol.* **2009**, *17*, 24–32. [CrossRef]
20. Palazzolo, E.; Laudicina, V.A.; Roccuzzo, G.; Allegra, M.; Torrisi, B.; Micalizzi, A.; Badalucco, L. Bioindicators and nutrient availability through whole soil profile under orange groves after long-term different organic fertilizations. *SN Appl. Sci.* **2019**, *1*, 468. [CrossRef]
21. Glab, T.; Zabinski, A.; Sadowska, U.; Gondek, K.; Kopec, M.; Mierzwa-Hersztek, M.; Tabor, S. Effects of co-composted maize, sewage sludge, and biochar mixtures on hydrological and physical qualities of sandy soil. *Geoderma* **2018**, *315*, 27–35. [CrossRef]
22. Rivier, P.-A.; Jamniczky, D.; Nemes, A.; Makó, A.; Barna, G.; Uzinger, N.; Rékási, M.; Farkas, C. Short-term effects of compost amendments to soil on soil structure, hydraulic properties, and water regime. *J. Hydrol. Hydromech.* **2022**, *70*, 74–88. [CrossRef]
23. Dexter, A.R. Soil physical quality Part I. Theory, effects of soil texture, density, and organic matter, and effects on root growth. *Geoderma* **2004**, *120*, 201–214. [CrossRef]
24. Van Genuchten, M.T. A Closed-form Equation for Predicting the Hydraulic Conductivity of Unsaturated Soils. *Soil Sci. Soc. Am. J.* **1980**, *44*, 892–898. [CrossRef]
25. Reynolds, W.D.; Drury, C.F.; Tan, C.S.; Fox, C.A.; Yang, X.M. Use of indicators and pore volume-function characteristics to quantify soil physical quality. *Geoderma* **2009**, *152*, 252–263. [CrossRef]
26. Reynolds, W.D.; Drury, C.F.; Tan, C.S.; Yang, X.M. Temporal effects of food waste compost on soil physical quality and productivity. *Can. J. Soil Sci.* **2015**, *95*, 251–268. [CrossRef]
27. Reynolds, W.D.; Bowman, B.T.; Drury, C.F.; Tan, C.S.; Lu, X. Indicators of good soil physical quality: Density and storage parameters. *Geoderma* **2002**, *110*, 131–146. [CrossRef]
28. Topp, G.C.; Reynolds, W.D.; Cook, F.J.; Kirby, J.M.; Carter, M.R. Physical attributes of soil quality. In *Development in Soil Science; Soil Quality for Crop Production and Ecosystem Health*; Gregorich, E.G., Carter, M.R., Eds.; Elsevier: New York, NY, USA, 1997; Volume 25, pp. 21–58.
29. Manici, L.M.; Castellini, M.; Caputo, F. Soil-inhabiting fungi can integrate soil physical indicators in multivariate analysis of Mediterranean agroecosystem dominated by old olive groves. *Ecol. Indic.* **2019**, *106*, 105490. [CrossRef]
30. Stellacci, A.M.; Castellini, M.; Diacono, M.; Rossi, R.; Gattullo, C.E. Assessment of soil quality under different soil management strategies: Combined use of statistical approaches to select the most informative soil physico-chemical indicators. *Appl. Sci.* **2021**, *11*, 5099. [CrossRef]
31. Agnese, C.; Bagarello, V.; Baiamonte, G.; Iovino, M. Comparing physical quality of forest and pasture soils in a Sicilian watershed. *Soil Sci. Soc. Am. J.* **2011**, *75*, 1958–1970. [CrossRef]
32. Iovino, M.; Castellini, M.; Bagarello, V.; Giordano, G. Using static and dynamic indicators to evaluate soil physical quality in a Sicilian area. *Land Degrad. Dev.* **2016**, *27*, 200–210. [CrossRef]
33. Cullotta, S.; Bagarello, V.; Baiamonte, G.; Gugliuzza, G.; Iovino, M.; La Mela Veca, D.S.; Maetzke, F.; Palmeri, V.; Sferlazza, S. Comparing Different Methods to Determine Soil Physical Quality in a Mediterranean Forest and Pasture Land. *Soil Sci. Soc. Am. J.* **2016**, *80*, 1038–1056. [CrossRef]
34. Shahab, H.; Emami, H.; Haghnia, G.H.; Karimi, A. Pore Size Distribution as a Soil Physical Quality Index for Agricultural and Pasture Soils in Northeastern Iran. *Pedosphere* **2013**, *23*, 312–320. [CrossRef]
35. Kutilek, M.; Nielsen, D.R. Soil Hydrology. In *GeoEcology Textbook*; Catena Verlag: Cremlingen, Germany, 1994; 370p, ISBN 3-923381-26-3.
36. Zhao, Y.; Wen, T.; Shao, L.; Chen, R.; Sun, X.; Huang, L.; Chen, X. Predicting hysteresis loops of the soil water characteristic curve from initial drying. *Soil Sci. Soc. Am. J.* **2020**, *84*, 1642–1649. [CrossRef]
37. Haverkamp, R.; Reggiani, P.; Ross, P.J.; Parlange, J.Y. Soil Water Hysteresis Prediction Model Based on Theory and Geometric Scaling. *Environ. Mech.* **2002**, *129*, 213–246.
38. Alagna, V.; Bagarello, V.; Cecere, N.; Concialdi, P.; Iovino, M. A test of water pouring height and run intermittence effects on single-ring infiltration rates. *Hydrol. Processes* **2018**, *32*, 3793–3804. [CrossRef]
39. Eden, M.; Gerke, H.H.; Houot, S. Organic waste recycling in agriculture and related effects on soil water retention and plant available water: A review. *Agron. Sustain. Dev.* **2017**, *37*, 11. [CrossRef]
40. Gugliuzza, G.; Verduci, A.; Iovino, M. Water retention characteristics of substrates containing biochar and compost as peat and perlite replacements for ornamental plant production. *Acta Hortic.* **2021**, *1305*, 507–511. [CrossRef]
41. Dane, J.H.; Hopmans, J.W. 3.3.2.2 Hanging water column. In *Methods of Soil Analysis, Part 4, Physical Methods, Number 5 in the Soil Science Society of America Book Series*; Dane, J.H., Topp, G.C., Eds.; Soil Science Society of America, Inc.: Madison, WI, USA, 2002; pp. 680–683.
42. Dane, J.H.; Hopmans, J.W. 3.3.2.4 Pressure plate extractor. In *Methods of Soil Analysis, Part 4, Physical Methods, Number 5 in the Soil Science Society of America Book Series*; Dane, J.H., Topp, G.C., Eds.; Soil Science Society of America, Inc.: Madison, WI, USA, 2002; pp. 688–690.
43. Reynolds, W.D.; Drury, C.F.; Yang, X.M.; Tan, C.S. Optimal soil physical quality inferred through structural regression and parameter interactions. *Geoderma* **2008**, *146*, 466–474. [CrossRef]

44. Seki, K. SWRC fit—A nonlinear fitting program with a water retention curve for soils having unimodal and bimodal pore structure. *Hydrol. Earth Syst. Sci.* **2007**, *4*, 407–437.
45. Castellini, M.; Iovino, M. Pedotransfer functions for estimating soil water retention curve of Sicilian soils. *Arch. Agron. Soil Sci.* **2019**, *65*, 1401–1416. [CrossRef]
46. Reynolds, W.D.; Yang, X.M.; Drury, C.F.; Zhang, T.Q.; Tan, C.S. Effects of selected conditioners and tillage on the physical quality of a clay of a clay loam soil. *Can. J. Soil Sci.* **2003**, *83*, 381–393. [CrossRef]
47. Dexter, A.R.; Czyz, E.A. Applications of s-theory in the study of soil physical degradation and its consequences. *Land Degrad. Dev.* **2007**, *18*, 369–381. [CrossRef]
48. Warrick, A.W. Appendix 1: Spatial variability. In *Environmental Soil Physics*; Hillel, D., Ed.; Academic Press: San Diego, USA, 1998; pp. 655–675.
49. Khaleel, R.; Reddy, R.; Overcash, M.R. Changes in Soil Physical Properties Due to Organic Waste Applications: A Review. *J. Environ. Qual.* **1981**, *10*, 133–141. [CrossRef]
50. Brown, S.; Cotton, M. Changes in soil properties and carbon content following compost application: Results of on-farm sampling. *Compost Sci. Util.* **2011**, *19*, 88–97. [CrossRef]
51. Carsel, R.F.; Parrish, R.S. Developing Joint Probability-Distributions of Soil-Water Retention Characteristics. *Water Resour. Res.* **1988**, *24*, 755–769. [CrossRef]
52. Haverkamp, R.; Leij, F.J.; Fuentes, C.; Sciortino, A.; Ross, P.J. Soil water retention: I. Introduction of a shape index. *Soil Sci. Soc. Am. J.* **2005**, *69*, 1881–1890. [CrossRef]
53. Al-Omran, A.; Ibrahim, A.; Alharbi, A. Effects of Biochar and Compost on Soil Physical Quality Indices. *Commun. Soil Sci. Plant Anal.* **2021**, *52*, 2482–2499. [CrossRef]
54. Doerr, S.H.; Shakesby, R.A.; Walsh, R.P.D. Soil water repellency: Its causes, characteristics and hydro-geomorphological significance. *Earth Sci. Rev.* **2000**, *51*, 33–65. [CrossRef]
55. Kaluarachchi, J.J.; Parker, J.C. Effects of hysteresis with air entrapment on water flow in the unsaturated zone. *Water Resour. Res.* **1987**, *23*, 1967–1976. [CrossRef]
56. DeBano, L.F. *Water Repellent Soils: A State-of-the-Art*; General Technical Report, PSW-46; United States Department of Agriculture, Forest Service: Berkeley, CA, USA, 1981; 21p.
57. Castellini, M.; Ventrella, D. Impact of conventional and minimum tillage on soil hydraulic conductivity in typical cropping system in southern Italy. *Soil Tillage Res.* **2012**, *124*, 47–56. [CrossRef]
58. Bagarello, V.; Castellini, M.; Iovino, M. Influence of the pressure head sequence on the soil hydraulic conductivity determined with the tension infiltrometer. *Appl. Eng. Agric.* **2005**, *21*, 383–391. [CrossRef]
59. Bagarello, V.; Castellini, M.; Iovino, M. Comparison of unconfined and confined unsaturated hydraulic conductivity. *Geoderma* **2007**, *137*, 394–400. [CrossRef]

 water

MDPI

Article

Water Use and Soil Water Balance of Mediterranean Vineyards under Rainfed and Drip Irrigation Management: Evapotranspiration Partition and Soil Management Modelling for Resource Conservation

Hanaa Darouich [1], Tiago B. Ramos [2], Luis S. Pereira [1], Danilo Rabino [3,*], Giorgia Bagagiolo [3], Giorgio Capello [3], Lucian Simionesei [2], Eugenio Cavallo [3] and Marcella Biddoccu [3]

1 Landscape Environment Agricultural and Food (LEAF), Institute of Agronomy, University of Lisbon, Tapada da Ajuda, 1349-017 Lisbon, Portugal; hdarouich@isa.ulisboa.pt (H.D.); luis.santospereira@gmail.com (L.S.P.)
2 Centro de Ciência e Tecnologia do Ambiente e do Mar (MARETEC), LARSyS, Instituto Superior Técnico, University of Lisbon, Av. Rovisco Pais, 1, 1049-001 Lisbon, Portugal; tiagobramos@tecnico.ulisboa.pt (T.B.R.); lucian.simionesei@tecnico.ulisboa.pt (L.S.)
3 Istituto di Scienze e Tecnologie per l'Energia e la Mobilità Sostenibili (STEMS), Consiglio Nazionale delle Ricerche (CNR), 10135 Torino, Italy; giorgia.bagagiolo@stems.cnr.it (G.B.); giorgio.capello@stems.cnr.it (G.C.); eugenio.cavallo@cnr.it (E.C.); marcella.biddoccu@cnr.it (M.B.)
* Correspondence: danilo.rabino@cnr.it

Abstract: Vineyards represent complex Mediterranean agrosystems that deliver significant ecosystem services to society. Yet, many vine-growers still need to assimilate the importance of crop and soil management to the conservation of soil and water resources. The main objective of this study was to evaluate water use and the water balance terms in rainfed and irrigated vineyards in Italy and Portugal, respectively, in both cases aiming at the sustainability of natural resources use. The SIMDualKc model is used for both sites after calibration and validation by fitting soil water content measurements. The Italian case study focused on the impacts of inter-row conservation management in hillslope vineyards while the Portuguese case study analyzed irrigation water management under scarcity in flat vineyards. For the Italian vineyards, the model results focused on the evapotranspiration fluxes and their partition, control of surface runoff, and soil water recharge provided by the inter-row soil management using cover crops. Model results of the Portuguese case study showed the need for improving irrigation water use and the terms of water balance, namely referring to percolation and soil water evaporation. Both case studies further demonstrated the advantages of using computational tools to better cope with climate variability in the Mediterranean region and made evident the benefits of improved crop and soil management practices in counteracting land degradation and valuing the use and conservation of natural resources.

Keywords: cover crops; inter-row management; evapotranspiration modeling and partition; FAO56 dual-K_c approach; soil water balance; viticulture

Citation: Darouich, H.; Ramos, T.B.; Pereira, L.S.; Rabino, D.; Bagagiolo, G.; Capello, G.; Simionesei, L.; Cavallo, E.; Biddoccu, M. Water Use and Soil Water Balance of Mediterranean Vineyards under Rainfed and Drip Irrigation Management: Evapotranspiration Partition and Soil Management Modelling for Resource Conservation. *Water* **2022**, *14*, 554. https://doi.org/10.3390/w14040554

Academic Editors: Leonardo V. Noto and Didier Orange

Received: 17 December 2021
Accepted: 10 February 2022
Published: 12 February 2022

1. Introduction

Viticulture is one of the most diffused cultivations in the world and has been practiced in the Mediterranean area for millennia [1]. In this region, the vineyard agricultural system is potentially well suited for delivering provisioning services such as grapes for table or wine production, the regulation of climate and the hydrologic cycle [2], and the preservation and enhancement of cultural heritage including landscape and aesthetic values [3,4]. Contrasting, vineyards are often associated with several environmental problems resulting from the intensification of production systems, which evidence the need for better management of soil and water resources.

There is abundant bibliography relative to soil management, namely in sloping fields in sub-humid to humid climates. Salomé et al. [5] provided a review based on 146 commercial plots and focused on three soil management practices: inter-row plant cover, weeding, and fertilization strategies. Fertilization contributes to improve yields and, therefore, to a higher water productivity while weeding has different impacts depending on whether it is performed mechanically or chemically. Both practices are influenced by the soil type. Mechanical weeding often contributes to runoff and erosion, mainly when intense rainfall occurs and when the soil structure is poor with unstable aggregates. Differently, a cover crop in the inter row, even temporary, benefits soil functioning whatever the soil type. The superiority of adopting a cover crop in the inter-row is reported by many authors, e.g., Biddoccu et al. [6,7] relative to control both soil erosion and runoff in sloping fields. Research results reported by Gómez et al. [8], Prosdocimi et al. [9], and Capello et al. [10] confirm those results. Plant cover contributes to retardation and control of runoff, water infiltration, reduction of soil erosion, increase in organic matter (OM), carbon sequestration, and nutrient supply and retention [5,11,12]. However, the ground cover plants generally compete for water with the vine plants and require special care in water scarce areas.

Inter-row management influences the response of the vineyard to different types of rainfall events, namely in terms of rainwater partition between infiltration and runoff, thus influencing the amount of soil erosion. It further impacts the whole field water balance as well as vines nutrition, growth, and productivity [13,14]. The importance of cover crops to control soil erosion is great, with decreasing soil losses up to 75%, but competition for water and nutrients by the cover crop may reduce yields by up to 54% [11]. Napoli et al. [12] reported for cover crops 68.5% less soil losses than in inter-rows with harrowed soil. Capello et al. [10] reported that grass cover reduced runoff by 65%, and soil erosion losses by 72%. In addition, the response of grass cover plot was less influenced by traffic conditions. In fact, the increased mechanization in vineyards is frequently associated with increased soil compaction due to repeated tractor passes on fixed paths [10,15], which increases runoff and soil erosion, mainly in sloping vineyards. This soil loss is particularly important when the soil is left bare in the inter-row and exposed to intense rainfall events [16,17]. Such question highlights the importance of inter-row management for the sustainability of the use of natural resources by the vineyard system. Biddoccu et al. [6] also reported on a 10-year field research in 15% sloping fields finding that the highest soil losses were observed for conventional and reduced tillage, respectively, 111.5 and 207.7 Mg ha^{-1} while losses were only 25.6 Mg ha^{-1} for the grass covered (GC) treatment. The worst soil management practice was reduced tillage. Keesstra et al. [18] reported that vegetation cover, soil moisture, and organic matter were significantly higher in covered plots than in tilled and herbicide treated plots, while sediment yield and soil erosion were significantly higher in herbicide treated plots. Nevertheless, as reported by Biddoccu et al. [7], adopting temporary cover crops may not produce the target effects on soil protection. In addition, the protection role of the cover crop depends on the grass type, e.g., with temporary natural vegetation being less efficient than saifoin [19].

Novara et al. [14] reported that the use of cover crops is a strategy that may positively influence water productivity by reducing excessive vines vigor in fertile soils and/or favoring deeper roots in layers. However, in low vigor vineyards, low fertile soils, and in dry environments, the competition for water needs to be considered to avoid negative impacts on yields. Moreover, for Mediterranean ecosystems, attention should be paid to their impact on water availability [14]. The use of cover crops in the inter-row provides various ecosystem services including reduction of runoff and erosion and improvement of water supply. Thus, permanent cover crops are commonly implemented in Europe where climate is not excessively dry [7,20–22]. Differently, in semi-arid Mediterranean regions, winegrowers are reluctant to use cover crops due to concerns over soil water competition [23,24]. The competition for soil water at critical crop stages can lead to excessive grapevine water stress, reducing the fruit set, causing premature defoliation, and negatively impacting growth, yields, and the quality of berries [25,26]. In this perspective, it is crucial to know the response of vineyards to rainfall distribution in terms of water availability to plants when considering the presence of cover crops in the inter-row, both in rainfed and irrigated vineyards, since the dynamics of evapotranspiration and water use is insufficiently known in both cases. Moreover, related measures and practices are among those required for adaptation and resilience to climate change impacts.

Costa et al. [27] assumed a long-term perspective when focusing on genetics, selection of varieties and plant amelioration of both the grapevines and the rootstocks, aimed at improving grapevine responses to heat stress and drought, thus in addition to the measures and practices reported. These authors largely referred to policies required to make effective the development of farm responses of vines to heat and drought and focused on water issues. Moreover, they assumed that future strategies to optimize the environmental performance of the wine sector in the Mediterranean must be focused on water and irrigation. Hannah et al. [28] also assumed that attempting to maintain wine grape productivity and quality under climate change implies increased water use for irrigation and to cool grapes through misting or sprinkling, which creates potential impacts on freshwater conservation. Thus, freshwater habitats may be affected where climate change undermines growing conditions for already established vineyards. Hannah et al. [28] concluded that climate change adaptation strategies are required for creating a positive future for producers, wine makers, and vine ecosystems.

In Italy and Portugal, which represent the 1st and 5th largest wine producers in Europe [29], vineyards were traditionally rainfed since, as in many European countries, irrigation was historically not used, forbidden by regulations for many quality wines. Nowadays, in Italy, many IGT, DOC and DOCG regulations admit only emergency irrigation. However, since the soil water stress strongly affects the growth and production of vines and the quality of berries [30], irrigation has become an increasingly frequent practice especially in dry areas of Southern Europe [31,32]. Yet, poor irrigation practices, namely the inadequacy of irrigation depths to crop water needs, climate, and soil and irrigation system characteristics may cause water percolation and leaching of fertilizers and pesticides, thus not promoting environmental friendliness [33–35], nor greater and more stable yields and quality of wine [16]. Better knowledge of the crop water requirements relative to the various crop growth stages are critical to better preserve the vineyards natural resources, namely soil and water.

Accurate knowledge of the soil water balance is fundamental for improving the soil and water management of the vineyard system and to further cope with the challenges resulting from climate change. In Piedmont, Italy, long-term simulations performed over 60 years (1950–2009) confirmed that the climate change is already influencing local vineyards since 1980 [36,37]. Furthermore, Fraga et al. [38] observed that mean phenological timings were projected to undergo significant advancements in Portugal (e.g., budburst and harvest may be anticipated by 1 month or more), with implications also in the corresponding pheno-phase intervals. Impacts of climate change on viticulture likely will be significant [39] in terms of increasing temperature combined with extreme events such as droughts and/or short-term storms [40]. Hence, there is the need for accurate tools for estimating crop water and irrigation requirements, namely soil water balance (SWB) models using weather data and soil water observations and/or other data such as eddy covariance and sap flow data [41,42]. In addition to local ground observations, also remote sensing data may be used in models. Pôças et al. [43] used hyperspectral reflectance data derived from a handheld spectroradiometer to estimate the predawn leaf water potential to assess the water status of grapevine cultivars in the Port wine region to be used in irrigation scheduling. A two-source model was used by Ortega-Farias et al. [44] to estimate vineyard energy balance using thermal images acquired by an unmanned aerial vehicle (UAV). Romero et al. [45] also reported on using multispectral imagery from an UAV platform for water status estimation.

Through modelling, farmers and/or farm advisers use models to support irrigation scheduling under diverse climates and management scenarios. Models may also be used to assess the impact of management decisions on soil and water resources, e.g., SIMDualKc model, which is able to both the partition of crop evapotranspiration (ET) into transpiration and soil evaporation, and the partition of transpiration into the fractions relative to the crop and the understory vegetation [46,47]. Another example is given by Cellete et al. [48] that developed the WaLIS model to simulate water resources partitioning and to estimate ET and the water use for both grapevine and the inter-row cover crop. Phogat et al. [49] applied a Richards-based mechanistic model for computing soil water fluxes and improving agricultural practices in irrigated vineyards, and Kustas et al. [50] and Kool et al. [51] applied the thermal-based two-source energy balance model for monitoring daily ET in vines and the inter-rows. A biosensing IoT platform for water management in vineyards is referred by Loddo et al. [52]. Moreover, there is abundant bibliography for ET studies with grapevines as reviewed by Rallo et al. [53], but not with a cover crop in the inter-row irrigated or rainfed.

The FAO 56 dual-K_c approach is a widely used method that refers to the determination of crop evapotranspiration (ET_c) as the product of a K_c value for a specific crop stage and the grass reference evapotranspiration (ET_o) computed with the FAO Penman-Monteith equation [54,55]. The K_c value is partitioned into the basal crop coefficient (K_{cb}) referring to transpiration and the soil evaporation coefficient (K_e), providing thus separate estimates of the ET_c components: crop transpiration (T_c) and soil evaporation (E_s). Examples of applications to vine and fruit crop systems can be found in the recent literature review provided by Rallo et al. [53] and Pereira et al. [56]. The SIMDualKc model [57] has adopted the dual-K_c approach for estimating daily ET fluxes of crops grown under different climatic regions and management practices, including grapevine [46,47,58], peach [59], and olive [60,61]. Yet, despite the growing use of the FAO56 dual-K_c approach for computing crop evapotranspiration fluxes and improving soil water management in complex agricultural systems, there is still the need for extending research focusing on the search for standard K_c and K_{cb} values as well the impact of active ground cover, cover crops, and mulches on crop evapotranspiration of vine systems [53].

Considering the insufficient knowledge on vineyards evapotranspiration as related to cover crops, the current study aims to estimate evapotranspiration fluxes and water use in two Mediterranean vineyards, a northern Italy sloping rainfed crop planted in a soil with large soil water holding capacity, and an irrigated one in southern Portugal cropped in a flat area with a sandy soil. Differences between vineyards allow perceiving differences in management and in water responses to both the rainfall regime and the ET. Objectives consist of (i) calibration and validation of the water balance model SIMDualKc [57], already proved for vine and tree crops; (ii) evaluation of the influence of the inter-row cover crop on the evapotranspiration dynamics and the soil water balance of the rainfed vineyard (2016–2019); (iii) evaluation of the evapotranspiration dynamics, the soil water balance, and the irrigation management applied to the drip irrigated vineyard (2018–2020); (iv) with support of the referred model, demonstrating the importance of improved soil and water conservation measures and practices for both Italian and Portuguese cases. The goal of the study focusses on the sustainable use of soil and water resources, thus on the sustainability and resilience of the vineyards production systems taking in account the challenges of climate change.

2. Materials and Methods

2.1. Description of the Study Areas

The Italian case study was located at the "Tenuta Cannona" Experimental Vine and Wine Center of Agrion Foundation (44°40′ N, 8°37′ E, 296 m a.s.l.), in the municipality of Carpeneto (AL), in the southern part of the Monferrato hilly area, known as "Alto Monferrato", North-West Italy. Data was collected from January 2016 to December 2019. The climate is alpine sublitoranean. According to records from the nearest weather station over the period 1951–1990 (Ovada, 187 m a.s.l.), the average annual precipitation is 965 mm, mainly concentrated in Autumn (October and November) and Spring (March), while the driest month is July [62]. At the experimental site, over the period 2000–2019, the average annual precipitation was slightly lower (881 mm), ranging from a maximum of 1455 mm (2019) to a minimum of 493 mm (2017). The mean annual air temperature was 13 °C (Figure 1). The Cannona vineyards lie on Pleistocenic fluvial terraces in the Tertiary Piedmont Basin, including highly altered gravel, sand, and silty clay deposits, with red alteration products [63]. The main physical characteristics of the studied soils are presented in Table 1. The soils had clay to clay-loam texture and were classified as Dystric Cambisols [64]. Particle size distribution, soil bulk density (ρ_b), and soil water contents at saturation (θ_s) and at field capacity (θ_{FC}) were obtained from undisturbed soil samples taken at different depths according to Blake and Hartge [65] and Cavazza [66], whereas soil water contents at the wilting point (θ_{WP}) were obtained using the Rosetta pedotransfer functions [67] and the particle size distribution and ρ_b as input (Table 1).

The Portuguese case study was located at Companhia das Lezírias, Samora Correia, southern Portugal (38.808° N, 8.900° W, 45 m a.s.l.). Data were collected from January 2018 to October 2020. The climate in the region is dry sub-humid, with mild winters and hot, dry summers. The mean annual precipitation is 669 mm, mainly concentrated between October and May, while the mean annual temperature is 16.8 °C. The weather data for the study area was obtained from the local weather station and is reported in Figure 2.

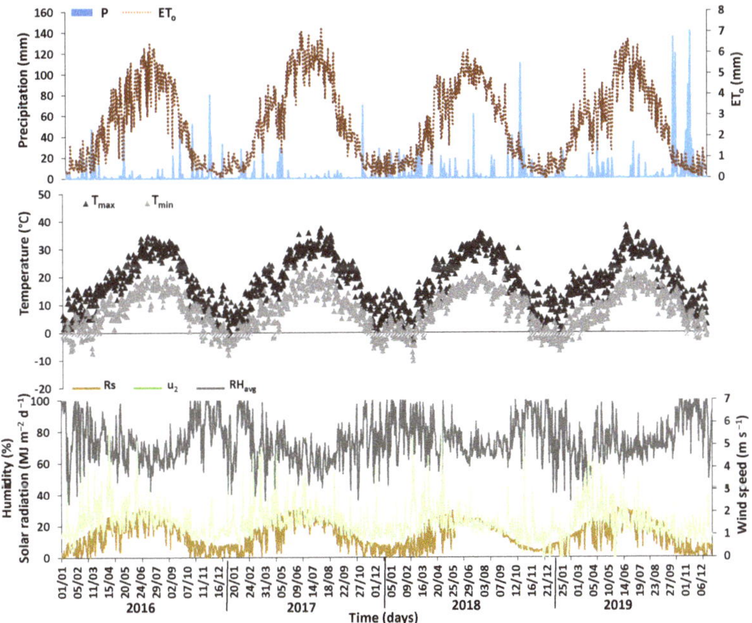

Figure 1. Weather data for Monferrato plots during the study period (P, precipitation, ET_o, reference evapotranspiration; T_{max} and T_{min}, maximum and minimum air temperatures, RH_{avg}, mean relative humidity; R_s, solar radiation, u_2, wind speed at 2 m height).

Table 1. Main soil physical characteristics in the different case studies (ρ_b, soil bulk density; θ_S, θ_{FC}, θ_{WP}, soil water contents at saturation, field capacity, and the wilting point, respectively; TAW, total available water).

Depth (cm)	Soil Texture (%)				ρ_b (Mg m^{-3})	Soil Hydraulic Properties (m^3 m^{-3})			TAW (mm)
	Coarse Sand (2000–200 μm)	Fine Sand (200–20 μm)	Silt (20–2 μm)	Clay (<2 μm)		θ_S	θ_{FC}	θ_{WP}	
Monferrato, Italy (conventional tillage plot)									
0–10	10.9	44.1	26.8	18.2	1.32	0.437	0.380	0.083	29.7
10–20	16.8	29.9	36.7	16.6	1.37	0.443	0.379	0.078	30.1
20–30	17.3	40.8	6.1	35.8	1.35	0.436	0.357	0.127	23.0
30–100	17.3	40.8	6.1	35.8	1.35	0.436	0.350	0.115	164.5
Monferrato, Italy (grass cover plot)									
0–10	11.1	27.6	29.8	31.5	1.33	0.453	0.375	0.115	26.0
10–20	9.5	27.3	30.4	32.8	1.32	0.448	0.398	0.119	27.9
20–30	13.8	28.4	26.8	31.0	1.28	0.454	0.372	0.115	25.7
30–100	13.8	28.4	26.8	31.0	1.28	0.454	0.350	0.135	150.5
Samora Correia, Portugal									
0–20	59.7	26.8	9.4	4.1	1.68	0.404	0.178	0.064	22.90
20–40	60.1	26.0	9.4	4.4	1.65	0.404	0.178	0.064	22.90
40–60	63.3	25.9	7.5	3.3	1.72	0.404	0.145	0.047	19.53
60–80	72.2	19.4	5.5	2.8	-	0.398	0.110	0.038	14.48
80–100	79.6	14.2	4.4	1.8	-	0.398	0.094	0.025	13.74

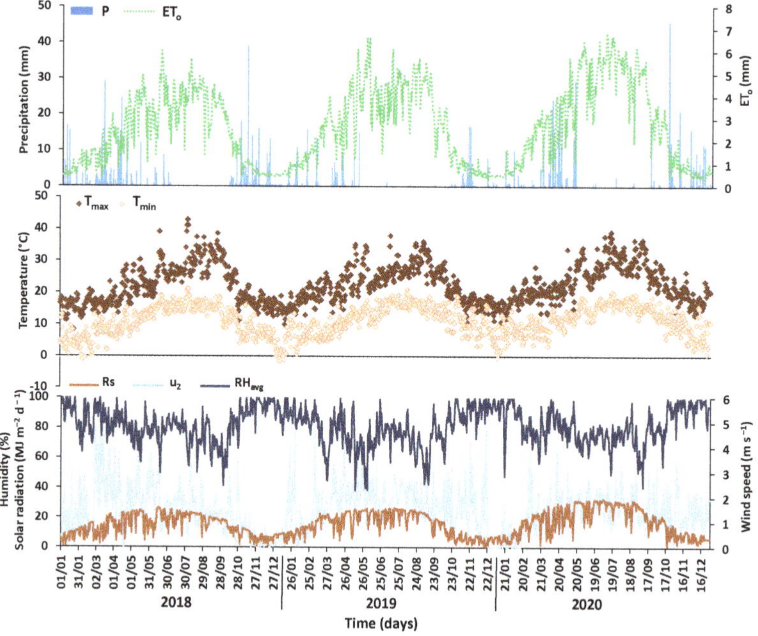

Figure 2. Weather data of Samora Correia case study (P, precipitation, ET_o, reference evapotranspiration; T_{max} and T_{min}, maximum and minimum air temperatures; RH_{avg}, mean relative humidity; R_s, solar radiation, u_2, wind speed at 2 m height).

The soil was classified as a Haplic Fluvisol [64], with main physical characteristics presented in Table 1. The particle size distribution was obtained using the pipette method for particles having diameters <2 μm (clay fraction) and between 2–20 μm (silt), and by sieving for particles between 20–200 µm (fine sand) and between 200–2000 µm (coarse sand). These textural classes follow the Portuguese classification system [68] and are based on international soil particle limits (Atterberg scale). The dry bulk density (ρ_b) was obtained by drying volumetric soil samples (100 cm^3) at 105 °C for 48 h. The soil water contents at saturation (θ_s), field capacity (θ_{FC}), and the wilting point (θ_{WP}) were obtained from pedotransfer functions using the particle size distribution as input [69,70].

2.2. Vineyards Management

As typical in the Monferrato area (Italy), vineyards were rainfed. The experiment was carried out in a vineyard planted in 1988 with Barbera vines, managed according to conventional farming for wine production. Two vineyard plots of 1221 m^2 (16.5 m wide and 74 m long) each were considered. Each plot was composed of 6 rows aligned along the slope (SE aspect, average slope 15%), spaced 2.75 m, where the vines were spaced 1.0 m along the row and grown on vertical shoot positioned trellis (VSP). Since 2000, the soil in the two plots has been managed with different techniques: (i) conventional tillage (CT, hereafter) cultivation with chisel (at a depth of about 0.25 m); and (ii) controlled grass cover (GC), i.e., mulching of the spontaneous grass cover. Both practices were usually carried out twice a year, in spring and autumn. Then, the grass cover in the inter-row was mowed additional times during the season, if necessary. Weeds under the rows were controlled with Glyphosate application in spring, on the surface, 0.6 m across the vine row. Wood pruning was carried out in winter, and residues were chipped in one inter-row out of two. Most of the farming operations in the vineyard were carried out using tracked or tyre tractors carrying or towing implements, with intensification of passages from spring to grape harvest time (from 14 to 27 passages per year). The dates of the main crop stages during the four growing seasons are reported in Table 2.

Table 2. Crop growth stage dates of vines in the various case studies and duration of growing seasons (in GDD).

Year	Crop Growth Stages							Total GDD
	Non-Growing	Initiation	Crop Development	Mid-Season	Late-Season	End-Season	Non-Growing	
				Monferrato, northern Italy				
2016	01/01	12/03	13/04	13/06	07/08	31/10	31/12	
GDD	-	44	325	698	684	-	-	1753
2017	01/01	15/03	30/04	18/06	01/08	31/10	31/12	
GDD	-	123	407	582	797	-	-	1910
2018	01/01	18/03	02/04	07/06	09/08	31/10	31/12	
GDD	-	5	391	845	742	-	-	1982
2019	01/01	24/03	08/04	15/06	19/08	31/10	31/12	
GDD	-	20	307	897	577	-	-	1801
				Samora Correia, southern Portugal				
2018	01/01	13/03	03/04	25/05	05/08	15/10	31/12	
GDD	-	54	308	799	904	-	-	2065
2019	01/01	10/03	13/04	11/06	13/08	20/10	31/12	
GDD	-	150	508	700	789	-	-	2147
2020	01/01	25/03	12/04	31/05	10/08	23/10	31/12	
GDD	-	86	423	906	822	-	-	2236

Note: GDD is the cumulative growing degree-days.

In the 4-years study period, i.e., from January 2016 to December 2019, weather variables, runoff amounts, soil losses, and hourly soil water content related to 89 runoff events were recorded for the two plots. Rainfall, air temperature, and humidity were recorded at 10-min intervals by a weather station placed near the plots. Daily values of solar radiation and wind speed and direction were obtained from stations located 10 km from the vineyard (Acqui Terme, Basaluzzo), belonging to the Regional Environmental Agency [71]. Each vineyard plot was hydraulically "isolated". The runoff generated by rainfall was collected separately for each plot by a channel connected with a sedimentation trap. A tipping bucket device measured the hourly volumes of runoff (RO, mm) in both CT e GC. Runoff samples were collected to obtain sediment concentration for erosive events, and, if sedimentation occurred in the channels and sediment trap, then the sediment yield was collected and weighted (see Biddoccu et al. [72] for details). Soil water contents were recorded every hour from an average of 1-min by indirect method [73] measurements of capacitance/frequency domain sensors (ECH_2O-5TM sensors, Decagon Devices Inc., Pullman, WA, USA), gravimetrically calibrated, placed at 0.1, 0.2, and 0.3 m depth, and stored by a Decagon EM50 datalogger. Soil water content measurements were carried out in the two plots, both in the track position (T), which is the portion of inter-row affected by the passage of tractor wheels or tracks, where the compressive effects tend to concentrate [74], and in the middle of the inter-row, identified as the no-track position (NT), which is not affected by direct contact with tractor wheels or tracks, and then averaged for comparison with model output.

The selected field in Samora Correia (Portugal), planted in 2008, was relatively flat, and part of a larger vineyard (130 ha). The field was a drip-irrigated plot, 5 ha in size, planted with different varieties of wine grapes but where Touriga Nacional was dominant. The plants were grown on VSP trellis, with wood pruning during the dormant period in winter. Plants were at a row distance of 1.0 m and a row spacing of 2.8 m, thus a plant density of approximately 3571 plants ha^{-1}, with an orientation in the east-west direction. The dates of the main crop stages during the three growing seasons can also be found in Table 2. Harvest of grapes in Samora Correia vineyards was performed during September. The inter-row was covered with spontaneous grass from autumn to spring. When necessary, weeds under the rows were controlled with Glyphosate application in spring. Irrigation was delivered through a drip system, with management practices performed according to the standard practices in the region and decided by the farmer. Drippers were spaced 1 m apart, and the drip line was placed on the trellis 0.5 m above the soil surface. The total water applied through irrigation summed 470, 625, and 465 mm in 2018, 2019, and 2020, respectively. The application depth during irrigation events varied from 1 to 12 mm. Soil water contents were continuously monitored in two locations at depths of 10, 20, 30, 40, 50, 60, 70, and 80 cm using EnviroPro MT (MAIT Industries, Australia) capacitance probes. Probes were installed in the crop row, distanced approximately 30 cm from emitters, and measurements were averaged for comparison with model output.

2.3. *The SIMDualKc Model*

2.3.1. The Soil Water Balance

The SIMDualKc model [57] computes the daily soil water balance at the field scale as follows:

$$D_{r,i} = D_{r,i-1} - (P - RO)_i - I_i - CR_i + DP_i + ET_{c\ act,i} \quad (1)$$

where D_r is the root zone depletion (mm), P is the rainfall (mm), RO is the runoff (mm), I is the net irrigation depth (mm), CR is the capillary rise from the groundwater table (mm), DP is the deep percolation (mm), and $ET_{c\ act}$ is the actual crop evapotranspiration (mm), all referring to day i or i−1. In this study, CR was not considered as the groundwater table was too deep (>5 m) in both case studies and could not contribute to crop evapotranspiration.

The SIMDualKc model adopts the FAO56 dual K_c approach for computing crop evapotranspiration [42,54,55]. In this approach, the components relative to crop transpiration (T_c, mm) and soil evaporation (E_s, mm) are computed separately as follows:

$$T_c = K_{cb}\, ET_o \quad (2)$$

$$E_s = K_e\, ET_o \quad (3)$$

where K_{cb} (-) is the standard basal crop coefficient that refers primarily to crop transpiration although some diffusive soil evaporation may also be included, particularly during the initial crop stage, K_e is the evaporation coefficient (-) that describes direct evaporation from the surface soil layer of depth Z_e (cm), and ET_o is the reference evapotranspiration (mm) computed with the FAO Penman-Monteith equation [54]. T_c values are reduced when water stress occurs through a multiplier stress coefficient (K_s) to K_{cb}:

$$T_{c\ act} = K_s\, K_{cb}\, ET_o = K_{cb\ act}\, ET_o \quad (4)$$

where $T_{c\ act}$ is the actual crop transpiration (mm) and $K_{cb\ act}$ is the actual basal crop coefficient (-). When only the matric potential is constraining T_c values, K_s is computed as follows [42,54]:

$$K_s = \frac{TAW - D_{r,i}}{TAW - RAW} \quad (5)$$

where TAW and RAW are, respectively, the total and readily available soil water relative to the rooting depth (mm). These are computed as:

$$TAW = 1000\ Z_r\ (\theta_{FC} - \theta_{WP}) \quad (6)$$

$$RAW = p\ TAW \quad (7)$$

where θ_{FC} and θ_{WP} are the soil water contents at field capacity and the wilting point ($m^3\ m^{-3}$), respectively, Z_r is the root depth (m), and p (-) is the soil water depletion fraction for no stress. When the depletion exceeds p, i.e., the soil water content drops below RAW, T_c values are reduced due to water stress ($K_s < 1.0$), otherwise $K_s = 1.0$.

Soil evaporation is limited by the amount of energy available at the soil surface in conjunction with the energy consumed by transpiration, and by water availability in the surface soil evaporation layer [54,55]. The evaporation coefficient (K_e) is then computed as:

$$K_e = K_r(K_{c\ max} - K_{cb\ min}) \leq f_{ew}\ K_{c\ max} \quad (8)$$

where K_r is the evaporation reduction coefficient (0–1), $K_{c\ max}$ is the maximum value of K_c (-) (i.e., $K_{cb} + K_e$) following a rain or an irrigation event (−), and f_{ew} is the fraction of the soil that is both exposed to radiation and wetted by rain or irrigation. f_{ew} depends upon the effective fraction of ground covered or shaded by vegetation near solar noon ($f_{c\ eff}$). K_r is calculated using the two-stage drying cycle approach where the first stage is the energy limited stage, and the second is the water limited stage [42,54,55,75]:

$$K_r = 1 \text{ for } D_{e,\ i-1} \leq REW \quad (9)$$

$$K_r = \frac{TEW - D_{e,i-1}}{TEW - REW} \text{ for } D_{e,\ i-1} > REW \quad (10)$$

where TEW is the maximum depth of water that can be evaporated from the evaporation soil layer when it has been completely wetted (mm), REW is the depth of water that can be easily evaporated without water availability restrictions (mm), and D_e is the evaporation layer depletion at the end of day i−1 (mm). D_e is computed through a daily water balance of the evaporation soil layer, with the evaporation decreasing as the evaporable soil water decreases in the evaporation soil layer beyond REW.

DP is calculated using a time decay function relating the soil water storage near saturation with the time after the occurrence of heavy rain or irrigation [76]:

$$W_a = a_D\ t^{b_D} \quad (11)$$

where W_a is the actual soil water storage in the root zone (mm), a_D is the soil water storage comprised between saturation (θ_S) and θ_{FC} (mm), b_D is an empirical dimensionless parameter (-), and t is the time after irrigation or rain that produces storage above field capacity (days). Further information on the estimation of a_D and b_D is provided by Liu et al. [76].

RO is estimated using the curve number (CN) approach [77], with the CN value depending on the soil type, vegetation type, and antecedent soil moisture. In SIMDualKc, CN is adjusted each day to reflect the impact of increasing or decreasing of the soil water content on soil infiltration properties by relating the CN value to the soil water depletion in the surface layer (D_e). A different CN value is thus computed according to the correspondence between D_e and the antecedent soil water conditions AWC I, AWC II, and AWC III (respectively, representing dry, average, and wet soil conditions) [77,78]:

$$CN = CN_I \rightarrow \text{for } D_e \geq D_{e-AWC\ I} \tag{12}$$

$$CN = CN_{III} \rightarrow \text{for } D_e \leq D_{e-AWC\ III} \tag{13}$$

$$CN = \frac{(D_e - 0.5\ REW)\ CN_I + (0.7\ REW + 0.3\ TEW - D_e)\ CN_{III}}{0.2\ REW + 0.3\ TEW} \rightarrow \text{for } D_{e-AWC\ III} < D_e < D_{e-AWC\ I} \tag{14}$$

with:

$$CN_I = \frac{CN_{II}}{2.281 - 0.01281 CN_{II}} \tag{15}$$

$$CN_{III} = \frac{CN_{II}}{0.427 - 0.00573 CN_{II}} \tag{16}$$

and:

$$D_{e-AWC\ III} = 0.5\ REW \tag{17}$$

$$D_{e-AWC\ I} = 0.7\ REW + 0.3\ TEW \tag{18}$$

where CN_I, CN_{II}, and CN_{III} are, respectively, the curve numbers associated with the antecedent soil water conditions AWC I (dry), AWC II (average), and AWC III (wet) (0–100), and $D_{e\text{-}AWC\ I}$, $D_{e\text{-}AWC\ II}$, and $D_{e\text{-}AWC\ III}$ are, respectively, the depletion of the evaporative layer at AWC I, AWC II, and AWC III conditions (mm). CN_{II} corresponds to tabulated values available in USDA-SCS [77] and Allen et al. [78].

The K_{cb} values are described for four growth stages: the initial stage or start of the crop season, the rapid growth or development stage, the mid-season stage, and the late-season stage. Yet, the K_{cb} curve is defined by three K_{cb} values corresponding to the initial ($K_{cb\ ini}$), mid-($K_{cb\ mid}$), and end-season ($K_{cb\ end}$). For tree crops and vineyards, these are computed based on crop characteristics [42,53,79,80]. When the inter-row has an active ground cover, the K_{cb} relative to the crop and the understory vegetation may be estimated with the following Equation [57,79]:

$$K_{cb} = K_{cb\ gcover} + K_d \left(\max \left(K_{cb\ full} - K_{cb\ gcover}, \frac{K_{cb\ full} - K_{cb\ gcover}}{2} \right) \right) \tag{19}$$

where $K_{cb\ gcover}$ is the K_{cb} of the ground cover vegetation in the absence of tree foliage, $K_{cb\ full}$ is the estimated basal K_c during peak plant growth for conditions having nearly full ground cover, and K_d is the crop density coefficient. The second term of the max function reduces the estimate for K_{cb} during the mid-season stage by half the difference between $K_{cb\ full}$ and $K_{cb\ cover}$ when this difference is negative. This accounts for impacts of the shading of the surface cover by overstory vegetation having a K_{cb} that is lower than that of the ground cover due to differences in stomatal conductance. When no ground cover exists or when the cover crop dries out becoming a less dense residuals mulch, the previous equation 19 is simplified by replacing $K_{cb\ cover}$ with the minimum K_c for bare soil ($K_{c\ min}$ = 0.15).

The $K_{cb\ full}$ is estimated primarily as a function of crop height and then adjusted for tree crops using a reduction factor (F_r; from Pereira et al. [56]) estimated from the mean leaf stomatal resistance, as follows [80]:

$$K_{cb\ full} = F_r \left(\min(1.0 + k_h\ h,\ 1.20) + [0.04(u_2 - 2) - 0.004(RH_{min} - 45)] \left(\frac{h}{3} \right)^{0.3} \right) \tag{20}$$

where u_2 is the average daily wind speed (m s^{-1}) at a height of 2 m above ground level during the crop growth period, RH_{min} (%) is the average daily minimum relative humidity during the growth period, and h is the mean plant height (m) during the mid-season.

Before climatic adjustment, an upper limit for $K_{cb\ full}$ is assumed 1.20. The effect of the crop height is considered through the sum $(1 + k_h\ h)$, with $k_h = 0.1$ for tree and vine crops [80]. Higher $K_{cb\ full}$ values are expected for taller crops and when the local climate is drier or windier than the standard climate conditions ($RH_{min} = 45\%$ and $u_2 = 2\ m\ s^{-1}$). When the vegetation shows more stomatal adjustment upon transpiration, the parameter F_r requires an empirical adjustment ($F_r < 1.0$), otherwise $F_r = 1.0$. For trees and vines, F_r is closer to 1.0 when crops exhibit great vegetative vigor and decreases under a limited water supply and due to pruning and training [80].

The density coefficient (K_d) is estimated from the fraction of ground cover as follows [79]:

$$K_d = \min\left(1,\ M_L\ f_{c\ eff},\ f_{c\ eff}^{\left(\frac{1}{1+h}\right)}\right) \quad (21)$$

where $f_{c\ eff}$ is the effective fraction of ground covered or shaded by vegetation near solar noon (-), M_L is a multiplier on $f_{c\ eff}$ (1.5–2.0) describing the effect of the canopy density on shading and on maximum relative evapotranspiration per fraction of ground shaded (to simulate the physical limits imposed on water flux through the plant root, stem, and leaf systems), and h is the mean height of vegetation (m).

The $K_{cb\ gcover}$ is computed through the following set of equations [57]:

$$K_{cb,\ h_{gcover}} = \min(1.0 + 0.1\ h_{gcover},\ 1.20) \quad (22)$$

$$K_{cb\ gcover\ full} = K_{cb,\ h_{gcover}} + [0.04(u_2 - 2) - 0.004(RH_{min} - 45)]\left(\frac{h_{gcover}}{3}\right)^{0.3} \quad (23)$$

$$f_{c\ eff\ gcover} = density.\ f_{c\ gcover} \quad (24)$$

$$K_{d\ gcover} = \min\left(1,\ M_L\ f_{c\ eff\ gcover},\ f_{c\ eff\ gcover}^{\left(\frac{1}{1+h_{gcover}}\right)}\right) \quad (25)$$

$$K_{cb\ gcover} = K_{c\ min} + K_{d\ gcover}\left(K_{cb\ gcover\ full} - K_{c\ min}\right) \quad (26)$$

where the variables shown are the same as defined earlier but refer to the active ground cover (subscript gcover). The equations 19 through 26 are not used when there is no need for partition of transpiration between the crop and the active ground cover.

2.3.2. Model Setup

The input data required by the SIMDualKc model were the following:

- Climatic data: daily weather data from local meteorological stations, namely the maximum and minimum temperatures (T_{min} and T_{max}, °C), minimum and maximum relative humidity (RH_{max} and RH_{min}, %), global solar radiation (R_s, MJ m^{-2} day^{-1}), wind speed at 2 m height (u_2, m s^{-1}), and rainfall (P, mm) (Figures 1 and 2).
- Soil data: θ_S, θ_{FC}, and θ_{WP} as well as the particle size distribution of the different soil layers defined in the rootzone domain (Table 1).
- Soil evaporation parameters: the evaporable layer depth (Z_e), with TEW and REW being then estimated using the textural and water holding characteristics of the Z_e [54,55].
- Initial conditions: initial (observed) values of the soil water content in both the root zone (% of TAW) and the evaporation layer (% of TEW) (Table 3).
- Crop data: the dates defining the different stages of vineyards during growing seasons (Table 2) and respective values of K_{cb}, p, h, the fraction of ground covered (f_c), M_L, and Z_r. Values of h and f_c were observed in the field and are presented in Table 4. Z_r was set to 1.0 m in both case studies. The M_L value is unique for all crop stages and was set in both case studies to 1.5 following Pereira et al. [56].
- Active ground cover: the density of the active ground cover (28–30% and 15–20% for the GC and CT plots in Italy, respectively, and 15–20% for the Portuguese plot); the initial values of active ground cover fraction ($f_{c\ gcover}$) and height (h_{gcover}), which

were then updated for each crop stage according to Table 4; and M_L, which was also assumed to be 1.5 following Fandiño et al. [58].

- Deep percolation: the deep percolation parameters a_D and b_D relative to the parametric equation of Liu et al. [76] were defined according to the soil texture data, and θ_S and θ_{FC} values in the soil profiles (Table 1).
- Runoff (more relevant for the Italian case studies): the CN value for each inter-row condition [77].
- Irrigation: the dates of irrigation events and irrigation depths, which were specified according to observations; and the fraction of the soil surface wetted by irrigation (f_w), which was measured in the field and set to 0.13.

Table 3. Initial soil water contents in the root zone (% of TAW) and evaporable soil layer (% of TEW) in each case study and growing season.

Growing Season	% of TAW			% of TEW		
	Monferrato (CT)	Monferrato (GC)	S. Correia	Monferrato (CT)	Monferrato (GC)	S. Correia
2016	67	34	-	67	34	-
2017	22	25	-	22	25	-
2018	45	33	30	45	33	30
2019	31	14	23	31	14	23
2020	-	-	17	-	-	17

Note: TAW, total available water; TEW, total evaporable water.

2.3.3. Model Calibration and Validation

The SIMDualKc model was calibrated by adjusting model parameters one at a time within reasonable ranges of values, using a trial-and-error procedure, until deviations between measured soil water contents and model predictions were minimized. For Monferrato plots, data from the 2019 growing season were used for calibration while for the Samora Correia plot the 2020 growing season data were used. The calibrated parameters were (i) the CN value (more relevant for the Italian plots); (ii) the K_{cb} and p values relative to the initial, mid-, and end-season; (iii) the depth of the evaporative soil layer (Z_e) and the TEW and REW values relative to soil evaporation; and (iv) the parameters relative to the deep percolation function, a_D and b_D.

The calibrated model parameters were validated using independent data sets corresponding to soil water contents measured in the Italian plots during the 2016–2018 growing seasons, and relative to the growing seasons 2018–2019 measured at Samora Correia. The resulting K_{cb} values are adjusted for climate. Following Pereira et al. [81], model performance was considered acceptable when the goodness-of-fit indicators relative to the validation were within a range of "20% variation" relative to calibration.

The goodness-of-fit indicators adopted for comparing SIMDualKc model simulations with soil water content observations were the coefficient of determination (R^2) of the ordinary least squares regression, the coefficient of regression (b_0) of the linear regression forced through the origin, the root mean square error (RMSE), the ratio of the RMSE to the standard deviation of observed data (NRMSE), the percent bias of estimation (PBIAS); and the modeling efficiency (EF), respectively, given as:

$$R^2 = \left\{ \frac{\sum_{i=1}^{n} (O_i - \overline{O})(P_i - \overline{P})}{\left[\sum_{i=1}^{n} (O_i - \overline{O})^2\right]^{0.5} \left[\sum_{i=1}^{n} (P_i - \overline{P})^2\right]^{0.5}} \right\}^2 \quad (27)$$

$$b_0 = \frac{\sum_{i=1}^{n} O_i P_i}{\sum_{i=1}^{n} O_i^2} \quad (28)$$

$$RMSE = \sqrt{\frac{\sum_{i=1}^{n} (O_i - P_i)^2}{n-1}} \quad (29)$$

$$\text{NRMSE} = \frac{\text{RMSE}}{\sqrt{\sum_{i=1}^{n}(O_i - P_i)^2}} \quad (30)$$

$$\text{PBIAS} = 100\,\frac{\sum_{i=1}^{n}(O_i - P_i)}{\sum_{i=1}^{n} O_i} \quad (31)$$

$$\text{EF} = 1 - \frac{\sum_{i=1}^{n}(O_i - P_i)^2}{\sum_{i=1}^{n}(O_i - \overline{O})^2} \quad (32)$$

where O_i and P_i are, respectively, the observed and model predicted values at time i, $\overline{O}$ and $\overline{P}$ are the respective mean values, and n is the number of observations. R^2 values close to 1 indicate that the model explains well the variance of observations. b_0 target is 1.0 with values $b_0 < 1.0$ indicating under-estimation of the predicted values and $b_0 < 1.0$ indicating over-estimation. RMSE and NRMSE close to zero indicate small estimation errors and good model predictions [82]. PBIAS values close to zero indicate that model simulations are accurate, while positive or negative values indicate under- or over-estimation bias, respectively. EF values close to one indicate that the residuals' variance is much smaller than the observed data variance, hence the model predictions are good. On the contrary, when EF is close to zero or negative, there is no gain in using the model [83].

Table 4. Fraction of ground cover f_c, crop height (h), fraction of the active ground cover ($f_{c\ gcover}$) and height of the active ground cover (h_{gcover}) in every plot and all growing seasons.

	Monferrato (CT)				Monferrato (GC)				Samora Correia			
	Vineyard		Inter-Row		Vineyard		Inter-Row		Vineyard		Inter-Row	
	f_c (-)	h (m)	$f_{c\ gcover}$ (-)	h_{gcover} (m)	f_c (-)	h (m)	$f_{c\ gcover}$ (-)	h_{gcover} (m)	f_c (-)	h (m)	$f_{c\ gcover}$ (-)	h_{gcover} (m)
2016												
Non-Growing	0.04	0.50	0. 20	0.20	0.05	0.50	0.82	0.18	-	-	-	-
Initiation	0.15	0.80	0.18	0.15	0.16	0.80	0.75	0.15	-	-	-	-
Mid-season	0.35	1.80	0.12	0.15	0.29	1.80	0.65	0.15	-	-	-	-
Late-season	0.27	1.20	0.18	0.15	0.23	1.20	0.77	0.15	-	-	-	-
Non-Growing	0.03	0.50	0.20	0.20	0.04	0.50	0.82	0.18	-	-	-	-
2017												
Non-Growing	0.04	0.50	0. 20	0.20	0.03	0.50	0.81	0.18	-	-	-	-
Initiation	0.15	0.80	0.17	0.15	0.14	0.80	0.70	0.15	-	-	-	-
Mid-season	0.37	1.80	0.12	0.15	0.25	1.70	0.61	0.15	-	-	-	-
Late-season	0.28	1.20	0.18	0.15	0.23	1.10	0.72	0.15	-	-	-	-
Non-Growing	0.03	0.50	0.20	0.20	0.04	0.50	0.81	0.18	-	-	-	-
2018												
Non-Growing	0.04	0.50	0.24	0.20	0.04	0.50	0.80	0.20	0.03	0.50	0.20	0.15
Initiation	0.16	0.80	0.18	0.15	0.15	0.80	0.71	0.18	0.13	0.80	0.18	0.15
Mid-season	0.37	1.90	0.13	0.15	0.29	1.80	0.59	0.17	0.36	1.70	0.07	0.10
Late-season	0.30	1.20	0.26	0.15	0.22	1.20	0.74	0.18	0.30	1.20	0.18	0.10
Non-Growing	0.03	0.50	0.26	0.20	0.04	0.50	0.80	0.20	0.04	0.50	0.20	0.15
2019												
Non-Growing	0.03	0.50	0.24	0.20	0.03	0.50	0.82	0.20	0.03	0.50	0.21	0.15
Initiation	0.16	0.80	0.19	0.15	0.15	0.80	0.72	0.18	0.13	0.80	0.19	0.15
Mid-season	0.39	1.90	0.13	0.15	0.29	1.80	0.60	0.18	0.35	1.70	0.08	0.10
Late-season	0.30	1.20	0.25	0.15	0.22	1.20	0.75	0.18	0.29	1.20	0.18	0.10
Non-Growing	0.03	0.50	0.25	0.20	0.04	0.50	0.82	0.20	0.03	0.50	0.20	0.15
2020												
Non-Growing	-	-	-	-	-	-	-	-	0.03	0.50	0.20	0.15
Initiation	-	-	-	-	-	-	-	-	0.17	0.80	0.18	0.15
Mid-season	-	-	-	-	-	-	-	-	0.36	1.70	0.08	0.10
Late-season	-	-	-	-	-	-	-	-	0.31	1.20	0.18	0.10
Non-Growing	-	-	-	-	-	-	-	-	0.03	0.50	0.20	0.15

3. Results and Discussion

3.1. Model Parametrization

Table 5 presents the model parameters calibrated for the Italian and Portuguese plots with data relative to 2019 and 2020, respectively. These parameters were then validated for 2016–2018 relative to the Monferrato vineyards and for 2017 and 2019 relative to the Samora Correia vineyard. The initial K_{cb} values were obtained from tabulated data provided by Pereira et al. [56]. For Monferrato vineyards in year 2019, the calibrated $K_{cb\ ini}$, $K_{cb\ mid}$, and $K_{cb\ end}$ values were, respectively, 0.20, 0.47, and 0.34 in the conventional tillage CT plot, and, respectively, 0.35, 0.47, and 0.40 in the ground cover GC plot. The active ground cover promoted an increase of the K_{cb} values in the initial and end season, while no differences were noticed in the mid-season. During this period, the inter-row grass was at the end of its annual cycle or already dried out. In the Samora Correia plot in year 2020, the calibrated $K_{cb\ ini}$, $K_{cb\ mid}$, and $K_{cb\ end}$ were 0.17, 0.47, and 0.39, with the same $K_{cb\ mid}$ as in Monferrato but with different $K_{cb\ ini}$ and $K_{cb\ end}$.

Table 5. Default and calibrated model parameters.

Parameters	Units	Default Values	Calibrated Values		
			Monferrato (CT)	Monferrato (GC)	Samora Correia
$K_{cb\ nongrowing}$	-	0.15	0.17	0.35	0.16
$K_{cb\ ini}$	-	0.15	0.20	0.35	0.17
$K_{cb\ mid}$	-	0.65	0.47	0.47	0.47
$K_{cb\ end}$	-	0.40	0.34	0.40	0.39
p_{ini}	-	0.45	0.45	0.35	0.40
p_{mid}	-	0.45	0.45	0.35	0.40
p_{end}	-	0.45	0.45	0.35	0.40
TEW	mm	34–15	30	31	15
REW	mm	9–7	9	11	7
Z_e	m	0.10	0.09	0.10	0.10
a_D	mm	-	360	370	152
b_D	-	−0.0173	−0.0150	−0.0170	−0.0173
CN	-	75–60	70	55	65

Note: CT, conventional tillage; GC, grass cover; K_{cb}, basal crop coefficient for the initial ($K_{cb\ ini}$), mid ($K_{cb\ mid}$), and end season ($K_{cb\ end}$); $K_{cb\ nongrowing}$, basal crop coefficient during the non-growing period; p, depletion fraction for no stress during the initial (p_{ini}), mid (p_{mid}), and end season (p_{end}); TEW, total evaporable water; REW, readily evaporable water; Z_e, depth of the soil evaporation layer; a_D and b_D, parameters of the deep percolation; CN, curve number.

Vineyards are heterogeneous, sparsely vegetated surfaces with complex canopies. For that, the K_{cb} values may vary among locations due to differences in varieties, age, training, irrigation, soil cover, and crop management as demonstrated in the recent literature review by Rallo et al. [53]. These authors updated the K_{cb} values proposed by Allen et al. [54] for vines and other perennial crops. Rallo et al. [53] provided standard K_{cb} for diverse training systems used in vineyards combined with the fraction of the ground cover (f_c) but not including active ground cover calculations. After reviewing literature relative to wine grapes, it was concluded that for the training system adopted in the all the studied vineyards (vertical shoot positioned trellis), the f_c values were reported to vary from 0.25 to 0.45, with $K_{cb\ mid}$ values ranging from 0.46 to 0.80 and $K_{cb\ end}$ values varying from 0.20 to 0.60. The K_{cb} values in both Monferrato and Samora Correia vineyards were therefore close to the lower limit of the range of variation of VSP trained vineyards reported by Rallo et al. [53]. The f_c values observed during the mid-season stage in the Monferrato CT plot (0.35−0.39) and in the GC plot (0.25–0.29), as well as in the Samora Correia plot (0.35−0.36) (Table 4) were closer to the lower f_c values reported by Rallo et al. [53], hence indicating that it is likely that the low $K_{cb\ mid}$ and $K_{cb\ end}$ values found in the current study are due to a small f_c observed in the vineyards under study, so in agreement with reported data of those authors.

The calibrated K_{cb} values listed in Table 5 correspond to the entire vineyard system (vine plants + active ground cover). Table 6 shows the partitions of those values based on formulations given in Section 2.3 and crop density data (f_c and h) given in Table 4 for each crop stage and growing season. Contribution of the active ground cover to the K_{cb} value was obviously higher in the Monferrato GC plot, but the K_{cb} values of vine plants were smaller, especially during the mid and end season stages since, as explained earlier, the cover crop may influence plant's vigor and reduce canopy leaf area and $T_{c\ act}$ rates of the grapevine in the earlier development stages which will then also affect the mid-season period. When looking at the partition of the K_{cb} values, it becomes also the higher $K_{cb\ mid}$ value at Samora Correia (Portugal), resulting from higher air temperature registered in that location.

Table 6. Basal crop coefficient (K_{cb}) partitioning between the main crop (vine) and active ground cover (gcover) during the nongrowing season and at the initial, mid, end season crop stages.

Parameters	Monferrato (CT)				Monferrato (GC)				Samora Correia
	2016	2017	2018	2019	2016	2017	2018	2019	2018–2020
$K_{cb\ gcover\ nongrowing}$	0.09	0.09	0.09	0.10	0.24	0.24	0.24	0.25	0.09
$K_{cb\ gcover\ ini}$	0.08	0.08	0.08	0.08	0.22	0.21	0.22	0.23	0.08
$K_{cb\ gcover\ mid}$	0.07	0.07	0.07	0.07	0.20	0.19	0.19	0.19	0.06
$K_{cb\ fcover\ end}$	0.08	0.08	0.09	0.09	0.22	0.21	0.23	0.22	0.07
$K_{cb\ vine\ nongrowing}$	0.09	0.09	0.10	0.10	0.24	0.24	0.24	0.24	0.09
$K_{cb\ vine\ ini}$	0.12	0.12	0.12	0.12	0.13	0.14	0.13	0.12	0.09
$K_{cb\ vine\ mid}$	0.40	0.40	0.40	0.40	0.27	0.28	0.28	0.28	0.41
$K_{cb\ vine\ end}$	0.26	0.26	0.25	0.25	0.18	0.19	0.17	0.18	0.32

The calibrated p_{ini}, p_{mid}, and p_{end} values (Table 5) matched those proposed by Allen et al. [54] in the Monferrato CT plot but were slightly smaller in the Monferrato GC and the Samora Correia vineyard. The GC plot showed the lowest p values, likely due to the competition for water by the active ground cover.

The Z_e, TEW, and REW, as well as the a_D and b_D calibrated values reflect the hydraulic properties of the soils of the studied vineyards, a loamy sand to sandy textured soil in Samora Correia vs. a clay to clay-loam soils at Monferrato. Lastly, the CN values were set to 70 and 55 in Monferrato CT and GC plots, respectively, with the inter-row grass cover clearly impacting the calibrated values of CN. Those values are lower than the CN reported by Gaudin et al. [84] from an analysis of three runoff events (CN = 82) for a bare soil. For olive orchards, Romero et al [85] reported CN = 89 for bare soil, CN = 82 and 70 for well-established cover crop strips with, respectively, 1 and 3 m wide, and CN = 88 and 87 for degraded cover crop strips with the same widths. For the Samora Correia vineyard, where runoff is less relevant due to the flat topography, the CN = 65 value is, as expected, smaller. Romero et al. [85] pointed out the need for improvements in the SCS-CN method aiming at a better understanding of the effects of soil management practices on runoff formation. As shown in Biddoccu et al. [7], different soil management implementations for the same treatment can lead to high variability in the soil roughness and coverage (namely the RUSLE C-factor), for the same region or among different countries.

3.2. Model Performance

Figure 3 compares the SIMDualKc simulated soil water content (SWC) values in the root zone with daily measured SWC of Monferrato CT and GC plots during the years 2016–2019. Likewise, Figure 4 shows the measured and simulated SWC in the root zone of the Samora Correia vineyard during 2018–2020. Both figures further provide the dates and depths of rainfall events and, for the Portuguese vineyard, also of the irrigation dates and depths. Both figures show that the measured values of SWC were kept between θ_{FC} and θ_p during most of the year. In Monferrato, this fact resulted from precipitation, with large or successive events occurring during the autumn and spring, leading to SWC values temporarily above θ_{FC}. However, in the summer dry season, the soil moisture dropped below θ_p for some extended periods when rainfall was lacking. The year 2017 had the lowest SWC values as annual precipitation was uncharacteristically low during that year, amounting to only 493 mm. Indeed, among the last 20 years, 2017 was the least rainy one, with only 56% of the mean annual precipitation, while summer was very dry, with less than 25 mm of precipitation from the beginning of June to the end of August (the average for this period in 20 years is about 100 mm).

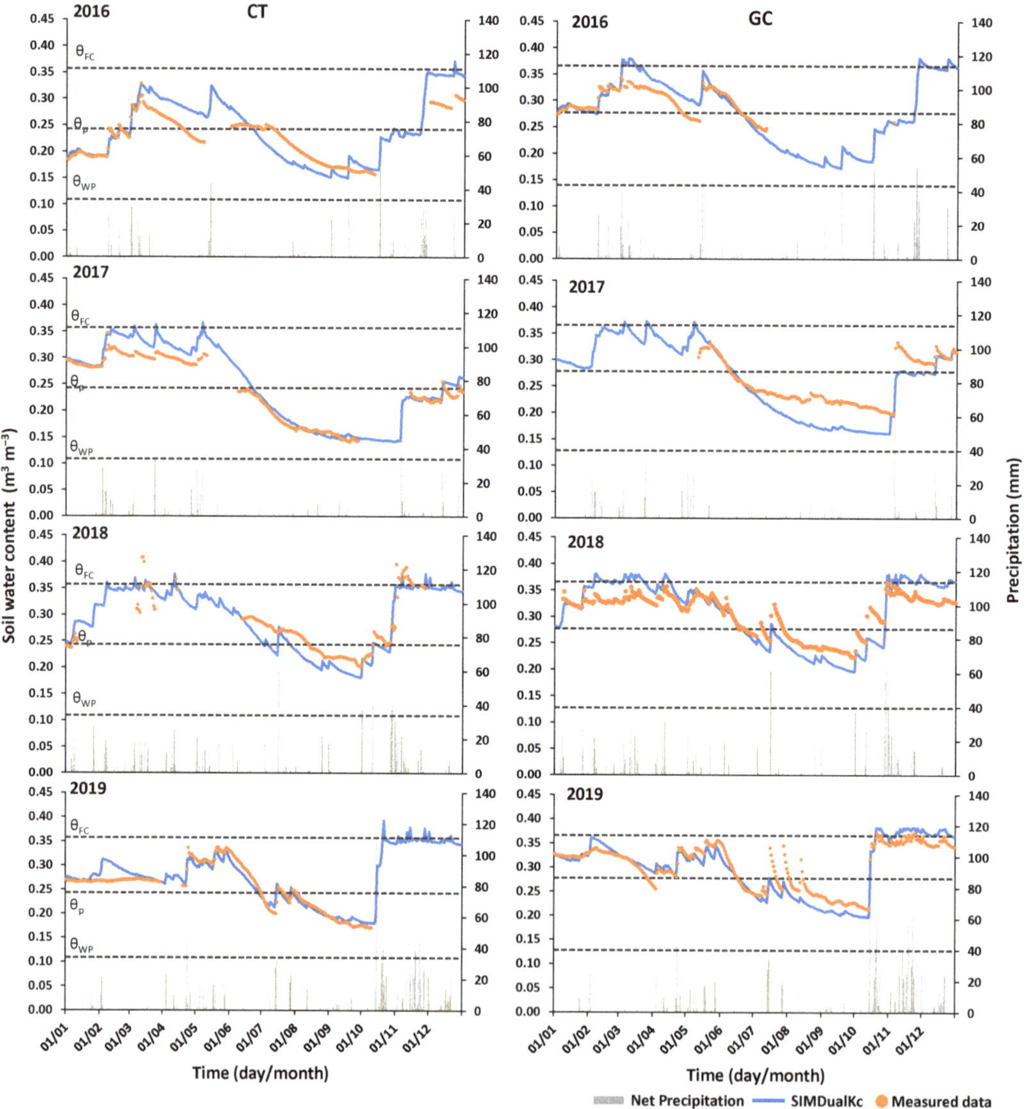

Figure 3. Measured and simulated soil water contents of the Monferrato conventional tillage (CT) and grass cover (GC) plots during the 2016–2019 growing seasons (θ_{FC}, θ_{WP}, and θ_p refer to soil water contents at field capacity, the wilting point, and at the depletion fraction for no stress, respectively).

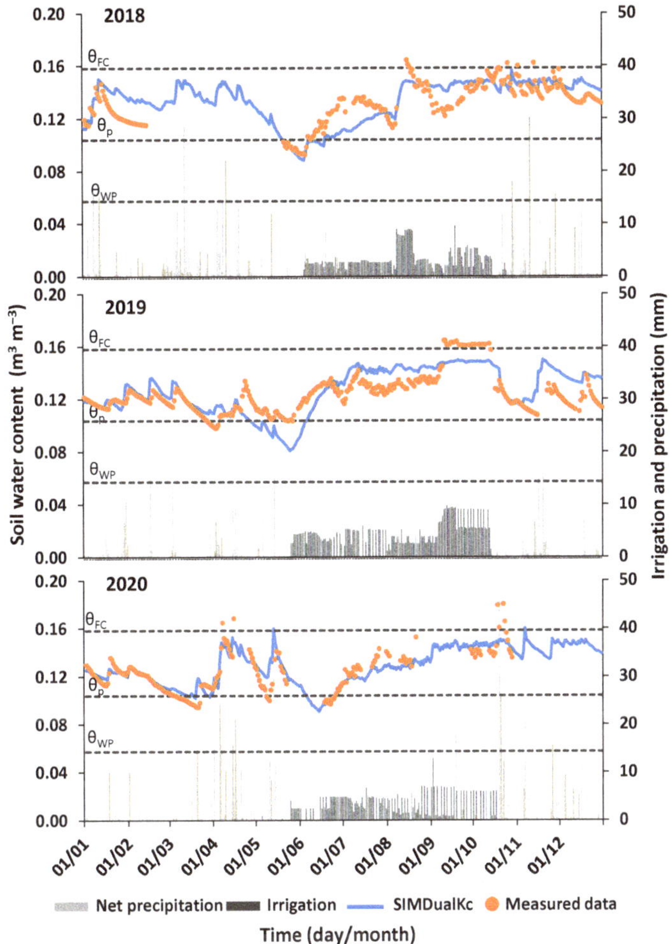

Figure 4. Measured and simulated soil water contents in the vineyard of Samora Correia during the 2018–2020 growing seasons (θ_{FC}, θ_{WP}, and θ_p refer to soil water contents at field capacity, the wilting point, and at the depletion fraction for no stress, respectively).

Simulated SWC were higher in GC plot relative to CT for all years, in agreement with the average of measured values. Likely, this would result from higher infiltration of rainfall water as already observed [5,11,12]. Nevertheless, Novara et al. [14], relative to a review of studies about the effects of a cover crop on SWC, reported that such effects often result in a decrease of soil moisture at a soil depth from 0–0.5 to 0–1.0 m, i.e., below the roots of the grass crop, between bloom and veraison, which can threaten yields in semiarid conditions.

In Samora Correia (Figure 4), drip irrigation was commonly applied at small depths (up to 12 mm) aimed at maintaining SWC above θ_p during the summer dry season. This type of irrigation management is uncommon in Portuguese vineyards of southern Portugal, where deficit irrigation is usually practiced during the summer dry period to better control shoot vigor and ripening, as well as the quality of fruit attributes [86,87]. Thus, the case observed at Samora Correia is rare and likely results from the very low water holding capacity of the soil, which calls the growers to avoid SWC < θ_p, at least during somewhat large periods.

The statistical indicators used to evaluate the agreement between measured and simulated SWC values are presented in Table 7 for both case studies. The SIMDualKc model performed well when simulating SWC in the Monferrato plots for calibration and for validation. The regression coefficient b_0 varied from 0.95 to 1.06 in the CT plot and from 0.91 to 1.03 in the GC plot, i.e., close to the 1.0 target indicating that simulated values are close to the observed ones. The value of R^2 was relatively high in both plots (0.76 to 0.90), showing that the model could explain most of the variability of the observed data. The errors of the estimates were small, resulting in a RMSE value ranging from 0.01 to 0.03 m^3 m^{-3} and a NRMSE value ranging between 0.32 and 0.81. In agreement with b_0, the PBIAS were small, not expressing a particular over- or under-estimation trend in simulating the measured data. The EF values were higher for CT than for GC (respectively, 0.53 to 0.90 and 0.34 to 0.80), indicating that the variance of the residuals was smaller than the measured data variance particularly for the calibration set. The goodness-of-fit indicators show better results for the CT plot relative to the GC case. This indicate that a supplemental effort for characterizing the active ground cover of the GC plots is required. Nevertheless, results obtained are reasonable enough for the pretended analysis.

Table 7. Goodness-of-fit indicators for the adjustment between measured and simulated values in all case studies.

Site	Year	Treatment	b_0 (-)	R^2 (-)	RMSE (m^3 m^{-3})	NRMSE (-)	PBIAS (%)	EF (-)
Monferrato	2016	CT	1.04	0.84	0.03	0.69	−3.30	0.53
		GC	1.03	0.76	0.02	0.78	−2.55	0.39
	2017	CT	1.06	0.97	0.02	0.34	−5.76	0.88
		GC	0.91	0.90	0.03	0.81	9.95	0.34
	2018	CT	0.95	0.86	0.03	0.51	5.60	0.74
		GC	1.02	0.89	0.03	0.74	−1.12	0.45
	2019	CT	1.01	0.90	0.01	0.32	−1.03	0.90
		GC	0.99	0.90	0.02	0.45	1.04	0.80
S. Correia	2018	-	1.01	0.59	0.01	0.74	−1.53	0.45
	2019	-	1.02	0.57	0.01	0.86	−1.84	0.30
	2020	-	1.00	0.75	0.01	0.50	−0.50	0.75

Note: b_0, regression coefficient; R^2, coefficient of determination; RMSE, root mean square error; NRMSE, ratio of the RMSE to the standard deviation of observed data; PBIAS, percent bias; NSE, model efficiency.

The performance of the SIMDualKc model during the calibration period (2018) for the Samora Correia vineyard was comparable to results in the Italian case studies, with the PBIAS values, close to zero, also indicating accuracy in predicting the measured SWC. The regression coefficient b_0 varied 1.00 to 1.02, thus leading to assume that there is no noticeable trends in the simulation, which is confirmed by the data. However, both the R^2 and EF values are not high, indicating that improvements in describing vineyards crop data are required, particularly relative to the ground cover that changes from active to dry residues. The errors of the estimate RMSE were small, but the NRMSE are high because soil water depletion by evapotranspiration are low.

Issues related to the quality of the measured dataset and the empirical one-dimensional modeling approach used here to describe the soil water balance in a three-dimensional drip irrigation system may also help explaining why the goodness-of-fit indicators are not as high as in former studies, e.g., Paço et al. [88]. Nevertheless, the reported goodness-of-fit indicators are within the range of values reported in the literature for soil water content simulations of field crops [89,90], vegetable crops [91,92], and perennial crops [58–61] using the same model.

3.3. Soil Water Balance in the Hillslope Vineyards of Monferrato, Northern Italy

In Figure 5 are presented the daily values of the potential non-stressed basal crop coefficients (K_{cb}), the actual basal crop coefficients ($K_{cb\ act}$), the soil evaporation coefficients

(K_e), and the actual crop coefficients ($K_{c\ act} = K_{cb\ act} + K_e$) computed by SIMDualKc. The rainfall events are also depicted in Figure 5. Table 8 presents then the components of the soil water balance in the Monferrato plots from 2016 to 2019.

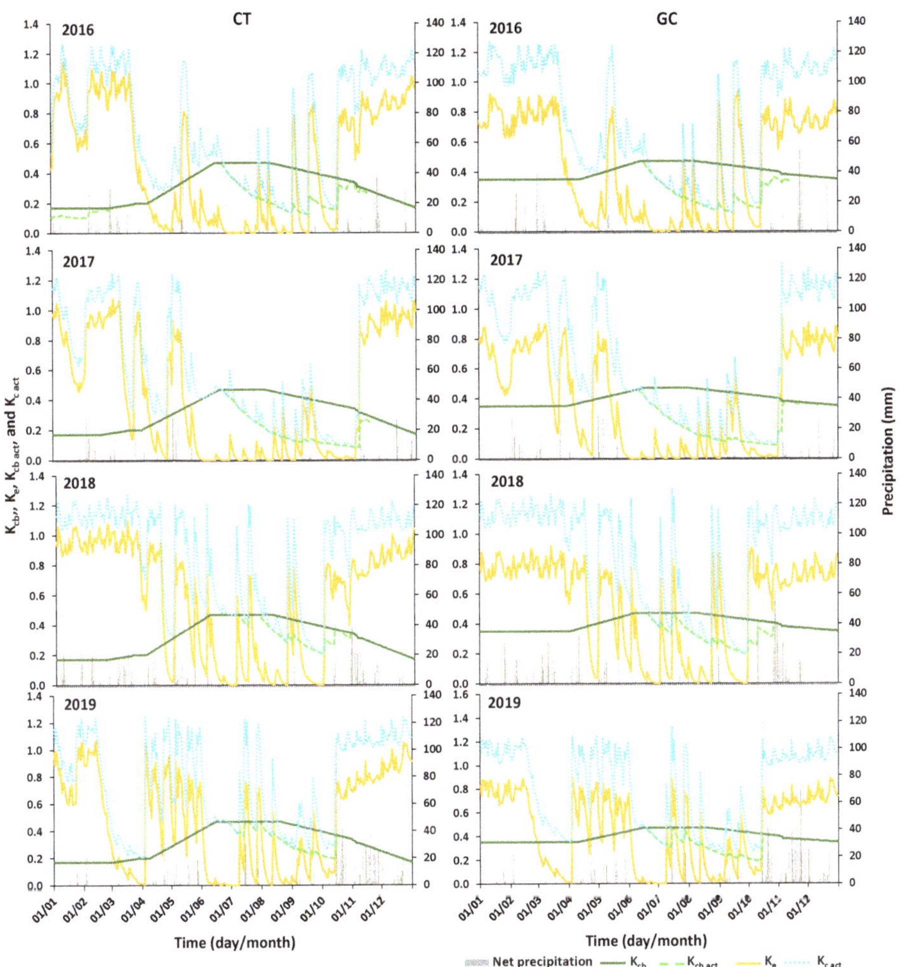

Figure 5. Seasonal evolution of the potential non-stressed basal crop coefficients (K_{cb}), the actual basal crop coefficients ($K_{cb\ act}$), the soil evaporation coefficients (K_e), and the actual crop coefficients ($K_{c\ act} = K_{cb\ act} + K_e$) computed by SIMDualKc for the Monferrato conventional tillage (CT) and grass cover (GC) plots during the 2016–2019 growing seasons.

The $K_{cb\ act}$ matched the K_{cb} values during most of crop season duration except for the dry summers (Figure 5). The resulting seasonal $T_{c\ act}$ values were estimated to vary from 262 to 313 mm in the CT plot and 298 to 339 mm in the GC plot (Table 8). $T_{c\ act}$ values were up to 14% higher in the GC plot because comprising the transpiration by the grass cover in the inter-row. The $T_{c\ act}$ reduction due to water stress started during mid-season and extended to the late-season due to low soil water availability during these dry months. The $T_{c\ act}/T_c$ ratios are always smaller during the late season (Table 9) and the lowest value was observed in the late season of 2017 (0.30) due to the exceptional lack of rainfall in this year. Model results even suggested the need for irrigation in such dry conditions despite northern Italy is a traditionally rainfed wine-growing region. The impact of water stress on transpiration rates as given by the $T_{c\,act}/T_c$ ratio was quite similar in both CT and GC plots with a general ratio value slightly smaller in the GC plot over three out of the four seasons. Despite the computed difference is small, it is noticeable in both the mid- and late-seasons, which explains the reluctance of winegrowers in using active ground cover in the inter-row due to concerns over soil water competition between the vines and the grass cover.

Table 8. Components of the soil water balance in the Monferrato plots.

Year	Plot	I (mm)	P (mm)	ΔSW (mm)	T_c (mm)	$T_{c\,act}$ (mm)	E_s (mm)	DP (mm)	RO (mm)
2016	CT	0	779	−156	372	262	234	27	90
	GC	0	779	−85	417	298	234	118	43
2017	CT	0	493	35	395	275	206	42	8
	GC	0	493	−17	445	300	203	8	2
2018	CT	0	1190	−100	347	313	307	310	178
	GC	0	1189	−86	386	326	306	315	92
2019	CT	0	1476	−68	353	313	274	467	355
	GC	0	1476	−40	403	339	284	600	198

Note: I, irrigation; P, precipitation; ΔSW, soil water storage variation; T_c, potential transpiration; $T_{c\,act}$, actual transpiration; E_s, evaporation; DP, deep percolation; RO, runoff.

Table 9. Evapotranspiration partitioning in the Monferrato vineyard plots during the different crop stages.

Period	Parameter	Units	2016		2017		2018		2019	
			CT	GC	CT	GC	CT	GC	CT	GC
Non-Growing	$T_{c\,act}$	(mm)	9	22	13	25	11	21	20	41
	$T_{c\,act\,gcover}$	(mm)	5	17	7	17	6	15	12	29
	$T_{c\,act}/T_c$	(-)	0.84	1.00	1.00	1.00	1.00	1.00	1.00	1.00
	E_s	(mm)	58	50	59	52	61	50	48	47
Initial stage	$T_{c\,act}$	(mm)	15	26	34	52	6	10	10	17
	$T_{c\,act\,gcover}$	(mm)	6	14	11	29	2	6	4	11
	$T_{c\,act}/T_c$	(-)	1.00	1.00	1.00	1.00	1.00	1.00	1.00	1.00
	E_s	(mm)	37	37	47	48	22	21	11	10
Rapid growth	$T_{c\,act}$	(mm)	85	99	92	99	72	86	79	94
	$T_{c\,act\,gcover}$	(mm)	18	30	17	44	16	43	17	47
	$T_{c\,act}/T_c$	(-)	1.00	1.00	1.00	0.99	1.00	1.00	1.00	1.00
	E_s	(mm)	41	45	39	41	94	95	107	111
Mid-season	$T_{c\,act}$	(mm)	98	93	85	72	149	135	149	132
	$T_{c\,act\,gcover}$	(mm)	15	16	13	29	22	55	22	53
	$T_{c\,act}/T_c$	(-)	0.69	0.66	0.72	0.61	0.96	0.87	0.91	0.81
	E_s	(mm)	19	21	7	8	47	54	51	61
Late-season	$T_{c\,act}$	(mm)	47	47	44	41	68	62	47	44
	$T_{c\,act\,gcover}$	(mm)	8	11	7	18	11	29	8	20
	$T_{c\,act}/T_c$	(-)	0.42	0.40	0.34	0.30	0.70	0.61	0.64	0.58
	E_s	(mm)	52	58	23	27	58	64	30	32
Non-Growing	$T_{c\,act}$	(mm)	7	11	7	12	7	11	7	11
	$T_{c\,act\,gcover}$	(mm)	1	8	2	9	2	7	2	8
	$T_{c\,act}/T_c$	(-)	0.97	0.95	0.80	0.93	1.00	1.00	1.00	1.00
	E_s	(mm)	26	23	32	28	24	21	26	23
Total	$T_{c\,act}$	(mm)	262	298	275	300	313	326	313	339
	$T_{c\,act\,gcover}$	(mm)	53	96	57	146	59	155	65	168
	$T_{c\,act}/T_c$	(-)	0.70	0.71	0.70	0.68	0.90	0.85	0.88	0.84
	E_s	(mm)	234	234	206	203	307	306	274	284

Note: T_c, potential transpiration; $T_{c\,act}$, actual transpiration; $T_{c\,act\,gcover}$, actual transpiration of the active ground cover; E_s, evaporation.

Table 9 further shows the $T_{c\,act}$ values relative to the active ground cover ($T_{c\,act\,gcover}$). In the GC plot, the weight of $T_{c\,act\,gcover}$ relative to $T_{c\,act}$ of the entire vineyard system was always high, amounting to a maximum of 63.6% to 77.3% of the $T_{c\ act}$ during the non-growing periods and to a minimum of 17.2% to 40.7% in the mid-season. In the CT plot, the ratios $T_{c\ act\ gcover}/T_{c\ act}$ were naturally always smaller than those reported for the GC plot. It should be noticed that the f_c values for the vineyard in the GC plot were always slightly lower than those in the CT plot (Table 4), which further exacerbates the importance of the cover crop to the soil water balance and preservation of soil and water resources in these hillslope areas.

The K_e values reached their maximum with the increase of soil moisture in the soil surface layer after rainfall events, rapidly decreasing in the following days as the soil dried out (Figure 5). For this reason, the K_e values have an enormous variability throughout the time. E_s values were relatively similar in the GC and CT plots, with the small differences resulting from the different Z_e set during calibration. The number of rainfall events influencing soil evaporation in each growing season is highly varied. E_s values in the CT plot varied from 206 to 307 mm while in the GC plot they varied from 203 to 306 mm. Soil evaporation occurred mostly during the non-growing periods and early crop stages (Table 9) and was slightly larger in the CT plot because a fraction of the potential E_s was used for ground cover transpiration.

In the CT plot, the seasonal simulated surface runoff ranged from 8 mm in 2017 to 355 mm in 2019. These values contrast with those estimated for the GC plot, which varied from 2 to 198 mm, i.e., 44% to 75% lower than in the CT plot. As shown in Figure 6, which compares the simulated and measured seasonal surface runoff in both plots during 2016–2019, the model provided a narrower range of runoff values when compared to field measurements. However, differences between CT and GC were well noticed. Computing runoff using a single CN value led the model to underestimate measured values during rainy years and overestimate them during drier seasons. Similarly, Celette et al. [48], when modelling the water balance in vineyards, observed that the use of a constant CN value for the surface runoff could not consider changes in the soil surface and their effects on runoff and water infiltration. Nevertheless, the SIMDualKc was able to catch the seasonal trends as well as the effect of inter-row management on runoff since measured values in the GC plot were also 50% to 78% lower than in the CT plot, thus very similar to the trend in model predictions. The model results are also in line with results of previous studies about cover crop and water conservation in olive orchards and vineyards in the Mediterranean area as reviewed by Novara et al. [14]. These authors showed that, on average, the annual runoff coefficient was reduced by 27% with the application of cover crop management. However, its effect on water conservation was higher in rainier regions (rainfall higher than 700 mm), whereas in drier regions it was not relevant. In sloping vineyards, the grass cover results not only in a reduction of runoff, but it contributes to decreasing significantly soil losses, up to 74% during very intense and erosive events [10,93]. On the one hand, improving the description of the soil coverage, either with the CT or the GC conditions, should improve the input data to the model and, therefore, its simulation capability; on the other hand, it is likely that simulations capabilities of the model may be enhanced. Nevertheless, current results are satisfactory and support the interpretation of the differences in behavior of vineyards, namely aspects that relate with the conservation of soil and water resources.

Lastly, the estimated percolation summed 27 to 467 mm in the CT plot and 8 to 600 mm in the GC plot. Except for 2017, which results were uncommon due to the very low precipitation, percolation was always higher 1.6% to 337% in the GC plot, with the grassed inter-row promoting higher infiltration rates, less runoff and soil erosion, thus higher groundwater recharge. Through direct measurements in vineyards of Montpellier, also with Mediterranean climate, Gaudin et al. [84] also observed that the presence of a permanent cover crop had a positive effect on water infiltration, favoring the refilling of the soil water profile. In any case, they observed that the soil water refilling at budbreak time was subjected to the rainfall pattern during the fall season.

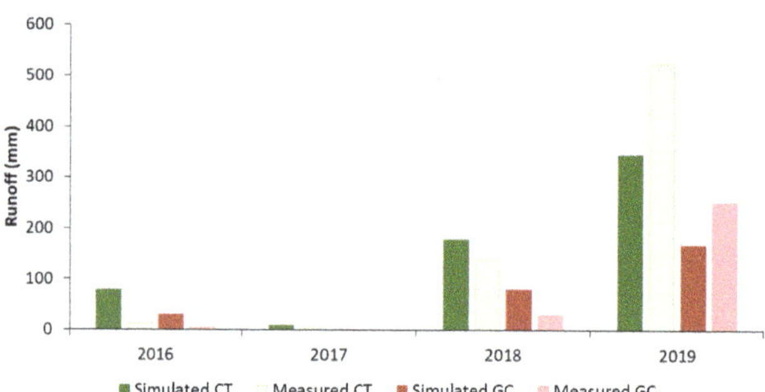

Figure 6. Simulated and measured seasonal runoff in the Monferrato plots with conventional tillage (CT) and grass cover (GC).

3.4. Soil Water Balance in the Irrigated Vineyard of Samora Correia, Southern Portugal

Figure 7 presents the daily values of the K_{cb}, $K_{cb\ act}$, K_e, and $K_{c\ act}$ computed with the SIMDualKc for the years of 2018–2020. Rainfall and irrigation events are also depicted. The $K_{cb\ act}$ matched the K_{cb} values throughout the growing seasons, meaning that the $T_{c\ act}$ was always close to its potential values during the different stages of crop development as shown by the $T_{c\ act}/T_c$ ratio always close to 1.0 in Table 10. The exception was only a small period during the rapid growth stage of the 2018 growing season where mild water stress was noticed. Irrigation was fundamental for such results, which otherwise would lead to a reduction of the $T_{c\ act}$ values due to the high atmospheric demand observed particularly during the mid- and late season stages. The cover crop had only a small impact on $T_{c\ act}$ values, with $T_{c\ act\ gcover}$ being only expressive during the non-growing period (Table 11).

Like in the Italian case study, the K_e values reached their maximum after rainfall events, to then drop with the depletion of the evaporation soil layer. During the irrigation season, K_e values were maintained at a much lower range, close to 0.2 (Figure 7), resulting from the low f_w value (0.13) describing evaporation from a small wetting bulb around emitters as opposite to rainfall events where the entire soil surface is wetted, and soil evaporation is maximized. The seasonal E_s values then ranged from 235 to 281 mm (Table 10), corresponding to 40% to 45% of the ET_c, and were distributed in relatively similar proportions throughout the different crop stages (Table 11).

The computed percolation was high, with values varying from 265 to 342 mm in the considered years (Table 10) which relates with the high infiltration and hydraulic conductivity of the sandy soil. In 2020, 23% (65 mm) of the seasonal percolation resulted from irrigation excess, corresponding to 14% of the water applied. In contrast, in 2019, seasonal percolation resulted mostly from irrigation surplus (92%; 243 mm), mainly during the late season, which corresponded to 39% of the total water applied through irrigation. These findings evidence the need for improving irrigation management in the considered field. Therefore, an optimization of irrigation schedules was considered by setting irrigation depths to 5 mm and the management allowed depletion (MAD) equal to the depletion fraction for no stress (p). This application revealed a substantial reduction of percolation losses during the monitored seasons without affecting the remaining components of the soil water balance (Table 10). This is visible in Figure 8, which compares the daily values of the different components of the soil water balance monitored in the field with those computed by the model when optimizing the irrigation schedules. Nonetheless, irrigation water management in vineyards is relatively more complex than the analysis shown herein since impacts on fruit quality are not considered. Hence, in future studies, it would be of interest to relate the SIMDualKc simulated irrigation schedules with both grapes yield and main characteristics of fruit quality for a better understanding of the impacts of irrigation management on crop development and yields.

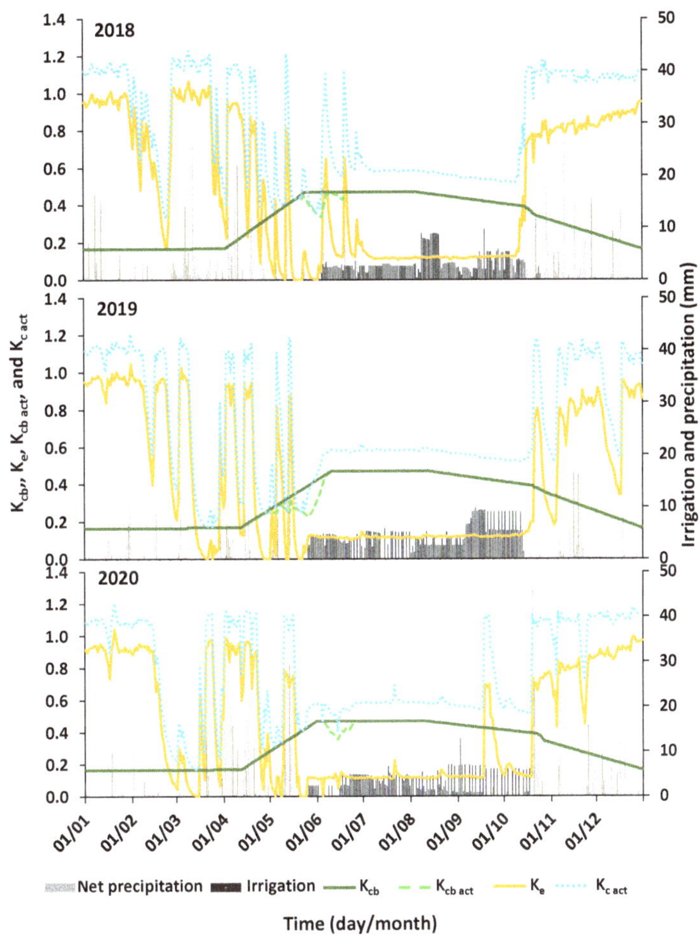

Figure 7. Seasonal evolution of the potential non-stressed basal crop coefficients (K_{cb}), the actual basal crop coefficients ($K_{cb\ act}$), the soil evaporation coefficients (K_e), and the actual crop coefficients ($K_{c\ act} = K_{cb\ act} + K_e$) computed by SIMDualKc for the Samora Correia vineyard during the 2018–2020.

Table 10. Components of the soil water balance in the Samora Correia vineyard.

Schedule	Year	I (mm)	P (mm)	ΔSW (mm)	T_c (mm)	$T_{c\ act}$ (mm)	E_s (mm)	DP (mm)	RO (mm)
Farmer	2018	454	499	−28	329	324	274	342	13
	2019	625	240	−17	350	337	235	265	1
	2020	465	506	−13	404	399	281	283	7
Model	2018	295	499	−19	329	327	278	169	13
	2019	375	240	−23	350	347	239	9	1
	2020	355	506	−14	404	399	274	170	7

Note: I, irrigation; P, precipitation; ΔSW, soil water storage variation; T_c, potential transpiration; $T_{c\ act}$, actual transpiration; E_s, evaporation; DP, deep percolation; RO, runoff.

Table 11. Evapotranspiration partitioning in the Samora Correia vineyard during the different crop stages.

Period	Parameter	Units	2018	2019	2020
Non-Growing	$T_{c\ act}$	(mm)	12	12	20
	$T_{c\ act\ gcover}$	(mm)	6	6	10
	$T_{c\ act}/T_c$	(-)	1.00	1.00	1.00
	E_s	(mm)	58	60	59
Initial stage	$T_{c\ act}$	(mm)	7	16	7
	$T_{c\ act\ gcover}$	(mm)	4	8	3
	$T_{c\ act}/T_c$	(-)	1.00	1.00	1.00
	E_s	(mm)	33	29	29
Vegetation growth	$T_{c\ act}$	(mm)	53	66	69
	$T_{c\ act\ gcover}$	(mm)	11	14	14
	$T_{c\ act}/T_c$	(-)	1.00	0.83	1.00
	E_s	(mm)	53	47	59
Mid-season	$T_{c\ act}$	(mm)	122	120	173
	$T_{c\ act\ gcover}$	(mm)	16	15	22
	$T_{c\ act}/T_c$	(-)	0.96	1.00	0.97
	E_s	(mm)	42	30	45
Late-season	$T_{c\ act}$	(mm)	110	106	115
	$T_{c\ act\ gcover}$	(mm)	21	20	21
	$T_{c\ act}/T_c$	(-)	1.00	1.00	1.00
	E_s	(mm)	33	30	45
Non-Growing	$T_{c\ act}$	(mm)	19	17	15
	$T_{c\ act\ gcover}$	(mm)	6	5	4
	$T_{c\ act}/T_c$	(-)	1.00	1.00	1.00
	E_s	(mm)	55	40	44
Total	$T_{c\ act}$	(mm)	324	337	399
	$T_{c\ act\ gcover}$	(mm)	63	68	76
	$T_{c\ act}/T_c$	(-)	0.99	0.96	0.99
	E_s	(mm)	274	235	281

Note: T_c, potential transpiration; $T_{c\ act}$, actual transpiration; $T_{c\ act\ gcover}$, actual transpiration of the active ground cover; E_s, evaporation from the soil.

4. Conclusions

This paper presents two study cases, one at Monferrato, northern Italy, analyzing alternative soil management issues, conventional tillage (CT) and grass cover (GC) as active ground cover, the other at Samora Correia, southern Portugal, drip irrigated. The vineyards are cultivated in different climates, humid at Monferrato, where soils have a large water holding capacity, and dry subhumid at Samora Correia, where the soil is sandy and therefore has a small water holding capacity. The SIMDualKc model was successful used for both cases to simulate water use (crop evapotranspiration) and the soil water balance. The model was calibrated and validated using, respectively, 4 and 3 years of soil moisture measurements, thus making it able to simulate soil and water management conditions observed in northern Italy and southern Portugal case studies.

The SIMDualKc model, adopting the dual Kc approach, was able to partitioning crop evapotranspiration into transpiration and evaporation from the soil. Moreover, it was able to compute distinctively the transpiration from the vine plants and the active ground cover. Results are satisfactory but it was observed that a better description of both the active ground cover and the mulch residues resulting from drying the former in summer is required, as well as improving related model simulation.

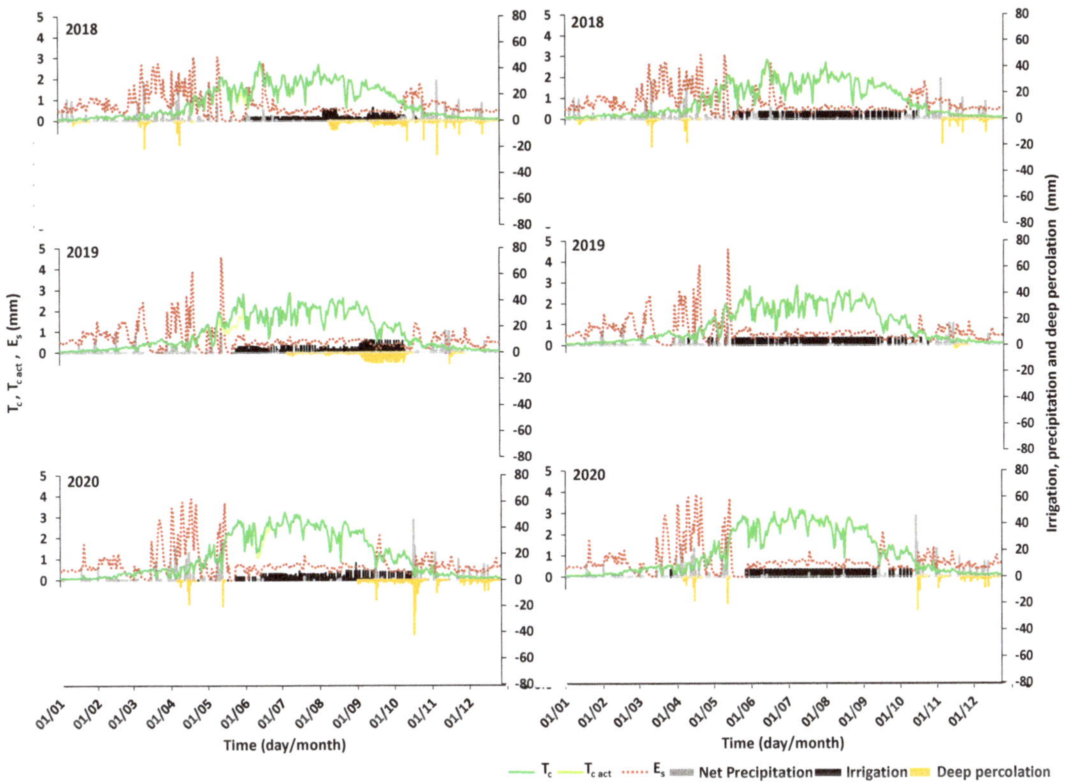

Figure 8. Daily evolution of the potential crop transpiration (T_c), actual crop transpiration ($T_{c\ act}$), soil evaporation (E_s), and deep percolation rates at Samora Correia vineyard as monitored in the field (left) and after optimization of the irrigation schedules (right) during 2018–2020.

The K_{cb} values estimated over the different crop stages in the plot having the inter-row grass cover (GC) were higher than in the tilled CT plot, indicating higher transpiration of both the vines and the grass cover. The use of grass cover in the inter-row did not affect the ratio between the actual and potential root water uptake rates, i.e., the cover crop did not cause additional water stress to the grapevine except in conditions of great weather dryness of a peculiar year. On the contrary, the cover crop promoted infiltration of surface water and groundwater recharge, thus decreasing the risk of soil erosion and land degradation in addition to increase soil water availability, which confirmed previous studies reported in literature. Adopting the SIMDualKc model may support further analysis of problems and respective issues for improvement. Long term economic gains for adopting active ground cover are important relative to resource conservation, water and soil, due to controlling runoff and erosion with small increases in water stress. Nevertheless, financial gains in the short term are likely small, which limits active ground cover adoption by growers. Thus, it is adequate to search innovative approaches to adopt an active ground cover in the inter-rows, including searching the most adequate grass to be used, as well as with better consideration of the impacts of field slopes, always focusing both the soil and water conservation and the economic and financial return of farming.

Relative to the drip irrigation of sandy soils, the computed components of the soil water balance evidenced the need for improving irrigation management, mainly irrigation schedules, and the drip irrigation design, despite not dealt in the current study but known as limiting the performance of the irrigation system. In fact, crop transpiration needs to be increased relative to the total water use, which refers to increase the wetted bulbs to favor roots development and water extraction from the soil in combination with controlling percolation losses, so also favoring resource conservation. Using SIMDualKc as a decision support system should be appropriate.

Author Contributions: Conceptualization, M.B. and T.B.R.; methodology, T.B.R.; software, H.D. and L.S.P.; validation, T.B.R., H.D. and M.B.; formal analysis, T.B.R. and E.C.; investigation, T.B.R., L.S., G.C. and G.B.; resources, D.R., M.B., E.C. and T.B.R.; data curation, G.C.; writing—original draft preparation, T.B.R., H.D. and M.B.; writing—review and editing, D.R., G.B. and L.S.; visualization, G.C. and G.B.; supervision, M.B. and L.S.; project administration, M.B. and T.B.R.; funding acquisition, D.R., M.B., E.C. and T.B.R. All authors have read and agreed to the published version of the manuscript.

Funding: This research was carried out within the framework of the WATER4EVER Project (Water-JPI/0010/2016), which is funded under the WaterWorks 2015 Call, with the contribution of EU, FCT, and MIUR (Ministero per l'Istruzione, l'Università e la Ricerca), contract agreement WaterJPI/0010/2016. T.B. Ramos and H. Darouich were supported by contracts CEECIND/01152/2017 and CEECIND/01153/2017, respectively, from the Portuguese Fundação para a Ciência e a Tecnologia (FCT).

Data Availability Statement: Data available in a publicly accessible repository that does not issue DOIs. Publicly available datasets were analyzed in this study. This data can be found here: https://ismn.geo.tuwien.ac.at/en/ (last accessed on 30 March 2020).

Acknowledgments: Authors thank the support of Paula Paredes in using the model.

Conflicts of Interest: The authors declare no conflict of interest.

References

1. Corti, G.; Cavallo, E.; Cocco, S.; Biddoccu, M.; Brecciaroli, G.; Agnelli, A. Evaluation of Erosion Intensity and Some of Its Consequences in Vineyards from Two Hilly Environments Under a Mediterranean Type of Climate, Italy. In *Soil Erosion Issues in Agriculture*; Godone, D., Stanchi, S., Eds.; InTechOpen: London, UK, 2011; pp. 113–160.
2. Aguilera, E.; Lassaletta, L.; Gattinger, A.; Gimeno, B.S. Managing soil carbon for climate change mitigation and adaptation in Mediterranean cropping systems: A meta-analysis. *Agr. Ecosyst. Environ.* **2013**, *168*, 25–36. [CrossRef]
3. Foronda-Robles, C. The territorial redefinition of the Vineyard Landscape in the sherry wine region (Spain). *Misc. Geogr.* **2018**, *22*, 95–101. [CrossRef]
4. UNESCO. Vineyard Landscape of Piedmont: Langhe-Roero and Monferrato. 2021. Available online: http://whc.unesco.org/en/list/1390 (accessed on 17 December 2021).
5. Salomé, C.; Coll, P.; Lardo, E.; Metay, A.; Villenave, C.; Marsden, C.; Blanchart, E.; Hinsinger, P.; Le Cadre, E. The soil quality concept as a framework to assess management practices in vulnerable agroecosystems: A case study in Mediterranean vineyards. *Ecol. Indic.* **2016**, *61*, 456–465. [CrossRef]
6. Biddoccu, M.; Opsi, F.; Cavallo, E. Relationship between runoff and soil losses with rainfall characteristics and long-term soil management practices in a hilly vineyard (Piedmont, NW Italy). *Soil Sci. Plant Nutr.* **2014**, *60*, 92–99. [CrossRef]
7. Biddoccu, M.; Guzman, G.; Capello, G.; Thielke, T.; Strauss, P.; Winter, S.; Zaller, J.G.; Nicolai, A.; Cluzeau, D.; Popescu, D.; et al. Evaluation of soil erosion risk and identification of soil cover and management factor (C) for RUSLE in European vineyards with different soil management. *Int. Soil Water Conserv. Res.* **2020**, *8*, 337–353. [CrossRef]
8. Gómez, J.A.; Llewellyn, C.; Basch, G.; Sutton, P.B.; Dyson, J.S.; Jones, C.A. The effects of cover crops and conventional tillage on soil and runoff loss in vineyards and olive groves in several Mediterranean countries. *Soil Use Manage.* **2011**, *27*, 502–514. [CrossRef]
9. Prosdocimi, M.; Cerdà, A.; Tarolli, P. Soil water erosion on Mediterranean vineyards: A review. *Catena* **2016**, *141*, 1–21. [CrossRef]
10. Capello, G.; Biddoccu, M.; Cavallo, E. Permanent cover for soil and water conservation in mechanized vineyards: A study case in Piedmont, NW Italy. *Ital. J. Agron.* **2020**, *15*, 323–331. [CrossRef]
11. Ruiz-Colmenero, M.; Bienes, R.; Marques, M.J. Soil and water conservation dilemmas associated with the use of green cover in steep vineyards. *Soil Till. Res.* **2011**, *117*, 211–223. [CrossRef]
12. Napoli, M.; Dalla Marta, A.; Zanchi, C.A.; Orlandini, S. Assessment of soil and nutrient losses by runoff under different soil management practices in an Italian hilly vineyard. *Soil Till. Res.* **2017**, *168*, 71–80. [CrossRef]
13. Medrano, H.; Tomás, M.; Martorell, S.; Escalona, J.-M.; Pou, A.; Fuentes, S.; Flexas, J.; Bota, J. Improving water use efficiency of vineyards in semi-arid regions. A review. *Agron. Sustain. Dev.* **2015**, *35*, 499–517. [CrossRef]
14. Novara, A.; Cerda, A.; Barone, E.; Gristina, L. Cover crop management and water conservation in vineyard and olive orchards. *Soil Till. Res.* **2021**, *208*, 104896. [CrossRef]

15. Pessina, D.; Galli, L.E.; Santoro, S.; Facchinetti, D. Sustainability of Machinery Traffic in Vineyard. *Sustainability* **2021**, *13*, 2475. [CrossRef]
16. Bagagiolo, G.; Biddoccu, M.; Rabino, D.; Cavallo, E. Effects of rows arrangement, soil management, and rainfall characteristics on water and soil losses in Italian sloping vineyards. *Environ. Res.* **2018**, *166*, 690–704. [CrossRef]
17. Rodrigo-Comino, J.; Senciales, J.M.; Ramos, M.A.; Martínez-Casasnovas, J.A.; Lasanta, T.; Brevik, E.C.; Ries, J.B.; Sinoga, J.R. Understanding soil erosion processes in Mediterranean sloping vineyards (Montes de Málaga, Spain). *Geoderma* **2017**, *296*, 47–59. [CrossRef]
18. Keesstra, S.; Pereira, P.; Novara, A.; Brevik, E.C.; Azorin-Molina, C.; Parras-Alcántara, L.; Jordán, A.; Cerdà, A. Effects of soil management techniques on soil water erosion in apricot orchards. *Sci. Total Environ.* **2016**, *551–552*, 357–366. [CrossRef]
19. Ben-Salem, N.; Álvarez, S.; López-Vicente, M. Soil and water conservation in rainfed vineyards with common sainfoin and spontaneous vegetation under different ground conditions. *Water* **2018**, *10*, 1058. [CrossRef]
20. Garcia, L.; Celette, F.; Gary, C.; Ripoche, A.; Valdés-Gómez, H.; Metay, A. Management of service crops for the provision of ecosystem services in vineyards: A review. *Agr. Ecosyst. Environ.* **2018**, *251*, 158–170. [CrossRef]
21. Winter, S.; Bauer, T.; Strauss, P.; Kratschmer, S.; Paredes, D.; Popescu, D.; Landa, B.; Guzmán, G.; Gómez, J.A.; Guernion, M.; et al. Effects of vegetation management intensity on biodiversity and ecosystem services in vineyards: A meta-analysis. *J. Appl. Ecol.* **2018**, *55*, 2484–2495. [CrossRef]
22. Guzmán, G.; Cabezas, J.M.; Sánchez-Cuesta, R.; Lora, Á.; Bauer, T.; Strauss, P.; Winter, S.; Zaller, J.G.; Gómez, J.A. A field evaluation of the impact of temporary cover crops on soil properties and vegetation communities in southern Spain vineyards. *Agr. Ecosyst. Environ.* **2019**, *272*, 135–145. [CrossRef]
23. Celette, F.; Gaudin, R.; Gary, C. Spatial and temporal changes to the water regime of a Mediterranean vineyard due to the adoption of cover cropping. *Eur. J. Agron.* **2008**, *29*, 153–162. [CrossRef]
24. Ruiz-Colmenero, M.; Bienes, R.; Eldridge, D.J.; Marques, M.J. Vegetation cover reduces erosion and enhances soil organic carbon in a vineyard in the central Spain. *Catena* **2013**, *104*, 153–160. [CrossRef]
25. Lopes, C.M.; Santos, T.P.; Monteiro, A.; Rodrigues, M.; Costa, J.M.; Chaves, M.M. Combining cover cropping with deficit irrigation in a Mediterranean low vigor vineyard. *Sci. Hortic.* **2011**, *129*, 603–612. [CrossRef]
26. Williams, L.E.; Matthews, M.A. Grapevine. In *Irrigation of Agricultural Crops*; Agronomy Monographs No. 30. ASA-CSSA-SSSA; Stewart, B.J., Nielsen, D.R., Eds.; Agronomy Monographs: Madison, WI, USA, 1990; pp. 1019–1055.
27. Costa, J.M.; Vaz, M.; Escalona, J.; Egipto, R.; Lopes, C.; Medrano, H.; Chaves, M.M. Modern viticulture in southern Europe: Vulnerabilities and strategies for adaptation to water scarcity. *Agric. Water Manage.* **2016**, *164*, 5–18. [CrossRef]
28. Hannah, L.; Roehrdanz, P.R.; Ikegami, M.; Shepard, A.V.; Shaw, M.R.; Tabor, G.; Zhi, L.; Marquet, P.A.; Hijmans, R.J. Climate change, wine, and conservation. *PNAS* **2013**, *110*, 6907–6912. [CrossRef]
29. Organisation Internationale de la Vigne et du Vin (OIV). State of the vitiviniculture world market 2021. 2021. Available online: https://www.oiv.int/en/technical-standards-and-documents/statistical-analysis/state-of-vitiviniculture (accessed on 17 December 2021).
30. Zarrouk, O.; Francisco, R.; Pinto-Marijuan, M.; Brossa, R.; Santos, R.R.; Pinheiro, C.; Costa, J.M.; Lopes, C.; Chaves, M.M. Impact of irrigation regime on berry development and flavonoids composition in Aragonez (Syn. Tempranillo) grapevine. *Agr. Water Manage.* **2012**, *114*, 18–29. [CrossRef]
31. Esteban, M.A.; Villanueva, M.J.; Lissarrague, J.R. Effect of irrigation on changes in the anthocyanin composition of the skin of cv Tempranillo (Vitis vinifera L) grape berries during ripening. *J. Sci. Food Agr.* **2001**, *81*, 409–420. [CrossRef]
32. Permanhani, M.; Costa, J.M.; Conceição, M.A.F.; De Souza, R.T.; Vasconcellos, M.A.S.; Chaves, M.M. Deficit irrigation in table grape: Eco physiological basis and potential use to save water and improve quality. *Theor. Exp. Plant Phys.* **2016**, *28*, 85–108. [CrossRef]
33. Whitmore, A.P.; Schröder, J.J. Intercropping reduces nitrate leaching from under field crops without loss of yield: A modelling study. *Eur. J. Agron.* **2007**, *27*, 81–88. [CrossRef]
34. Clothier, B.E.; Green, S.R. The leaching and runoff of nutrients from vineyards. In *Science and Policy: Nutrient Management Challenges for the Next Generation*; Occasional Report No. 30; Currie, L.D., Hedley, M.J., Eds.; Fertilizer and Lime Research Centre, Massey University: Palmerston North, New Zealand, 2017; 7p.
35. Steenwerth, K.L.; Belina, K.M. Vineyard weed management practices influence nitrate leaching and nitrous oxide emissions. *Agric. Ecosyst. Environ.* **2010**, *138*, 127–131. [CrossRef]
36. Andreoli, V.; Cassardo, C.; La Iacona, T.; Spanna, F. Description and preliminary simulations with the Italian vineyard integrated numerical model for estimating physiological values (IVINE). *Agron. J.* **2019**, *9*, 94. [CrossRef]
37. Bagagiolo, G.; Rabino, D.; Biddoccu, M.; Nigrelli, G.; Cat Berro, D.; Mercalli, L.; Spanna, F.; Capello, G.; Cavallo, E. Effects of inter-annual climate variability on grape harvest timing in rainfed hilly vineyards of Piedmont (NW Italy). *Ital. J. Agrometeorol.* **2021**, *1*, 37–49. [CrossRef]
38. Fraga, H.; De Cortázar Atauri, I.G.; Malheiro, A.C.; Moutinho-Pereira, J.; Santos, J.A. Viticulture in Portugal: A review of recent trends and climate change projections. *Oeno One* **2017**, *51*, 61–69. [CrossRef]
39. Scherrer, S.C.; Begert, M.; Croci Maspoli, M.; Appenzeller, C. Long series of Swiss seasonal precipitation: Regionalization trends and influence of large scale flow. *Int. J. Climatol.* **2016**, *36*, 3673–3689. [CrossRef]

40. Masson-Delmotte, V.; Zhai, P.; Pörtner, H.-O.; Roberts, D.; Skea, J.; Shukla, P.R.; Pirani, A.; Moufouma-Okia, W.; Péan, C.; Pidcock, R.; et al. Impacts of Global Warming of 1.5 °C above Pre-industrial Levels and Related Global Greenhouse Gas Emission Pathways, in the Context of Strengthening the Global Response to the Threat of Climate Change, Sustainable Development, and Efforts to Eradicate Poverty. Global Warming of 1.5 °C., an IPCC Special Report, IPCC. 2018. Available online: https://www.ipcc.ch/sr15/ (accessed on 17 December 2021).
41. Allen, R.G.; Pereira, L.S.; Howell, T.A.; Jensen, M.E. Evapotranspiration information reporting: I. Factors governing measurement accuracy. *Agric. Water Manag.* **2011**, *98*, 899–920. [CrossRef]
42. Pereira, L.S.; Paredes, P.; Jovanovic, N. Soil water balance models for determining crop water and irrigation requirements and irrigation scheduling focusing on the FAO56 method and the dual Kc approach. *Agric. Water Manage.* **2020**, *241*, 106357. [CrossRef]
43. Pôças, I.; Gonçalves, J.; Costa, P.M.; Gonçalves, I.; Pereira, L.S.; Cunha, M. Hyperspectral-based predictive modelling of grapevine water status in the Portuguese Douro wine region. *Int. J. Appl. Earth Observ. Geoinf.* **2017**, *58*, 177–190. [CrossRef]
44. Ortega-Farias, S.; Condori, W.E.; Riveros-Burgos, C.; Fuentes-Peñailillo, F.; Bardeen, M. Evaluation of a two-source patch model to estimate vineyard energy balance using high-resolution thermal images acquired by an unmanned aerial vehicle (UAV). *Agric. For. Meteo.* **2021**, *304–305*, 108433. [CrossRef]
45. Romero, M.; Luo, Y.; Su, B.; Fuentes, S. Vineyard water status estimation using multispectral imagery from an UAV platform and machine learning algorithms for irrigation scheduling management. *Comput. Electron. Agric.* **2018**, *147*, 109–117. [CrossRef]
46. Cancela, J.J.; Fandiño, M.; Rey, B.J.; Martinez, E.M. Automatic irrigation system based on dual crop coefficient, soil and plant water status for *Vitis vinifera* (cv Godello and cv Mencía). *Agric. Water Manage.* **2015**, *151*, 52–63. [CrossRef]
47. Silva, S.P.; Valín, M.I.; Mendes, S.; Araujo-Paredes, C.; Cancela, J. Dual crop coefficient approach in *Vitis vinifera* L. cv. Loureiro. *Agronomy* **2021**, *11*, 2062. [CrossRef]
48. Celette, F.; Ripoche, A.; Gary, C. WaLIS—A simple model to simulate water partitioning in a crop association: The example of an intercropped vineyard. *Agr. Water Manage.* **2010**, *97*, 1749–1759. [CrossRef]
49. Phogat, V.; Pitt, T.; Stevens, R.M.; Cox, J.W.; Šimůnek, J.; Petrie, P.R. Assessing the role of rainfall redirection techniques for arresting the land degradation under drip irrigated grapevines. *J. Hydrol.* **2020**, *587*, 125000. [CrossRef]
50. Kustas, W.P.; Alfieri, J.G.; Nieto, H.; Wilson, T.G.; Gao, F.; Anderson, M.C. Utility of the two-source energy balance (TSEB) model in vine and interrow flux partitioning over the growing season. *Irrig. Sci.* **2019**, *37*, 375–388. [CrossRef]
51. Kool, D.; Kustas, W.P.; Ben-Gal, A.; Agam, N. Energy partitioning between plant canopy and soil, performance of the two-source energy balance model in a vineyard. *Agr. For. Meteorol.* **2021**, *300*, 108328. [CrossRef]
52. Loddo, S.; Soccol, M.; Perra, A.; Ucchesu, M.; Meloni, P.; Barbaro, M.; Lo Cascio, M.; Sirca, C. Biosensing IoT platform for water management in vineyards. In Proceedings of the 2020 IEEE International Symposium on Circuits and Systems (ISCAS), Seville, Spain, 12–14 October 2020. [CrossRef]
53. Rallo, G.; Paço, T.A.; Paredes, P.; Puig-Sirera, À.; Massai, R.; Provenzano, G.; Pereira, L.S. Updated single and dual crop coefficients for tree and vine fruit crops. *Agric. Water Manage.* **2021**, *250*, 106645. [CrossRef]
54. Allen, R.G.; Pereira, L.S.; Raes, D.; Smith, M. *Crop Evapotranspiration—Guidelines for Computing Crop Water Requirements*; Irrigation & Drainage Paper 56; FAO: Rome, Italy, 1998.
55. Allen, R.G.; Pereira, L.S.; Smith, M.; Raes, D.; Wright, J.L. FAO-56 dual crop coefficient method for estimating evaporation from soil and application extensions. *J. Irrig. Drain. Eng.* **2005**, *131*, 2–13. [CrossRef]
56. Pereira, L.S.; Paredes, P.; Melton, F.; Johnson, L.; Mota, M.; Wang, T. Prediction of crop coefficients from fraction of ground cover and height: Pratical application to vegetable, field, and fruit crops with focus on parameterization. *Agric. Water Manage.* **2021**, *252*, 106663. [CrossRef]
57. Rosa, R.D.; Paredes, P.; Rodrigues, G.C.; Alves, I.; Fernando, R.M.; Pereira, L.S.; Allen, R.G. Implementing the dual crop coefficient approach in interactive software. 1. Background and computational strategy. *Agric. Water Manage.* **2012**, *103*, 8–24. [CrossRef]
58. Fandiño, M.; Cancela, J.J.; Rey, B.J.; Martínez, E.M.; Rosa, R.G.; Pereira, L. Using the dual-Kc approach to model evapotranspiration of Albariño vineyards (Vitis vinifera L. cv. Albariño) with consideration of active ground cover. *Agric. Water Manage.* **2012**, *112*, 75–87. [CrossRef]
59. Paço, T.; Ferreira, M.; Rosa, R.; Paredes, P.; Rodrigues, G.; Conceição, N.; Pacheco, C.; Pereira, L. The dual crop coefficient approach using a density factor to simulate the evapotranspiration of a peach orchard: SIMDualKc model versus eddy covariance measurements. *Irrig. Sci.* **2012**, *30*, 115–126. [CrossRef]
60. Paço, T.A.; Pôças, I.; Cunha, M.; Silvestre, J.C.; Santos, F.L.; Paredes, P.; Pereira, L.S. Evapotranspiration and crop coefficients for a super intensive olive orchard. An application of SIMDualKc and METRIC models using ground and satellite observations. *J. Hydrol.* **2014**, *519*, 2067–2080. [CrossRef]
61. Puig-Sirera, À.; Rallo, G.; Paredes, P.; Paço, T.A.; Minacapilli, M.; Provenzano, G.; Pereira, L.S.S. Transpiration and water use of an irrigated traditional olive grove with Sap-Flow observations and the FAO56 dual crop coefficient approach. *Water* **2021**, *13*, 2466. [CrossRef]
62. Biancotti, A.; Bellardone, G.; Bovo, S.; Cagnazzi, B.; Giacomelli, L.; Marchisio, C. *Distribuzione Regionale di Piogge e Temperature. Collana Studi Climatologici del Piemonte, Vol.1*; Regione Piemonte: Torino, Italy, 1998.
63. Servizio Geologico Italiano. Carta Geologica d'Italia alla scala 1:100.000. 1969. Available online: http://193.206.192.231/carta_geologica_italia/cartageologica.htm (accessed on 17 December 2021).

64. IUSS Working Group. *World Reference Base for Soil Resources 2014: International Soil Classification System for Naming Soils and Creating Legends for Soil Maps*; World Soil Resources Reports No. 106; Food and Agriculture Organization of the United Nations (FAO): Rome, Italy, 2014.
65. Blake, G.R.; Hartge, K.H. Bulk density. In *Methods of Soil Analysis. Part 1. Physical and Mineralogical Methods*, 2nd ed.; Klute, A., Ed.; American Society of Agronomy–Soil Science Society of America: Madison, WI, USA, 1986; pp. 363–375.
66. Cavazza, L. Fisica del Terreno Agrario. UTET: Torino, Italy, 1981.
67. Schaap, M.G.; Leij, F.J.; van Genuchten, M.T. ROSETTA: A computer program for estimating soil hydraulic parameters with hierarchical pedotransfer functions. *J. Hydrol.* **2001**, *251*, 163–176. [CrossRef]
68. Gomes, M.P.; Silva, A.A. Um novo diagrama triangular para a classificação básica da textura do solo. *Garcia Orta* **1962**, *10*, 171–179.
69. Ramos, T.B.; Gonçalves, M.C.; Brito, D.; Martins, J.C.; Pereira, L.S. Development of class pedotransfer functions for integrating water retention properties into Portuguese soil maps. *Soil Res.* **2013**, *51*, 262–277. [CrossRef]
70. Ramos, T.B.; Horta, A.; Gonçalves, M.C.; Martins, J.C.; Pereira, L.S. Development of ternary diagrams for estimating water retention properties using geostatistical approaches. *Geoderma* **2014**, *230*, 229–242. [CrossRef]
71. ARPA Piemonte—Meteorologia e Clima. Available online: https://www.arpa.piemonte.it/dati-ambientali/dati-meteoidrografici-giornalieri-richiesta-automatica (accessed on 17 December 2021).
72. Biddoccu, M.; Ferraris, S.; Opsi, F.; Cavallo, E. Long-term monitoring of soil management effects on runoff and soil erosion in sloping vineyards in Alto Monferrato (North–West Italy). *Soil Till. Res.* **2016**, *155*, 176–189. [CrossRef]
73. Raffelli, G.; Previati, M.; Canone, D.; Gisolo, G.; Bevilacqua, I.; Capello, G.; Biddoccu, M.; Cavallo, E.; Deiana, R.; Cassiani, G.; et al. 2017. Local and plot-scale measurements of soil moisture: Time and spatially resolved field techniques in plain, hill and mountain sites. *Water* **2017**, *9*, 706. [CrossRef]
74. Sohne, W. Druckverteilung im Boden und Boden-verformung unter Schlepper Reifen. *Grundl. Der Landtech.* **1953**, *5*, 49–63.
75. Ritchie, J.T. Model for predicting evaporation from a row crop within complete cover. *Water Resour. Res.* **1972**, *8*, 1204–1213. [CrossRef]
76. Liu, Y.; Pereira, L.S.; Fernando, R.M. Fluxes through the bottom boundary of the root zone in silty soils: Parametric approaches to estimate groundwater contribution and percolation. *Agric. Water Manage.* **2006**, *84*, 27–40. [CrossRef]
77. USDA-SCS. *National Engineering Handbook*; Section 4, Table 10.1; 1972. Available online: https://directives.sc.egov.usda.gov/OpenNonWebContent.aspx?content=17752.wba (accessed on 17 December 2021).
78. Allen, R.G.; Wright, J.L.; Pruitt, W.O.; Pereira, L.S.; Jensen, M.E. Water requirements. In *Design and Operation of Farm Irrigation Systems*, 2nd ed.; Hoffman, G.J., Evans, R.G., Jensen, M.E., Martin, D.L., Elliot, R.L., Eds.; ASABE: St. Joseph, MI, USA, 2007; pp. 208–288.
79. Allen, R.G.; Pereira, L.S. Estimating crop coefficients from fraction of ground cover and height. *Irrig. Sci.* **2009**, *28*, 17–34. [CrossRef]
80. Pereira, L.S.; Paredes, P.; Melton, F.; Johnson, L.; Wang, T.; López-Urrea, R.; Cancela, J.J.; Allen, R. Prediction of crop coefficients from fraction of ground cover and height. Background and validation using ground and remote sensing data. *Agric. Water Manage.* **2020**, *240*, 106197. [CrossRef]
81. Pereira, L.S.; Paredes, P.; Rodrigues, G.C.; Neves, M. Modeling malt barley water use and evapotranspiration partitioning in two contrasting rainfall years. Assessing AquaCrop and SIMDualKc models. *Agric. Water Manage.* **2015**, *159*, 239–254. [CrossRef]
82. Moriasi, D.N.; Arnold, J.G.; Van Liew, M.W.; Bingner, R.L.; Harmel, R.D.; Veith, T.L. Model Evaluation Guidelines for Systematic Quantification of Accuracy in Watershed Simulations. *Trans. ASABE* **2007**, *50*, 885–900. [CrossRef]
83. Nash, J.E.; Sutcliffe, J.V. River flow forecasting through conceptual models part I—A discussion of principles. *J. Hydrol.* **1970**, *10*, 282–290. [CrossRef]
84. Gaudin, R.; Celette, F.; Gary, C. Contribution of runoff to incomplete off season soil water refilling in a Mediterranean vineyard. *Agric. Water Manage.* **2010**, *97*, 1534–1540. [CrossRef]
85. Romero, P.; Castro, G.; Gomez, J.A.; Fereres, E. Curve number values for olive orchards under different soil management. *Soil Sci. Soc. Am. J.* **2007**, *71*, 1758–1769. [CrossRef]
86. Ferreira, M.I.; Silvestre, J.; Conceição, N.; Malheiro, A.C. Crop and stress coefficients in rainfed and deficit irrigation vineyards using sap flow techniques. *Irrig. Sci.* **2012**, *30*, 433–447. [CrossRef]
87. Zarrouk, O.; Brunetti, C.; Egipto, R.; Pinheiro, C.; Genebra, T.; Gori, A.; Lopes, C.M.; Tattini, M.; Chaves, M.M. Grape ripening is regulated by deficit irrigation/elevated temperatures according to cluster position in the canopy. *Front. Plant Sci.* **2016**, *7*, 1640. [CrossRef]
88. Paço, T.A.; Paredes, P.; Pereira, L.S.; Silvestre, J.; Santos, F.L. Crop Coefficients and Transpiration of a Super Intensive Arbequina Olive Orchard using the Dual K_c Approach and the K_{cb} Computation with the Fraction of Ground Cover and Height. *Water* **2019**, *11*, 383. [CrossRef]
89. Paredes, P.; Rodrigues, G.C.; Alves, I.; Pereira, L.S. Partitioning evapotranspiration, yield prediction and economic returns of maize under various irrigation management strategies. *Agric. Water Manage.* **2014**, *135*, 27–39. [CrossRef]
90. Rosa, R.D.; Ramos, T.B.; Pereira, L.S. The dual Kc approach to assess maize and sweet sorghum transpiration and soil evaporation under saline conditions: Application of the SIMDualKc model. *Agric. Water Manage.* **2016**, *115*, 291–310. [CrossRef]

91. Darouich, H.; Karfoul, R.; Eid, H.; Ramos, T.B.; Baddour, N.; Moustafa, A.; Assaad, M.I. Modeling zucchini squash irrigation requirements in the Syrian Akkar region using the FAO56 dual-Kc approach. *Agric Water Manage.* **2020**, *229*, 105927. [CrossRef]
92. Darouich, H.; Karfoul, R.; Ramos, T.B.; Moustafa, A.; Shaheen, B.; Pereira, L.S. Crop water requirements and crop coefficients for jute mallow (*Corchorus olitorius* L.) using the SIMDualKc model and assessing irrigation strategies for the Syrian Akkar region. *Agric. Water Manage.* **2021**, *255*, 107038. [CrossRef]
93. Novara, A.; Cerdà, A.; Gristina, L. Sustainable vineyard floor management: An equilibrium between water consumption and soil conservation. *Curr. Opin. Environ. Sci. Health.* **2018**, *5*, 33–37. [CrossRef]

 water

Article

Spatiotemporal Analysis of Evapotranspiration and Effects of Water and Heat on Water Use Efficiency

Yuan-Yuan Tang [1,2], Jian-Ping Chen [1,*], Feng Zhang [3] and Shi-Song Yuan [3]

1 School of Earth Science and Resouces, China University of Geosciences (Beijing), Beijing 100083, China; 3001160121@cugb.edu.cn
2 Changsha Natural Resources Comprehensive Survey Center, China Geological Survey, Changsha 410000, China
3 Geophysical Survey Center, China Geological Survey, Langfang 065000, China; zhangfeng01@mail.cgs.gov.cn (F.Z.); yuanshisong@mail.cgs.gov.cn (S.-S.Y.)
* Correspondence: 3s@cugb.edu.cn

Abstract: Water Use Efficiency (WUE) is an important indicator of the carbon cycle in the hydrological and ecological system. It is of great significance to study the response of different hydrological processes to climate and to understand ecosystem carbon sink. However, little is known about the effects and mechanisms of precipitation and temperature on the WUE of different hydrological processes. Thus, three kinds of WUEs (GPP/E (eWUE), GPP/Et (tWUE), and GPP/P (pWUE)) are defined for three different hydrological indicators in semi-arid areas in this study in order to reveal the variation pattern of WUEs based on hydrological indicators and their response to climate. We found that in the past 15 years, the seasonal fluctuation of evapotranspiration in arid areas was large, and the spatial difference of WUE of different hydrological processes was obvious. In semi-arid areas, temperature had a significant effect on WUE (about 68–81%). However, precipitation had a lag effect on WUEs, and the negative impact of precipitation has a great influence (about 84–100%). Secondly, the threshold values of precipitation to WUEs (200 or 300 mm) and temperature to WUEs (2 or 7 °C) are also different from previous studies. This study advances our understanding of the influence of different hydrological processes on ecosystem carbon and climate.

Keywords: evapotranspiration; spatial and temporal; water use efficiency; Gleam Product

Citation: Tang, Y.-Y.; Chen, J.-P.; Zhang, F.; Yuan, S.-S. Spatiotemporal Analysis of Evapotranspiration and Effects of Water and Heat on Water Use Efficiency. *Water* **2021**, *13*, 3019. https://doi.org/10.3390/w13213019

Academic Editors: Mirko Castellini, Simone Di Prima, Ryan Stewart, Marcella Biddoccu, Mehdi Rahmati and Vincenzo Alagna

Received: 15 August 2021
Accepted: 21 October 2021
Published: 27 October 2021

1. Introduction

Terrestrial ecosystems are a major carbon sink in the global carbon cycle [1]. The issue of how the carbon and water cycles can be balanced is of great interest to the scientific community [2]. WUE is an important indicator of the coupling of carbon and water cycles in ecosystems, which closely links biological processes (photosynthesis, transpiration) and physical processes (evaporation). It is a key parameter for ecosystem response to climate change [3].

WUE controls the coupling between plant water use and carbon assimilation and has been defined in many previous studies as the ratio of GPP (gross primary productivity) to evapotranspiration [4,5]. In some GPP simulation models, WUE is also a critical metric [6]. Recently, other studies have determined the effects of changes in vegetation, climate, land use, and atmospheric CO_2 on changes in WUE [7–9]. Responses of ecosystem WUE's to climate are generally primarily based on spatial or temporal gradient analysis [10]. Due to the uncertain interactions of many environmental factors on WUE, it is still yet to be tackled how to accurately predict the regional impact of climate change on carbon water resources [11]. The application of the mechanism model makes it possible for WUE to expand from the blade or regional to the global level, so as to explore the impact of single and multiple environmental factors on WUE greatly. Climate factors such as precipitation and temperature can sensitively reflect climate change in arid and semi-arid areas [11]. Evaporation, transpiration, and precipitation, which regulate the growth of plants, are important parts of the water cycle. Huang et al. (2016) defined the GPP/p and GPP/Et (transpiration) ratio as a new indicator of water efficiency and have been widely applied [12]. Yang et al. discussed the ecosystem response mechanism to drought by comparing the responses of different terrestrial ecosystems around the global to drought [13].

WUE not only exhibits strong spatial heterogeneity but also has significant interannual changes. Spatially, areas with low WUE values are mainly located in Central Asia, the Sahel in Africa, the western coasts of North and South America, and areas such as Oceania, where evaporation is strong and vegetation is scarce [14]. Global WUEs have generally increased over the 30 years from 1982 to 2008, but they vary significantly from region to region, depending on the direction and magnitude of changes between the GPP and *ET* [14]. Among them, meteorological factors such as precipitation, temperature, and solar radiation were crucial factors driving WUE change [12]. However, due to the different geographical environments effect, the results obtained from different studies often vary significantly [15]. For example, in the water-scarce Black River basin, precipitation plays a leading role in WUE [16]. However, in the more humid alpine meadows, temperature is a more important factor [17]. There are very few studies of WUE in arid areas, especially based on newly defined indicators of water efficiency, and studies of climate responses have not been reported.

Xinjiang belongs to an arid and semi-arid area with a shortage of water resources and scarce surface vegetation. It has extremely fragile ecosystems, and the impact of climate change is severe [18–20]. The process of small water circulation in inland Xinjiang is accelerating, and the rapid melting of alpine ice and snow directly changes the composition of runoff recharge, affecting the water consumption of vegetation in mountains, oases, and deserts, and the growth of vegetation and spatial distribution patterns [21]. The total amount of water resources consumed in arid areas of Xinjiang is relatively large, leading to an increasing depletion of water resources. It has become the core restriction on the development and regional stability of Xinjiang [22]. Therefore, based on previous research, this study analyzed the multi-year spatial–temporal changes of Xinjiang evapotranspiration and WUE, and it quantitatively evaluated WUE's sensitivity to temperature and precipitation, providing data support for ecological restoration and water resource utilization and management in the region.

In summary, the objectives of this study are as follows: (1) monthly and annual scale changes and spatial changes of Xinjiang actual evapotranspiration and potential evapotranspiration (*Ep*) were analyzed; then, *E* and *Ep* trend changes can be predicted according to the Hurst index; (2) The spatiotemporal variation trend of based on different hydrological indicators of water use efficiency; (3) The response and threshold of different hydrological processes of WUE indexes to climate change were analyzed.

2. Materials and Methods

2.1. Study Area

Xinjiang is located in the hinterland of Eurasia and the northwestern border of China, with a geographical position of 34°15′ N to 48°10′ N, 73°20′ E to 96°25′ E (Figure 1). It has the unique landscape featuring three mountains. The aera is rich in glaciers, which is 46.9% of China's total glacial reserves. Rivers are mainly supplied with ice and snow melting water, the annual runoff is 8.79×10^{10} m^3, and the lake area exceeds 1.29×104 km^2 [23]. As a result of its particular geographical location, topography, climate and other factors, the coverage rate of vegetation in the territory is low. Desert, gobi, and bare land are widely distributed, and the ecosystem is fragile [24].

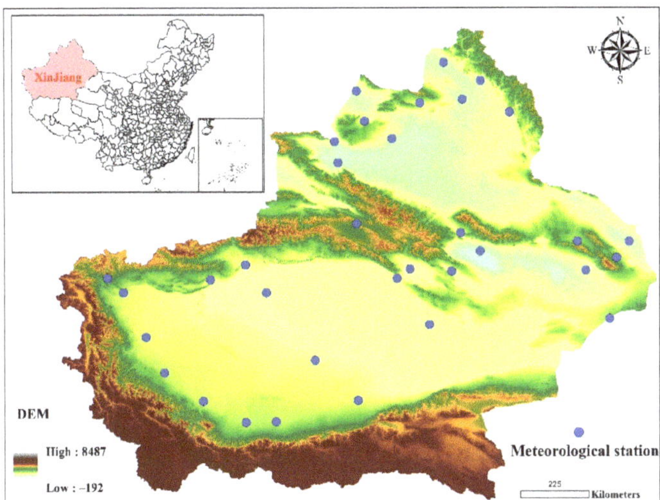

Figure 1. Location of the study area (the small picture inserted shows the geographical location of Xin-jiang).

2.2. Evapotranspiration and Drought Index Data

Gleam v3.3a (Global Land Evapotranspiration Amsterdam Model 3.3a) is a set of data estimating the different components of land surface evapotranspiration independently: actual evaporation (*E*), potential evapotranspiration (*Ep*), transpiration (*Et*), evaporation stress factor, root zone soil moisture, surface soil moisture, etc. [25]. This data product is based on the surface radiation and near-surface temperature of multiple reanalysis datasets, and the Gleam Ep dataset with a spatial resolution of $0.25° \times 0.25°$ is obtained by using the Priestley–Taylor formula. Evapotranspiration data are a monthly dataset (https: //www.gleam.eu/) (accessed on 14 October 2021). GLEAM evapotranspiration products have been widely used in the field of hydrology and meteorology since their development [26]. These data has been applied by scholars to the study of the arid region of northwest China, and the data quality has been verified accordingly [27,28]. Therefore, monthly data of GLEAM V3.3A, *E*, *Ep*, and *Et* from 2000 to 2014 are used for data analysis.

Two drought indices, SPEI and PDSI, were selected in this paper. The SPEI dataset is provided by the National Cryosphere Desert Data Center. (http://www.ncdc.ac.cn) (accessed on 14 October 2021). The spatial resolution is 1°, and the temporal resolution is the monthly average from 2000 to 2014. The PDSI is the first comprehensive drought index developed and is most commonly used to assess the impact of drought on Agriculture in the United States. It was obtained from KNMI Climate Explorer (http://climexp.knmi.nl/) (accessed on 14 October 2021), has a spatial resolution 1°, and the temporal resolution is the monthly average from 2000 to 2014.

Data Availability

To verify the accuracy of GLEAM-E in Xinjiang region, the daily datasets from 34 sites of the China Meteorological Data Network (http://data.cma.cn/) (accessed on 14 October 2021) are used. Site scale assessment of surface evapotranspiration products was carried out. Due to the absence of observation data in time series, 132 valid observations were screened. The evaporating dish is the maximum evaporation condition of free water in the region under the condition of sufficient moisture [29]. In arid and semi-arid areas, the actual evapotranspiration is mainly determined by water and energy. At the same time, the evaporating dish evapotranspiration is not affected by moisture, and the main factor determining its size is solar energy. Therefore, the evaporating dish evapotranspiration has an inverse relationship with the actual evapotranspiration, and potential evapotranspiration represents the maximum evapotranspiration of an underlying solid surface under sufficient water supply [30]. It can be seen that the evapotranspiration in the evaporating dish is close to the potential evapotranspiration in the GLEAM product, and the feasibility of the GLEAM-*E* product in the Xinjiang region can be verified by the correlation between the evapotranspiration and GLEAM-*Ep*. The results (Figure 2) show that there was a high correlation between the observations evapotranspiration and GLEAM-*Ep*. The correlation coefficient R^2 = 0.8654, indicating that the GLEAM-*E* product had good verification accuracy in the Xinjiang region, and the verification results meet the requirements.

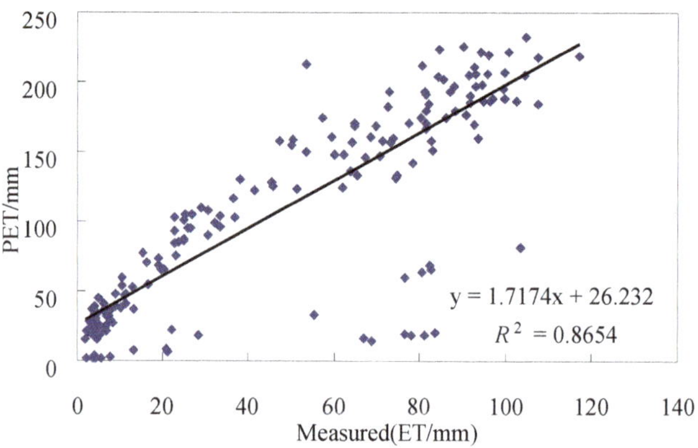

Figure 2. Relationship between GLEAM *Ep* product and observations evapotranspiration (measured represents observations of evapotranspiration; PET represents GLEAM *Ep* at meteorological stations.

2.3. Precipitation and Temperature Data

The monthly precipitation and temperature (2 m air temperature) data from 2000 to 2014 used in this study come from a dataset with a spatial resolution of 1 km × 1 km [31]. The dataset shows the data of January 1901 to December 2017 in a time series from the National Tibetan Plateau Data Center (NTPDC) (http://data.tpdc.ac.cn) (accessed on 14 October 2021). The accuracy of this dataset has been verified and tested for good efforts [32,33]. Precipitation and temperature data spatial resolution was resampled to be 0.25° × 0.25°. The precipitation data were used to calculate WUE and temperature to analyze the response of WUE to climate.

2.4. Gross Primary Production (GPP) Data

The MOD17 product is the first continuous satellite-driven dataset that monitors global surface vegetation productivity and provides 8-day composite GPP products, net photosynthesis production products, and annual total GPP and NPP numerical products [34]. In MODIS satellite products, the standard product mod17a2h GPP of version 6 is a cumulative 8-day composite product with a spatial resolution of 1 km. The GPP of this product was estimated according to the BIOME-BGC model of radiation utilization efficiency [35]. In addition, the algorithm of this product was based on the relationship between the total photosynthesis and the amount of photosynthetically active radiation absorbed by the plant canopy. For detailed calculation, please refer to Heinsch's literature [36]. In this study, GPP data from 2000 to 2014 were selected, and the spatial resolution was resampled to 0.25° × 0.25°. The GPP product was applied to calculate WUE.

2.5. Land-Use/Land-Cover (LULC) Datasets

The LULC dataset is collected from the resource and environmental data cloud platform released by the Chinese Academy of Sciences (http://www.resdc.cn/Data/) (accessed on 14 October 2021). Then, we downloaded the LULC datasets in 2000 and 2015 and used ArcGIS 10.1 to extract the study area by mask. There are 25 types of raster datasets with a spatial resolution of 1 km. According to the actual situation in Xinjiang, LULC types are reclassified into 6 types, including cultivated land, forest, grassland, construction and residential land, water, and unused land.

2.6. Coefficient of Variation

The coefficient of variation (CV) measures the degree of dispersion of random variables. It is used to analyze the spatial pattern and spatial differentiation of *ET* [37].We calculated the coefficient of variation of evapotranspiration and potential evapotranspiration from 2000 to 2014 by pixels and analyzed the stability of evapotranspiration and potential evapotranspiration in time series. The higher the CV, the more discrete the data distribution is across years, which means the time series fluctuates and is unstable. Conversely, it indicates that the data distribution is relatively concentrated, and the time series is relatively stable.

2.7. Hurst Index

The Hurst index is used to describe the sustainability of time-series data quantitatively and has extensive applications in hydrology, economics, and climatology [38]. Consider the *ET* (*ET* represents variables) time series $ET(t)$ (t = 1,2,3,4,. . . ,n), and define the mean sequence of the time series:

$$\overline{ET}_{(\tau)} = \frac{1}{\tau}\sum_{t=1}^{\tau} ET_{(\tau)} \quad \tau = 1,2,\ldots,n. \tag{1}$$

Calculate the cumulative dispersion as:

$$X_{(t,\tau)} = \sum_{t=1}^{t}\left(ET_{(t)} - \overline{ET}_{(\tau)}\right) \quad 1 \leq t \leq \tau. \tag{2}$$

The sequence of the range:

$$R_{(\tau)} = \underset{1\leq t\leq \tau}{max} X_{(t,\tau)} - \underset{1\leq t\leq \tau}{\min} X_{(t,\tau)} \quad \tau = 1,2,\ldots,n. \tag{3}$$

The standard deviation sequence is:

$$S_{(\tau)} = \left[\frac{1}{\tau}\sum_{t=1}^{\tau}\left(ET_{(t)} - ET_{(\tau)}\right)^2\right]^{\frac{1}{2}} \quad \tau = 1,2,\ldots,n. \tag{4}$$

Calculate the Hurst index:

$$\frac{R_{(\tau)}}{S_{(\tau)}} = (c\tau)^H \tag{5}$$

where c is a constant. The Hurst empirical formula is obtained by using the logarithm of both sides of the method (5). Based on the time series and using Hurst's index, we get a cluster of H values for least-squares fitted line, the slope of which is the modified Hurst index (H), revealing the fractal characteristics of the time series.

2.8. Calculation of WUEs

WUE is defined as the ratio of the total primary productivity of terrestrial ecosystems to the actual vapor emplacement of *ET* in a unit of time [39]:

$$\text{WUEs} = \frac{\text{GPP}}{H} \tag{6}$$

where WUEs (g C·mm^{-1}·m^{-2}) represents the water use efficiency, *ET* is the evapotranspiration, and GPP is the gross primary productivity. Here, we define three similar water use efficiencies; eWUE represents the ecosystem water use efficiency (*H* stands for evapotranspiration); tWUE is the vegetation transpiration water use efficiency (*H* stands for transpiration); and pWUE is the precipitation of water use efficiency (*H* stands for precipitation). Here, GPP and precipitation data with a spatial resolution of 1 km and a time scale of month are resampled with a spatial resolution of 0.25° × 0.25°. Finally, monthly WUEs are calculated.

This study used a linear trend estimation method to analyze the spatiotemporal change trend of WUEs. The *x* axis is the year (time series), and the *y* axis is the slope $y = a + bx$ of the WUEs, the linear regression equation representing the trend rate. When the slope is greater than zero, the WUEs are growing, and vice versa [40,41]. We have divide the trend change into five levels, as shown in Table 1.

Table 1. Scale of trend change.

$Slope_{WUE}$	Increase Degree	Abbreviations
≥0.40	High Increase	HI
$0.2 \leq Slope_{WUE} < 0.4$	Relatively High Increase	RHI
$0.1 \leq Slope_{WUE} < 0.2$	Medium Increase	MI
$0 \leq Slope_{WUE} < 0.1$	Relatively Low Increase	RI
$-0.1 \leq Slope_{WUE} < 0$	Low Decrease	LD

2.9. WUEs Response to Climate Change

We used the relative sensitivity of WUE to temperature and precipitation to estimate the impact of terrestrial ecosystem processes on climate response [13]. Firstly, we used the dplR package tool to detrend all variables. Second, we define sensitivity analysis with WUE as the dependent variable and temperature and precipitation as the independent variables. Finally, the estimated regression coefficient represents the sensitivity of each independent variable to the dependent variable [42]. The sensitivity coefficient calculation equation is as follows:

$$S_p = \frac{\partial \mathrm{WUE}}{\partial P} \tag{7}$$

where S_p is the sensitivity coefficient; WUE is water use efficiency; and *P* is air temperature or precipitation, respectively.

The correlation technique was used to calculate the coefficients (R) of each relationship between variables in the study, and the *p* test was used to test R significance. The Rstudio Tool (https://rstudio.com/) (accessed on 14 October 2021) and ArcGIS 10.2 Python (ArcGIS 10.2 and Python, ESRI, Redlands, CA, USA) were employed to calculate and model using all datasets.

3. Results and Discussion

3.1. Temporal Variation in E and Ep

Considering the fact that *E* represents the actual evapotranspiration and *Ep* represents the potential evapotranspiration under sufficient water supply conditions, the difference between *Ep* and *E* can reflect the drought situation in the study area [43,44]. From Figure 3a, it is shown that the difference between *Ep* and *E* in Xinjiang is relatively large in 2000–2014. Among these years, the difference between *Ep* and *E* was the largest in 2008, reaching around 1200 mm. The difference in 2003 was the wettest and is about 1000 mm. Hence, the drought degree was the biggest in 2008 and the wettest in 2003. We can also see from Figure 3c,d that the year 2008 was the driest and 2003 was the wettest. It can be seen from Figure 3c,d that the drought situation in Xinjiang gradually alleviated, which is consistent with the findings of previous studies that precipitation is increasing in Xinjiang. Figure 3b shows the monthly trend of *E* and *Ep*. Specifically, the difference between *E* and *Ep* is the largest in July and the smallest in January, and the differences are around

170 mm and 15 mm respectively. This indicates that the drought degree is the largest in July and the smallest in January. Overall, in spring (March–May), both *E* and *Ep* were in an increasing state. As the temperature rises as well as the melting of snow and ice increases, the transpiration of vegetation increases and evaporation increases. In summer (June–August), *E* and *Ep* reached the peak, and the temperature is the highest at this time. Under the influence of westerly wind, water vapors, rainfall, and vegetation are abundant. In autumn (September–November), the temperature drops, vegetation withers, and transpiration weakens, leading to the gradual decrease in *E* and *Ep*. The temperature in winter (December–February) reached the lowest, with less precipitation and radiation, and *E* and *Ep* reduced to the lowest. From 2000 to 2014, the *E* and *Ep* seasonal changes are significant.

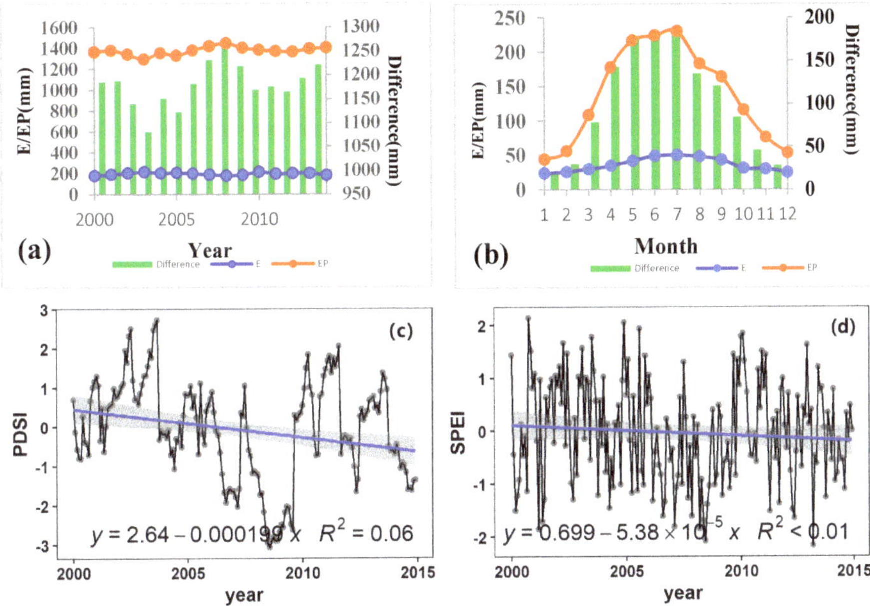

Figure 3. Annual and monthly variations of *E* and *Ep* in Xinjiang during 2000–2014. (**a**) Annual mean changes; (**b**) Monthly mean changes; (**c**) Trends of PDSI monthly scale variation; (**d**) Trends of SPEI monthly scale variation).

3.2. Stability Analysis of Evapotranspiration

In order to further understand the spatial stability pattern of *E* and *Ep* in Xinjiang, we used the coefficient of variation method for analysis. The CV of *E* and *Ep* is divided into five levels, with CV less than 0.05 defined as low fluctuation, 0.05 to 0.10 defined as low to moderate fluctuation, 0.10 to 0.15 defined as moderate fluctuation, 0.15 to 0.20 defined as moderate to strong fluctuation, and greater than 0.20 defined as very strong fluctuation.

According to Figure 4a, the CV value of *E* near Yili River Valley, Tacheng, and Kunlun Mountain is significantly larger than 0.2, which indicates that the interannual *E* in this region has large fluctuations.

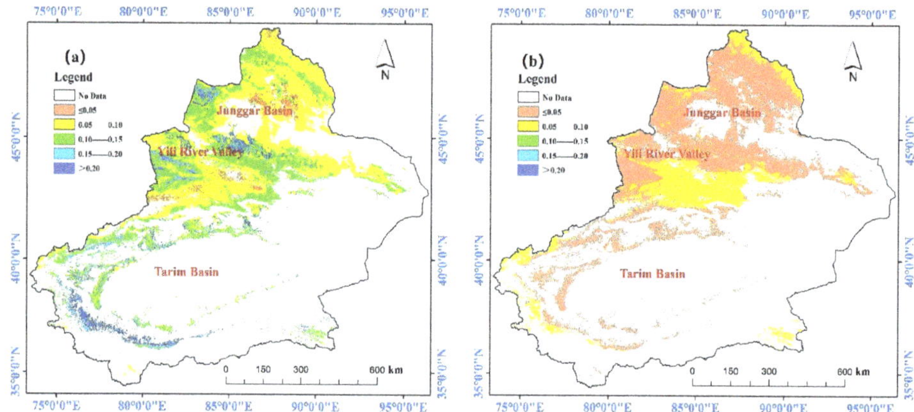

Figure 4. Spatial distribution of variation coefficient of E and Ep in Xinjiang. (**a**) *E*; (**b**) *Ep*.

The *E* is the most unstable, while the CV of the Junggar Basin and the vicinity of the Tianshan Mountains is the lowest (less than 0.1), indicating that the interannual *E* fluctuation in this region is small. The time-series fluctuation is the most stable and the time-series fluctuation of *E* in northern Xinjiang (to the north of the Tianshan Mountains) is more durable than that in southern Xinjiang. This may be due to the continuous "warm and wet" conditions in the northwest [45]. From Figure 4b, we can see that the CV values of *Ep* in Xinjiang are all low (less than 0.1), and the minimum CV values ($\leq$0.05) are mainly distributed in the vicinity of the Junggar Basin. The larger CV (0.05–0.1) is primarily distributed near the Tianshan Mountains; it shows that the Ep's space sequence in Xinjiang is relatively stable, and the stability of the Junggar Basin is most significant. To sum up, the regions with large fluctuations in evapotranspiration in Xinjiang are mainly located around the Yili River Valley and the Tarim Basin. The possible reason is that the vegetation in these areas has poor water conservation capacity, causing the interannual evapotranspiration to vary greatly.

LULC also affected the spatial pattern of *E* and *Ep*. As can be seen from Figure A1, the regions with strong fluctuations of *E* are mainly distributed in cultivated land, grassland and forest land in mountainous areas such as Tianshan mountains, Altai mountains, Qilian mountains and Yili river valley. The regions with low *ET* fluctuation are mainly in the margin of Tarim Basin in southern Xinjiang and Junggar Basin in northern Xinjiang. Meanwhile, the *ET* of agricultural land in oasis is higher than that of unused land at oasis edge. According to the changes of different land cover types (Figure A1C), cultivated land increased by 1.08%, grassland decreased by 0.66%, and woodland decreased by 0.08%. From the perspective of spatial differences, the difference in dynamic and thermal properties of different land covers leads to the redistribution of energy in the land-air interaction [46]. Secondly, due to artificial planting and irrigation, vegetation coverage and soil moisture in cultivated areas are higher than those in other areas, which makes evapotranspiration show significant spatial heterogeneity.

3.3. Hurst Index

The Hurst index method is used to analyze the sustainability of the time series of *E* and *Ep* in Xinjiang from 2000 to 2014 (Figure 5a). It can be seen that the regions with an *E* Hurst index less than 0.5 are mainly distributed in the northern Junggar Basin region, which shows that the future *E* variations in this region are contrary to the past. As northern Xinjiang is mainly an area with more agricultural land and greater land use change, the future trend of evapotranspiration in this region is opposite to that in the past. Most of the rest of the areas have a Hurst index of more than 0.5, indicating that the regional distribution of *E* in the vast majority of regions in Xinjiang is consistent with the trend of the past. The change of *E* Hurst index from 2000 to 2014 shows continuity. The most persistence is near the Tianshan Mountains, and the *E* Hurst index in this region is mostly above 0.8 from 2000 to 2014.

The region with an *Ep* Hurst index less than or equal to 0.5 is mainly distributed in the Yili River valley in northern Xinjiang, indicating that the future *Ep* changes in this area are contrary to the past. The areas with a Hurst index greater than 0.5 for *Ep* are mainly distributed in the Junggar Basin, the nearby Tianshan Mountains, and near Tarim River Basin in northern Xinjiang. The Hurst index indicates that the *Ep* in these regions is consistent with past trends, and the entire process is continuous. The Tianshan Mountains and the Junggar Basin have the most robust persistence, and the Hurst index in the eastern part of the Tianshan Mountains is above 0.9. In conclusion, *E* and *Ep* in Xinjiang will continue for a period of time in line with the past trend with no turning point, and the volatility is relatively unstable.

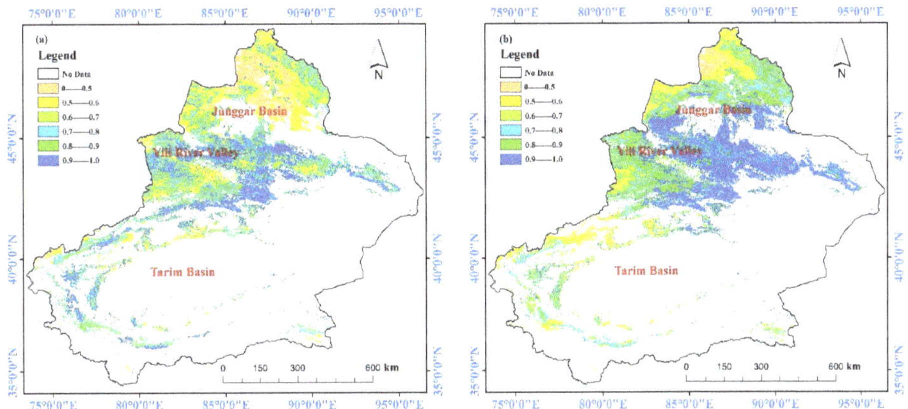

Figure 5. Spatial distribution of Hurst index in Xinjiang from 2000 to 2014.(**a**) *E*; (**b**) *Ep*.

3.4. Spatiotemporal Characteristic of WUEs Defined by GPP/E, GPP/Et and GPP/P

There are obvious spatial differences between different WUEs in which the spatial distribution patterns of eWUE and pWUE are generally consistent. Spatiotemporal WUE (2000–2014 average) consists of GPP/E (eWUE), GPP/Et (tWUE), and GPP/P (pWUE) (Figure 6). During the period of 2000–2014, the value of WUE showed slight changes, and the spatial distribution of WUEs was uneven (the white area means no value). The spatial distribution of eWUE presents a high–low interval distribution. The eWUE around the Taklimakan Desert was relatively high, indicating that the WUE of the desert ecosystem is generally high. However, the tWUE around the Taklimakan Desert is higher than the eWUE. Secondly, the spatial distribution of pWUE and eWUE is basically the same (Figure 6a,c), but they are different from the trend of tWUE. It shows that precipitation and evapotranspiration have consistent changes in semi-arid areas. In arid areas, 90% of the surface precipitation returns to the atmosphere in the form of vapor [45,46]; therefore, the precipitation and evaporation show the same pattern. The range of WUE was 0–5 g C·mm^{-1}·m^{-2}, and Zou Jie et al. analyzed the range of nearly 15 ecosystems in Central Asia and Xinjiang, confirming the accuracy of the calculation results [41]. We normalized data from 0–5, mapped them to values between 0–1, and divided WUE into five levels for statistical analysis (Table 2): When 0 < WUE < 0.2, the proportion of eWUE and pWUE is similar (65.25% and 67.11%), but the proportion of tWUE (23.66%) is less than eWUE and pWUE. When 0.2 < WUE < 0.4, overall, the values of eWUE and pWUE range mainly from 0–0.4, which is more than 85% of the total area. However, the tWUE at 0–0.2, 0.2–0.4, and >0.8 levels accounted for 23.66%, 26.78%, and 25.66%, respectively. In summary, the mean annual value of tWUE is larger than that of eWUE and pWUE, and the spatial difference is obvious. According to previous studies [47,48], there has been an increase in temperature and precipitation since the mid-1980s, with the increase in precipitation in Xinjiang being particularly significant compared with other regions. However, according to the results of this study, WUEs has a negative effect on precipitation. We think the increase in precipitation is bad news for the vegetation that has adapted to the arid environment. It takes a while for the vegetation to adapt. Such a research phenomenon may be the lag effect. Firstly, we found that there was a high degree of spatial similarity between pWUE and NDVI in the growing season [20]. Secondly, pWUE values were also higher in the regions with higher NDVI values in the growing season, except for sparse vegetation around the desert. Such a phenomenon occurs around the desert. We believe that the reason why pWUE is high around the desert is that mainly coniferous vegetation around the desert leads to the reduction of transpiration, which leads to a high WUE. Finally, we believe that there is a difference between NDVI in the growing season and pWUE, which may be caused by the difference in time scale.

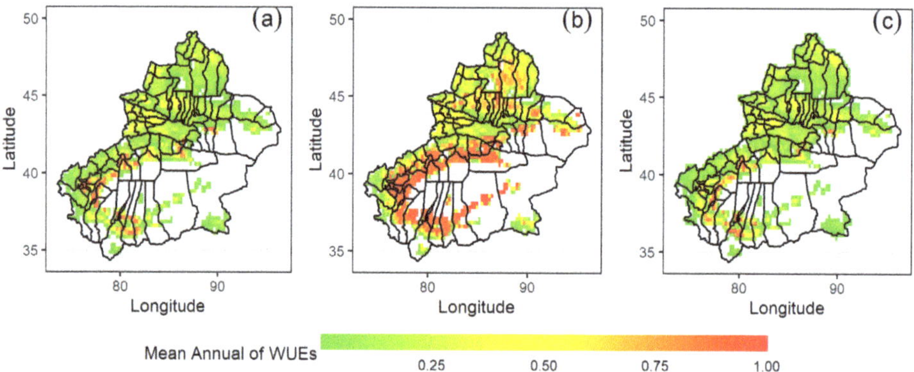

Figure 6. Spatial distribution of WUEs during 2000–2014; (**a**) eWUE; (**b**) tWUE; (**c**) pWUE.

We divided the change trends (Slope) into five categories (Table 1). The regions with increasing eWUE and pWUE accounted for 68.18% and 81.47%, respectively (Table 2), while the regions with decreasing trends were mainly distributed in the north, central, and a few southern regions of Xinjiang (Figure 7a,c). Secondly, the area with a slight increasing trend of TWUE accounted for 41.32%. The spatial distribution is scattered and irregular, which is mainly because the vegetation cover distribution and precipitation distribution are jointly determined. Surprisingly, the regions with high and relatively high increasing trends in southern Xinjiang accounted for 5.74% of the total. This phenomenon is mainly due to the relatively large fluctuations caused by farmland and irrigation in southern Xinjiang (Figure 7b). The tWUE change greatly compared with eWUE and pWUE, which indicates that the tWUE of arid and semi-arid areas changes significantly due to the strong fluctuation of the transpiration of vegetation [49].

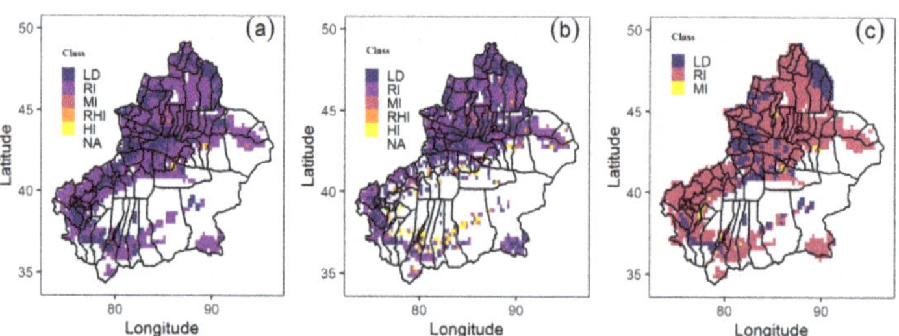

Figure 7. Latitude change trend of WUEs; (**a**) eWUE; (**b**) tWUE; (**c**) pWUE.

Table 2. Trend change and mean annual of WUEs percentage (units: %).

	Trend of WUEs					Mean Annual of WUEs				
Class	LD	RI	MI	RHI	HI	0–0.2	0.2–0.4	0.4–0.6	0.6–0.8	>0.8
eWUE	31.57	66.87	0.81	0.31	0.19	65.25	23.64	5.18	1.68	4.24
tWUE	34.77	41.32	3.00	2.43	3.31	23.66	26.78	15.04	8.86	25.66
pWUE	18.54	80.13	1.34	0	0	67.11	22.72	5.81	1.92	2.44

As shown in Figure 7c, pWUE has only three categories, and most regions show a slight increase, accounting for 80.13%. There are no areas of high increase. In conclusion, eWUE and pWUE have basically the same variation pattern, and the tWUE spatial variation trend is not uniform (Figure 7b). Most areas in northern Xinjiang show an upward trend in tWUE, while the changes around the desert are dramatic. This is due to the rising temperature in northern Xinjiang with increasing precipitation, resulting in increased vegetation photosynthesis and a steaming effect [12,50]. However, the drastic diurnal temperature difference in the desert leads to the unstable changes of vegetation photosynthesis and water loss, resulting in drastic changes of tWUE.

The increase in atmospheric CO_2 emissions caused global warming, which in theory could result in an increase in WUE. High temperatures inhibit photosynthesis and promote an increase in actual steaming emissions, which ultimately led to a decrease in WUE [51,52].

3.5. WUEs Responds to Temperature and Precipitation Thresholds

Numerous studies have shown that climatic condition is an important environmental factor that affects the water use efficiency of terrestrial ecosystems [52,53]. There is a threshold effect of both temperature and precipitation on WUEs. Xue et al. (2015) found that WUE was positively correlated with air temperature when the temperature was below 18.5 °C ($p < 0.05$), and it significantly decreased when the temperature was between 18.5 and 25.7 °C ($p < 0.05$) [54]. However, this study indicates that the response of diverse types of water use efficiency to temperature fluctuation is different in semi-arid area. Figure 8b,d,f show that water use efficiency is proportional to temperature. Due to the increased stomatal conductance of vegetation leaves, the increase in photosynthetic rate is greater than that the transpiration rate, resulting in an increase in WUE. The temperature thresholds of WUEs in different hydrological processes are different. Figure 8b,d,f show that the temperature, when around −10–2 °C, is proportional to eWUE. However, when the temperature is greater than 2 °C, the correlation between temperature and eWUE fluctuates. Temperature is also proportional to tWUE when it is within the range of −10–7 °C. Similarly, the pWUE and eWUE thresholds are basically the same.

The thresholds of this paper differ from those of Xue et al. [54], which is mainly because different regional climates influence the changes in thresholds. The study by Liu et al. (2015) on Chinese terrestrial ecosystems shows that most of the vegetation WUE has a positive response threshold of 500 mm to precipitation [55]; however, our study does not find such a pattern, and this may be due to the influence of arid climate characteristics. There is a positive correlation between eWUE and precipitation, which indicates that precipitation has a positive response to eWUE. From Figure 8a,c,e, it can be seen that for eWUE, when precipitation is less than 300 mm, precipitation is proportional to eWUE; when precipitation is greater than 300 mm, precipitation is negatively correlated with eWUE. For tWUE, when precipitation is less than 200, precipitation has a positive relationship with tWUE; when precipitation is greater than 200 mm, precipitation and eWUE exhibit a negative correlation. The pWUE and eWUE threshold are basically the same. The standard precipitation threshold in this paper is different from that of Liu et al. [55]. This is mainly because the small average annual precipitation in Xinjiang affects the change of threshold value.

The sensitive coefficient in Figure 9 shows the response of WUEs to climate factors. From Figure 9a,e it can be seen that tWUE has the largest response to temperature (Sp = 0.4) (Figure 9a); tWUE has the greatest response to precipitation (Sp = 1.0) (Figure 9e). tWUE responds more quickly to precipitation but responds more slowly to temperature. This shows that the effect of temperature on vegetation has a lag effect.

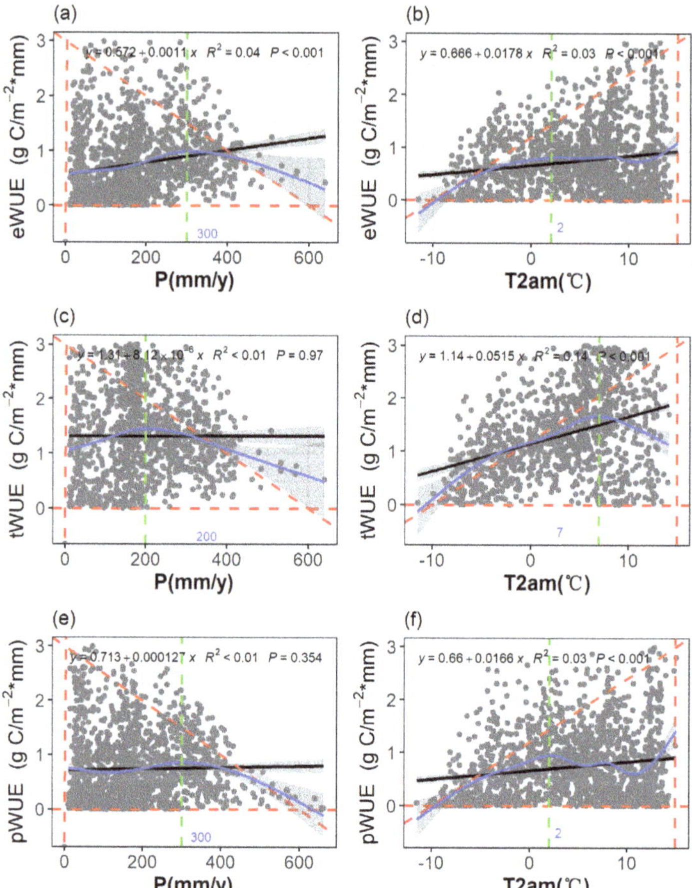

Figure 8. Relationship between precipitation, temperature, and WUEs. (The black line represents linear fitting; the blue line represents the variation trend of the fitting points. The gray is the 95% confidence interval, and the dashed green line is the threshold.)

The red line in Figure 8 shows the constraint analysis of all metadata in the semi-arid region. It was surprising to find two symmetrical "triangular" relationships between precipitation, temperature, and water use efficiency in the point distribution. From Figure 8a,c,e, we can see that the relationship between precipitation and WUEs shows a "left triangle shape". This indicates that in the semi-arid zone, an increase in precipitation does not in the meantime increase the water use efficiency. Due to the low fluctuation in the response of semi-arid vegetation to water, there is a lagged effect, which is consistent with the findings of Zou et al. (2018) [56], who suggested that there is a lagged effect of drought on ecosystem WUE. However, the relationship between air temperature and WUEs showed a "right triangle" shape, indicating that air temperature contributes to WUEs. This pattern is currently seen to be specific to semi-arid regions.

3.6. WUEs Responds to Temperature and Precipitation

From 2000 to 2014, a significant spatial difference in the sensitivity of WUEs to precipitation and temperature in Xinjiang can be observed (Figure 9), while the negative region of precipitation to WUEs was significantly larger than the positive regions. Precipitation is positively correlated with eWUE and tWUE Figure 9d,e in central Xinjiang, while the vast majority of the rest is negatively correlated. Among them, the negative correlation between precipitation and EWUE accounts for 88.8% of the region (Figure 9d); the negative correlation between precipitation and tWUE accounted for 84.8% of the region (Figure 9e); and the negative correlation between precipitation and pWUE prevailed throughout the region (Figure 9f).

It is shown that the sensitivity of vegetation transpiration to precipitation is relatively negative compared with evapotranspiration, and the difference is estimated in part because ecosystem evapotranspiration includes evaporation from exposed soils. The increase in precipitation in arid areas leads to the decrease in WUE, since the increase in ET is greater than the increase in photosynthesis. On the contrary [57], the decrease in precipitation leads to the increase in WUE in the region, because the decrease in ET is greater than the decrease in GPP [58], and the threshold effect on WUE may be related to the effective precipitation in the region.

From the spatial perspective, there is a positive correlation between temperature and WUEs, and the proportion of positive correlation is 70.5% (eWUE), 68.3% (tWUE), and 81.3% (pWUE), respectively (Figure 9a,b,c). The positive correlation area is mainly distributed in the north of Xinjiang, because the north of Xinjiang is the main agricultural area in Xinjiang, and the response of irrigation and farmland to temperature is more positive [45]. However, the southern region of Xinjiang is more scattered and has both positive and negative relationships, which may be related to the local geographical location and temperature inequality caused by deserts. The negative values are mainly distributed in the central and southern regions of Xinjiang, as the negative values are located near low latitude and have sufficient thermal energy. This region is highly sensitive to precipitation (Figure 9d–f).

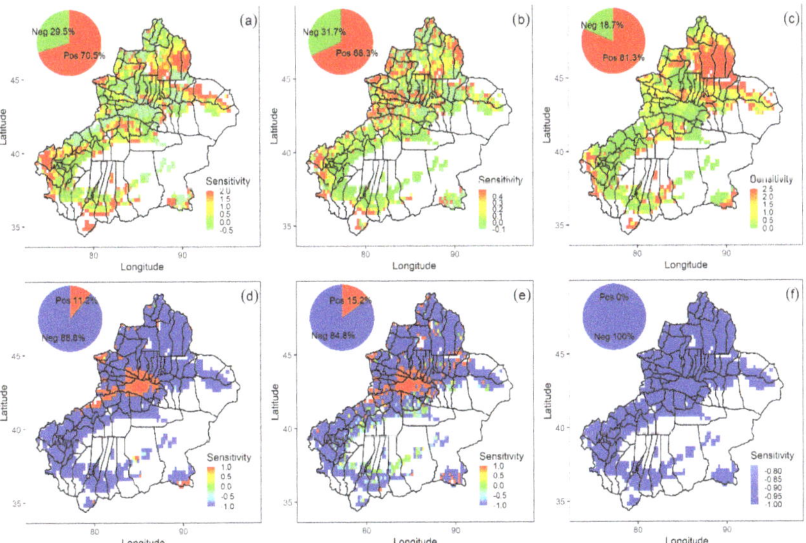

Figure 9. WUEs sensitivity coefficient to precipitation and temperature (**a**), (**b**), and (**c**) respectively represent the sensitivity coefficient of temperature to eWUE, tWUE, and pWUE; (**d**), (**e**), and (**f**) respectively represent the sensitivity coefficient of precipitation to eWUE, tWUE, and pWUE; the upper left illustration shows the proportion of positive (*Pos*) and negative (*Neg*) sensitivities.

4. Conclusions

This study of water use efficiency (WUE) of different hydrological processes helps to shed new light on the response of hydrological processes and ecosystems to climate change and provides valuable approaches for further research. In particular, we found that in the past 15 years (2000–2014), the seasonal fluctuation of evapotranspiration in arid areas was large and the interannual variation was unstable, which was mainly due to the continuous "warm and wet" phenomenon in the northwest of China. The spatial variations of WUE in different hydrological processes were significant, and the trend of WUE in most regions showed an upward trend (eWUE (68.43%); tWUE (65.23%); pWUE (81.46%)). We also found that in arid areas, temperature has a positive effect on WUE, while precipitation has a lagging effect on WUE. Secondly, the threshold value of precipitation to WUEs (200 or 300 mm) and temperature to WUEs (2 or 7 °C) are also different from previous studies. The causes of the differences require further studies.

Author Contributions: Data curation, F.Z. and S.-S.Y.; Investigation, F.Z.; Supervision, Y.-Y.T.; Writing—original draft, Y.-Y.T.; Writing—review and editing, J.-P.C. All authors have read and agreed to the published version of the manuscript.

Funding: This research was funded by the by the China-Geological Survey Bureau Coastal Zone and Island Reef Military-civilian Integration Geological Survey Project [Grant No. DD20208017].

Institutional Review Board Statement: Not applicable.

Informed Consent Statement: Not applicable.

Data Availability Statement: The data presented in this study are available on request from the corresponding author.

Conflicts of Interest: The authors declare that they have no known competing financial interests or personal relationships that would influence the work reported in this paper.

Appendix A

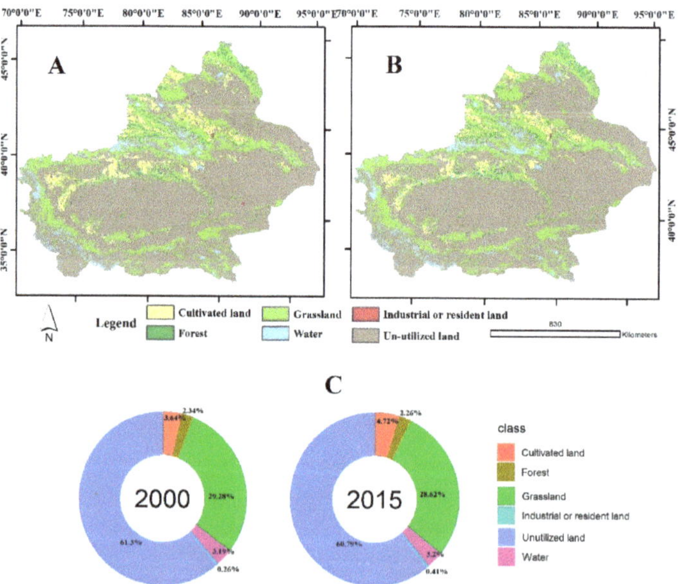

Figure A1. The LULC of Xinjiang in 2000 (**A**) and 2015 (**B**,**C**) Pie chart showing the percentage of land type.

References

1. Zhao, M.; Running, S.W. Drought-Induced Reduction in Global Terrestrial Net Primary Production from 2000 Through 2009. *Science* **2010**, *329*, 940–943. [CrossRef]
2. Zhang, Z.; Jiang, H.; Liu, J.; Zhou, G.; Liu, S.; Zhang, X. Assessment on water use efficiency under climate change and heterogeneous carbon dioxide in China terrestrial ecosystems. *Procedia Environ. Sci.* **2012**, *13*, 2031–2044. [CrossRef]
3. Ponce-Campos, G.E.; Moran, M.S.; Huete, A.; Zhang, Y.G.; Bresloff, C.; Huxman, T.E.; Eamus, D.; Bosch, D.D.; Buda, A.R.; Gunter, S.A.; et al. Ecosystem resilience despite large-scale altered hydro-climatic conditions. *Nature* **2013**, *494*, 349–352. [CrossRef] [PubMed]
4. Huang, M.; Piao, S.; Sun, Y.; Ciais, P.; Cheng, L.; Mao, J.; Poulter, B.; Shi, X.; Zeng, Z.; Wang, Y. Change in terrestrial eco-sys-tem water-use efficiency over the last three decades. *Glob. Chang. Biol.* **2015**, *21*, 2366–2378. [CrossRef] [PubMed]
5. Reichstein, M.; Ciais, P.; Papale, D.; Valentini, R.; Running, S.; Viovy, N.; Cramer, W.; Granier, A.; Ogée, J.; Allard, V.; et al. Reduction of ecosystem productivity and respiration during the European summer 2003 climate anomaly: A joint flux tower, remote sensing and modelling analysis. *Glob. Chang. Biol.* **2007**, *13*, 634–651. [CrossRef]
6. Keenan, T.F.; Hollinger, D.Y.; Bohrer, G.; Dragoni, D.; Munger, J.W.; Schmid, H.P.; Richardson, A.D. Increase in forest water-use efficiency as atmospheric carbon dioxide concentrations rise. *Nat. Cell Biol.* **2013**, *499*, 324–327. [CrossRef]
7. Liu, X.; Hu, B.; Ren, Z. Spatiotemporal variation of water use efficiency and its driving forces on the Loess Plateau during 2000–2014. *Sci. Agric. Sin.* **2018**, *51*, 302–314.
8. Zhou, S.; Yu, B.; Schwalm, C.R.; Ciais, P.; Zhang, Y.; Fisher, J.B.; Wang, G. Response of water use efficiency to global environmental change based on output from terrestrial biosphere models. *Glob. Biogeochem. Cycles* **2017**, *31*, 1639–1655. [CrossRef]
9. Woodward, C.; Shulmeister, J.; Larsen, J.; Jacobsen, G.E.; Zawadzki, A. The hydrological legacy of deforestation on global wetlands. *Science* **2014**, *346*, 844–847. [CrossRef]
10. Zheng, H.; Lin, H.; Zhou, W.; Bao, H.; Zhu, X.; Jin, Z.; Song, Y.; Wang, Y.; Liu, W.; Tang, Y. Revegetation has increased ecosystem water-use efficiency during 2000–2014 in the Chinese Loess Plateau: Evidence from satellite data. *Ecol. Indic.* **2019**, *102*, 507–518. [CrossRef]
11. Tian, H.; Chen, G.; Liu, M.; Zhang, C.; Sun, G.; Lu, C.; Chappelka, A. Model estimates of net primary productivity, evapotranspiration, and water use efficiency in the terrestrial ecosystems of the southern United States during 1895–2007. *For. Ecol. Manag.* **2010**, *259*, 1311–1327. [CrossRef]
12. Huang, M.; Piao, S.; Zeng, Z.; Peng, S.; Ciais, P.; Cheng, L.; Wang, Y. Seasonal responses of terrestrial ecosystem wateruse efficiency to climate change. *Glob. Chang. Biol.* **2016**, *22*, 2165–2177. [CrossRef]
13. Yang, Y.; Guan, H.; Batelaan, O.; McVicar, T.; Long, D.; Piao, S.; Liang, W.; Liu, B.; Jin, Z.; Simmons, C.T. Contrasting responses of water use efficiency to drought across global terrestrial ecosystems. *Sci. Rep.* **2016**, *6*, 23284. [CrossRef]
14. Sun, Y.; Piao, S.; Huang, M.; Ciais, P.; Zeng, Z.; Cheng, L.; Zeng, H. Global patterns and climate drivers of water-use efficiency in terrestrial ecosystems deduced from satellite-based datasets and carbon cycle models. *Glob. Ecol. Biogeogr.* **2016**, *25*, 311–323. [CrossRef]
15. Li, S.C.; Cai, Y.L. Some scaling issues of geography. *Geogr. Res.* **2005**, *24*, 11–18.
16. Li, X.J.; Zhang, F.P.; Wang, H.W.; Veroustraete, F. Analysis of Spatio-Temporal Characteristics of water use efficiency of Vegetation and its Relationship with Climate in the Heihe River Basin. *J. Desert Res.* **2017**, *37*, 733–741.
17. Hu, Z.; Yu, G.; Fu, Y.; Sun, X.; Li, Y.; Shi, P.; Wang, Y.; Zheng, Z. Effects of vegetation control on ecosystem water use efficiency within and among four grassland ecosystems in China. *Glob. Chang. Biol.* **2008**, *14*, 1609–1619. [CrossRef]
18. Yao, J.; Zhao, Y.; Chen, Y.; Yu, X.; Zhang, R. Multi-scale assessments of droughts: A case study in Xinjiang, China. *Sci. Total Environ.* **2018**, *630*, 444–452. [CrossRef]
19. Xu, C.; Chen, Y.; Yang, Y.; Hao, X.; Shen, Y. Hydrology and water resources variation and its response to regional climate change in Xinjiang. *J. Geogr. Sci.* **2010**, *20*, 599–612. [CrossRef]
20. Yu, H.; Bian, Z.; Mu, S.; Yuan, J.; Chen, F. Effects of Climate Change on Land Cover Change and Vegetation Dynamics in Xinjiang, China. *Int. J. Environ. Res. Public Health* **2020**, *17*, 4865. [CrossRef]
21. White, M.A.; Brunsell, N.; Schwartz, M.D. Vegetation phenology in global change studies. *Phenol. Integr. Environ. Ment. Sci.* **2003**, *39*, 453–466.
22. Chen, Y.; Li, Z.; Fang, G.; Li, W. Large Hydrological Processes Changes in the Transboundary Rivers of Central Asia. *J. Geophys. Res. Atmos.* **2018**, *123*, 5059–5069. [CrossRef]
23. Shen, T.; Su, H.C.; Wang, G.Y.; Mao, W.M.; Wang, S.D.; Han, P.; Wang, N.; Li, Z.Q. The Responses of Glaciers and Snow Cover to Climate Change in Xinjiang (I): Hydrological Effects. *J. Glaciol. Geocryol.* **2013**, *35*, 513–527.
24. Xu, C.; Li, J.; Zhao, J.; Gao, S.; Chen, Y. Climate variations in northern Xinjiang of China over the past 50 years under global warming. *Quat. Int.* **2015**, *358*, 83–92. [CrossRef]
25. Miralles, D.G.; Holmes, T.R.H.; De Jeu, R.A.M.; Gash, J.H.; Meesters, A.G.C.A.; Dolman, A.J. Global land-surface evaporation estimated from satellite-based observations. *Hydrol. Earth Syst. Sci.* **2011**, *15*, 453–469. [CrossRef]
26. Vicente-Serrano, S.; Miralles, D.; Domínguez-Castro, F.; Azorin-Molina, C.; ElKenawy, A.; McVicar, T.; Tomas-Burguera, M.; Beguería, S.; Maneta, M.; Peña-Gallardo, M. Global assessment of the standardized Evapotranspiration Deficit Index (SEDI) for drought analysis and monitoring. *J. Clim.* **2018**, *31*, 5371–5393. [CrossRef]

27. Long, B.; Zhang, B.; He, C.; Shao, R.; Tian, W. Is There a Change From a Warm-Dry to a Warm-Wet Climate in the Inland River Area of China? Interpretation and Analysis Through Surface Water Balance. *J. Geophys. Res. Atmos.* **2018**, *123*, 7114–7131. [CrossRef]
28. Li, S.; Wang, G.; Sun, S.; Chen, H.; Bai, P.; Zhou, S.; Huang, Y.; Wang, J.; Deng, P. Assessment of Multi-Source Evapotranspiration Products over China Using Eddy Covariance Observations. *Remote. Sens.* **2018**, *10*, 1692. [CrossRef]
29. Allen, R.G.; Pereira, L.; Raes, D.; Smith, M. Crop evapotranspiration: Guidelines for computing crop water requirements. In *FAO Irrigation and Drainage Paper 56*; FAO: Rome, Italy, 1998.
30. Junde, W. Variations of Land Evapotranspiration in the Plain of the Middle Reaches of Heihe River in the Recent 35 Years. *J. Glaciol. Geocryol.* **2007**, *29*, 406–412.
31. Peng, S.; Ding, Y.; Liu, W.; Li, Z. 1 km monthly temperature and precipitation dataset for China from 1901 to 2017. *Earth Syst. Sci. Data* **2019**, *11*, 1931–1946. [CrossRef]
32. Peng, S.; Gang, C.; Cao, Y.; Chen, Y. Assessment of climate change trends over the Loess Plateau in China from 1901 to 2100. *Int. J. Clim.* **2018**, *38*, 2250–2264. [CrossRef]
33. Ding, Y.; Peng, S. Spatiotemporal Trends and Attribution of Drought across China from 1901–2100. *Sustainability* **2020**, *12*, 477. [CrossRef]
34. Zhao, M.; Running, S.W.; Nemani, R.R. Sensitivity of moderate resolution imaging spectroradiometer (modis) terrestri-al primary production to the accuracy of meteorological reanalyses. *J. Geophys. Res. Biogeoences* **2015**, *111*, 1–13. [CrossRef]
35. White, M.A.; Thornton, P.E.; Running, S.W.; Nemani, R.R. Parameterization and sensitivity analysis of the BIOME-BGC terrestrial ecosystem model: Net primary production controls. *Earth Interact.* **2000**, *4*, 1–84. [CrossRef]
36. Heinsch, F.; Zhao, M.; Running, S.; Kimball, J.; Nemani, R.; Davis, K.; Bolstad, P.; Cook, B.; Desai, A.; Ricciuto, D.; et al. Evaluation of remote sensing based terrestrial productivity from MODIS using regional tower eddy flux network observations. *IEEE Trans. Geosci. Remote. Sens.* **2006**, *44*, 1908–1925. [CrossRef]
37. Milich, L.; Weiss, E. GAC NDVI inter-annual coefficient of variation (CoV) images: Ground truth sampling of the Sahel along north-south transects. *Int. J. Remote. Sens.* **2000**, *21*, 235–260. [CrossRef]
38. Guli·jiapaer, G.; Liang, S.; Yi, Q.; Liu, J. Vegetation dynamics and responses to recent climate change in Xinjiang using leaf area index as an indicator. *Ecol. Indic.* **2015**, *58*, 64–76. [CrossRef]
39. Yang, S.; Zhang, J.; Zhang, S.; Wang, J.; Bai, Y.; Yao, F.; Guo, H. The potential of remote sensing-based models on global wateruse efficiency estimation: An evaluation and inter comparison of an ecosystem model (BESS) and algorithm (MODIS) using site level and upscaled eddy covariance data. *Agric. For. Meteorol.* **2020**, *287*, 107959. [CrossRef]
40. Wang, D.; Zhang, F.; Yang, S.; Xia, N.; Ariken, M. Exploring the spatial-temporal characteristics of the aerosol optical depth (AOD) in Central Asia based on the moderate resolution imaging spectroradiometer (MODIS). *Environ. Monit. Assess.* **2020**, *192*, 1–15. [CrossRef] [PubMed]
41. Zou, J.; Ding, J.; Welp, M.; Huang, S.; Liu, B. Using MODIS data to analyse the ecosystem water use efficiency spa-tial-temporal variations across Central Asia from 2000 to 2014. *Environ. Res.* **2020**, *182*, 108985. [CrossRef] [PubMed]
42. Gu, C.; Tang, Q.; Zhu, G.; Ma, J.; Gu, C.; Zhang, K.; Niu, S. Discrepant responses between evapotranspiration-and tran-spiration based ecosystem water use efficiency to inter annual precipitation fluctuations. *Agric. For. Meteorol.* **2021**, *303*, 108385. [CrossRef]
43. Kim, H.W.; Hwang, K.; Mu, Q.; Lee, S.O.; Choi, M. Validation of MODIS 16 global terrestrial evapotranspiration products in various climates and land cover types in Asia. *Ksce J. Civ. Eng.* **2012**, *16*, 229–238. [CrossRef]
44. Yin, Y.; Wu, S.; Zheng, D.; Yang, Q. Regional difference of aridity/ humidity conditions change over China during the last thirty years. *Chin. Sci. Bull.* **2005**, *50*, 2226–2233. [CrossRef]
45. Zhang, H.; Song, J.; Wang, G.; Wu, X.; Li, J. Spatiotemporal characteristic and forecast of drought in northern Xinjiang, China. *Ecol. Indic.* **2021**, *127*, 107712. [CrossRef]
46. He, Y.B.; Su, Z.; Li, J.; Wang, S.L. Regional evapotranspiration of different land covers based on remotesensing. *Chin. J. Appl. Ecol.* **2007**, *18*, 288–296.
47. Wang, Q.; Zhai, P.; Qin, D. New perspectives on «warming-wetting» trend in Xinjiang, China. *Adv. Clim. Chang. Res.* **2020**, *11*, 252–260. [CrossRef]
48. Song, L.L.; Yin, Y.H.; Wu, S.H. Advancements of the metrics of evapotranspiration. *Prog. Geogr.* **2012**, *31*, 1186–1195.
49. Kaufmann, R.K.; Kauppi, H.; Mann, M.L.; Stock, J.H. Reconciling anthropogenic climate change with observed temper-ature 1998–2008. *Sch. Artic.* **2011**, *108*, 11790–11793.
50. Gong, T.; Lei, H.; Yang, D.; Liu, T.; Duan, L. Assessing impacts of extreme water and temperature conditions on carbon fluxes in two desert shrub lands. *J. Hydroelectr. Eng.* **2018**, *37*, 32–46.
51. DeLucia, E.H.; Drake, J.; Thomas, R.B.; Gonzalez-Meler, M. Forest carbon use efficiency: Is respiration a constant fraction of gross primary production? *Glob. Chang. Biol.* **2007**, *13*, 1157–1167. [CrossRef]
52. Xia, L.; Wang, F.; Mu, X.; Jin, K.; Sun, W.; Gao, P.; Zhao, G. Water use efficiency of net primary production in global terres-trial ecosystems. *J. Earth Syst. Sci.* **2015**, *124*, 921–931. [CrossRef]
53. Zhang, F.; Ju, W.; Shen, S.; Wang, S.; Yu, G.; Han, S. How recent climate change influences water use efficiency in East Asia. *Theor. Appl. Clim.* **2013**, *116*, 359–370. [CrossRef]
54. Xue, B.-L.; Guo, Q.; Otto, A.; Xiao, J.; Tao, S.; Li, L. Global patterns, trends, and drivers of water use efficiency from 2000 to 2013. *Ecosphere* **2015**, *6*, art174. [CrossRef]

55. Liu, Y.; Xiao, J.; Ju, W.; Zhou, Y.; Wang, S.; Wu, X. Water use efficiency of China's terrestrial ecosystems and responses to drought. *Sci. Rep.* **2015**, *5*, 13799. [CrossRef] [PubMed]
56. Zou, J.; Ding, J.L.; Qin, Y.; Wang, F. Response of water use efficiency of Central Asia ecosystem to drought based on remote sensing data. *Trans. CSAE.* **2018**, *34*, 145–152. (In Chinese)
57. Zhang, Y.; Yu, G.; Yang, J.; Wimberly, M.; Zhang, X.; Tao, J.; Jiang, Y.; Zhu, J. Climate-driven global changes in carbon use efficiency. *Glob. Ecol. Biogeogr.* **2013**, *23*, 144–155. [CrossRef]
58. Salve, R.; Sudderth, E.A.; Clair, S.B.S.; Torn, M.S. Effect of grassland vegetation type on the responses of hydrological processes to seasonal precipitation patterns. *J. Hydrol.* **2011**, *410*, 51–61. [CrossRef]

 water

Article

Estimation of Groundwater Evapotranspiration of Different Dominant Phreatophytes in the Mu Us Sandy Region

Wuhui Jia [1], Lihe Yin [2,*], Maosheng Zhang [3], Kun Yu [2], Luchen Wang [1,2] and Fusheng Hu [1]

[1] School of Water Resources and Environment, China University of Geosciences, Beijing 100083, China; jiawh@cugb.edu.cn (W.J.); wangluchen@cugb.edu.cn (L.W.); 1991010439@cugb.edu.cn (F.H.)
[2] Xi'an Center of Geological Survey, China Geological Survey, Xi'an 710054, China; yukun01@mail.cgs.gov.cn
[3] Institute of Disaster Prevention and Ecological Restoration, Xi'an Jiaotong University, Xi'an 710049, China; xjtzms@xjtu.edu.cn
* Correspondence: ylihe@cgs.cn; Tel.: +86-29-8782-1680

Abstract: Groundwater evapotranspiration (ET_G) estimation is an important issue in semiarid areas for groundwater resources management and environmental protection. It is widely estimated by diurnal water table fluctuations. In this study, the ET_G at four sites with different plants was estimated using both diurnal water table and soil moisture fluctuations in the northeastern Mu Us sandy region, in order to identify the groundwater utilization strategy by different dominant phreatophytes. Groundwater level was monitored by ventilatory pressure transducers (Solinst LevelVent, Solinst Canada Ltd.; accuracy ±3 mm), while soil moisture was monitored using EM50 loggers (Decagon Devices Inc., Pullman, USA) in K1 and K14 and simulated by Hydrus-1D in other observation wells. A significant spatial variation of ET_G was found within a limited area, indicating a poor representativeness of site ET_G for regional estimation. The mean values of ET_G are 4.01 mm/d, 6.03 mm/d, 8.96 mm/d, and 12.26 mm/d at the *Achnatherum splendens* site, *Carex stenophylla* site, *Salix psammophila* site and *Populus alba* site, respectively, for the whole growing season. ET_G is more sensitive to depth to water table (DWT) in the *Carex stenophylla* site than in the *Achnatherum splendens* site for grass-dominated areas and more sensitive to DWT in the *Populus alba* site than in *Salix psammophila* site for tree-dominated areas. Groundwater extinction depths are estimated at 4.1 m, 2.4 m, 7.1 m, and 2.9 m in the *Achnatherum splendens* site, *Carex stenophylla* site, *Salix psammophila* site and *Populus alba* site, respectively.

Keywords: diurnal fluctuations; Phreatophyte; semiarid; wetland; vegetation restoration

Citation: Jia, W.; Yin, L.; Zhang, M.; Yu, K.; Wang, L.; Hu, F. Estimation of Groundwater Evapotranspiration of Different Dominant Phreatophytes in the Mu Us Sandy Region. *Water* **2021**, *13*, 440. https://doi.org/10.3390/w13040440

Academic Editors: Mirko Castellini, Simone Di Prima, Ryan Stewart, Marcella Biddoccu, Mehdi Rahmati and Vincenzo Alagna

Received: 7 January 2021
Accepted: 4 February 2021
Published: 8 February 2021

1. Introduction

In arid and semi-arid areas, restoring and protecting the environment is a huge project for the benefit of mankind [1]. While groundwater is a crucial source of ecological restoration because of the lower precipitation and higher evapotranspiration in arid regions. However, these areas also depend heavily on groundwater for industry, drinking water supply and agriculture [2,3]. Reasonable allocation of groundwater resources in all sectors is thus a major environmental issue.

Groundwater is a main water source for phreatophytes in arid and semiarid areas. Evapotranspiration partitioning shows that transpiration accounts for the majority of evapotranspiration [4,5]. Over the past few decades, vegetation coverage has displayed a large increase because of the multiple ecological restoration programs including the grain for green and afforestation in some arid areas of China [6]. However, ecological restoration by afforestation may increase groundwater depth and create potentially large ecological and water costs in arid and semiarid [7]. For example, Gribovszki et al. [8] found that the water table beneath forest was 0.4–0.5 m lower than that beneath grassland due to the afforestation in the Hungarian Great Plain. So, the groundwater consumption strategy of each phreatophyte should be considered in the process of afforestation.

In addition, groundwater evapotranspiration (ET_G) is a vital factor for groundwater balance analysis in arid and simi-arid regions. For example, almost 60% of natural groundwater discharge is through ET_G in the Ordos Plateau in NW China [9]. Many studies usually estimate ET_G indirectly, i.e., by the pan coefficient method [10], the remote sensing method [11] and the eddy covariance method [12]. The pan coefficient method is simple, but coefficients mainly are empirical with low precision. The remote sensing method is usually used to calculate evapotranspiration at a regional scale and the time step depends on remote sensing data. It cannot analyze the temporal and spatial variation of evapotranspiration at small scale. The eddy covariance method depends on expensive instruments, so it is not widely used. In addition, all of these methods estimate rather total evapotranspiration than ET_G. Evapotranspiration partitioning is needed to exclude the contribution of soil water, which is a complicated process. Therefore, over the past few decades, the method of estimating evapotranspiration based on diurnal water table fluctuations, developed by White [13], has attracted more and more attention [14–18], because it can estimate ET_G directly with fewer parameters and lower costs.

The Mu Us sandy region is a part of the Ordos Plateau in NW China and is used to be one of the most seriously desertified areas in China [19]. It has also undergone a significant vegetation restoration in recent decades [20]. While Cheng et al. [21] estimated the ET_G at four sites in the center of the Mu Us sandy region, using only one observation at each site. However, the spatial variation of ET_G at point scale under each vegetation is unclear and the representativeness of the ET_G for one observation well is also uncertain. Multiple wells are expected to improve the accuracy of ET_G estimation in each vegetation site [22]. Therefore, the objectives of this study are (1) to estimate the ET_G from multiple observation wells based on the water table and soil moisture fluctuations at different sites; (2) to study the temporal and spatial variations of ET_G; and (3) to analyze the sensitivity of ET_G of different vegetation to water table depth.

2. Materials and Methods

2.1. Site Description

The study site (39°16′54″~39°18′48″ N, 108°47′40″~108°50′03″ E) is located in the northeastern Mu Us sandy region, within a wetland of the Mukai Lake in Ordos City, Inner Mongolia, NW China. The geomorphology is mainly lake basin that contains some microrelief, such as wetlands, crescent sand dunes, and highlands (Figure 1). Vegetation is distributed unevenly in the various microrelief. Vegetation coverage is higher in wetlands, up to about 70%. While, there is only sparse vegetation on sand dunes. The climate is semi-arid characterized by a low rainfall and high evapotranspiration. According to the long-term meteorological data (2000–2017) from the Wushenzhao meteorological station near the study area, the mean annual precipitation is ~330 mm and over 70% of the annual precipitation falls from June to September. Average reference evapotranspiration (ET_0) calculated by the Penman–Monteith equation [23] is ~1150mm/a. Average annual air temperature is about 7.0 °C, with a maximum of 22.4 °C in July and a minimum of −11.0 °C in January. Population density is low and arable lands near the study area are scarce, so the consumption of groundwater by human activities is negligible.

The main species of vegetation are *Achnatherum splendens* (Figure 2a), *Carex stenophylla* (Figure 2b), *Populus alba* (Figure 2c), *Salix psammophila* (Figure 2d), and *Caragana korshinskii* (Figure 2e). The grass sites (*Achnatherum splendens* and *Carex stenophylla*) are distributed around the lake, while the tree sites (*Populus alba, Salix psammophila* and *Caragana korshinskii*) are located adjacent and upstream of the grass sites.

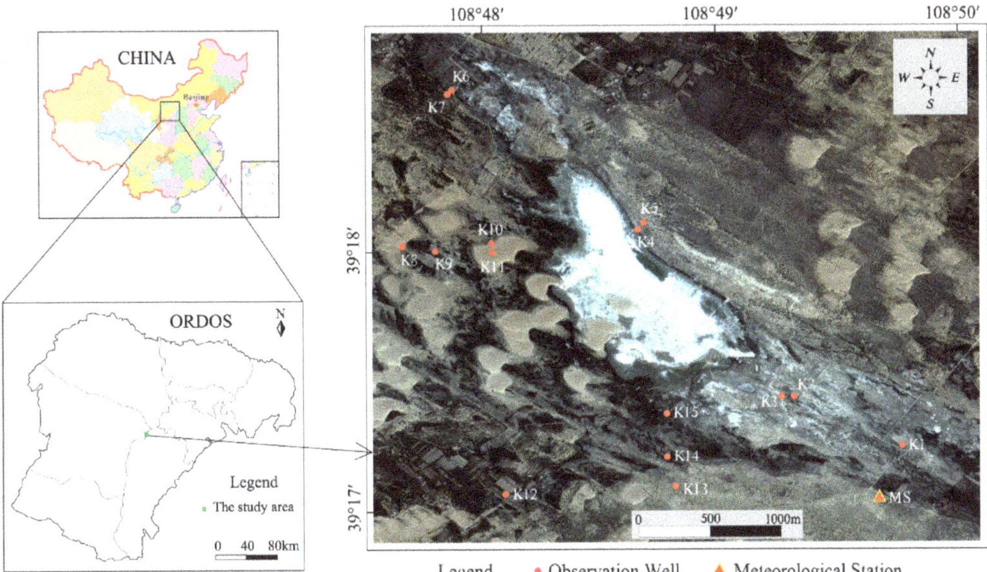

Figure 1. Location map of the study site showing the observation wells and meteorological station.

Figure 2. The dominate vegetations, i.e., (**a**) *Achnatherum splendens*, (**b**) *Carex stenophylla*, (**c**) *Populus alba*, (**d**) *Salix psammophila* and (**e**) *Caragana korshinskii*, in the study site.

2.2. Monitoring Design

Fifteen shallow observation wells (K1 to K15) were drilled by a hand auger around the lake (Figure 1). K1, K2, and K3 are in the *Achnatherum splendens* site; K4, K9, K10, K14, and K15 are in the *Carex stenophylla* site; K5 and K13 are in the *Caragana korshinskii* site; K6, K7 are in the *Salix psammophila* site; K8 and K11 are located in sand dunes; K12 is in the *Populus alba* site. The depth of the wells varies between 2 m and 4 m, depending on the depth to water table (DWT) at each site. Each was constructed using a 50 mm PVC pipe screened over the entire subsurface length. Augured soil was backfilled around the PVC pipe and granular bentonite was then packed around land surface to avoid preferential flow.

Groundwater level was continuously monitored from 1 April 2019 to 31 October 2019 at one-hour intervals by ventilatory pressure transducers with data logging capability (Solinst LevelVent, Solinst Canada Ltd.; accuracy ±3 mm) that automatically eliminated the effect of air pressure fluctuations by a vented cable to the wellhead. The well cap assembly of the transducers serves to prevent direct rainfall input and evaporation loss. In order to remove noise, a five-point moving average was used to smoothen groundwater level data [18]. The data were also calibrated frequently in the field with manual measurements of DWT.

Soil samples were collected from different depths and the particle sizes were determined by the sieving method for the sand fractions and by a soil hydrometer for the silt and clay fractions. Results showed that the soil texture of unsaturated zone at the study sites was mainly fine sand and homogeneous in the vertical direction at each well (Table 1). The soil textures of all observation wells were divided artificially into two categories, with sand content higher than 90%, and with sand content less than 90%. Then, two observation sites, K1 and K14, were thought to be representative of the two soil textures. Soil moisture (θ_s) was monitored once an hour from 20 cm to 100 cm below ground surface with 20 cm intervals at K1 and from 20 cm to 60 cm with 10 cm intervals at K14 using EM50 loggers (Decagon Devices Inc., Pullman, USA).

Table 1. Particle size distribution at different depths in the Observation wells.

Wells	Soil Depth (cm)	Sand (%)	Silt (%)	Clay (%)
K1	40	82.9	17.1	0
	60	83.4	16.6	0
	80	86.3	13.7	0
K3	50	85.0	15.0	0
	70	85.1	14.9	0
K4	20	85.5	14.5	0
	50	83.2	16.8	0
	70	81.9	18.1	0
K6	50	91.0	9.0	0
	80	91.1	8.9	0
K9	20	85.6	14.4	0
	50	85.2	14.8	0
K10	20	92.7	7.3	0
	50	94.7	5.3	0
K12	50	92.4	7.6	0
	80	92.9	7.1	0
	100	95.7	4.3	0
K14	30	98.0	2.0	0
	60	98.4	1.6	0
	80	92.6	7.4	0
K15	20	87.2	12.8	0

Meteorological data (i.e., solar radiation, air temperature, air humidity, and wind speed) were monitored at one-hour intervals by a meteorological station (HOBO-U30, Onset, USA) at the southeast of study site. Hourly reference evapotranspiration (ET_0) was calculated by the Penman–Monteith Equation (Equation (1)) using the meteorological data [23].

$$ET_0 = [0.408\Delta(R_n - G) + 37\gamma\mu(e^{\circ}(T_h) - e_a)/(T_h + 273)]/[\Delta + \gamma(1 + 0.34\mu)]. \quad (1)$$

where ET_0 is reference evapotranspiration ($mm \cdot h^{-1}$); R_n is net radiation ($MJ \cdot m^{-2} \cdot h^{-1}$); G is soil heat flux density ($MJ \cdot m^{-2} \cdot h^{-1}$); T_h is mean hourly air temperature (°C); Δ is saturation slope vapor pressure curve at T_h ($kPa \cdot {}^{\circ}C^{-1}$); γ is psychrometric constant ($kPa \cdot {}^{\circ}C^{-1}$); $e^{\circ}(T_h)$ is saturation vapor pressure at air temperature T_h (kPa); e_a is average hourly actual vapor pressure (kPa); μ is average hourly wind speed ($m \cdot s^{-1}$). It is worth noting that ET_0 is the evapotranspiration from the reference surface which is under a hypothetical grass reference crop with an assumed crop height of 0.12 m, a fixed surface resistance of 70 $s \cdot m^{-1}$ and an albedo of 0.23. The reference surface closely resembles an extensive surface of green, well-watered grass of uniform height, actively growing and completely shading the ground. The fixed surface resistance of 70 s·m^{-1} implies a moderately dry soil surface resulting from about a weekly irrigation frequency [23].

2.3. Numerical Simulations

2.3.1. Soil Hydraulic Parameters Estimated by Inverse Modeling

Soil hydraulic parameters were obtained through inverse modeling by Hydrus-1D, which was widely used by previous studies [24–26]. For K1 and K14, as the representative of the two soils, both groundwater level and soil moisture were monitored. Therefore, the meteorological data, water table and soil moisture data were used to inverse soil hydraulic parameters of the two soil textures in the study site.

For the inverse modeling, a 150 cm homogeneous soil profile was divided uniformly into 150 elements, with 151 nodes and two observations corresponding to the soil moisture monitoring depths (20 cm and 100 cm at K1; 20 cm and 40 cm at K14). The maximum number of iterations was set to 50. The van Genuchten–Mualem hydraulic model was selected and the soil hydraulic parameters of loamy sand in the model were used as the initial parameters because the soil texture is classified as loamy sand based on particle size analysis. The maximum and minimum values of each parameter, i.e., θ_s, θ_r, φ, n, and Ks, are set to limit the range of inversion. According to the particle analysis of soils in the study site (Table 1), the maximum values of φ, n and Ks were set to the default parameters of sand in Hydrus-1D code and the minimum values equaled to the default parameters of silt, as shown in Table 2. Data for inverse solution were soil moisture data measured in the field at the two observations depths. Time discretization was in hours and initial time step was set to 0.01.

Table 2. The initial, minimum and maximum values of the soil hydraulic parameters in inverse solution.

	θ_r (cm^3/cm^3)	θ_s (cm^3/cm^3)	φ (1/cm)	n	K_s (cm/h)
Initial Value	0.057	0.41	0.124	2.28	14.5917
Minimum	0	0	0.016	1.37	0.25
Maximum	0.078	0.43	0.145	2.68	29.7

The upper boundary condition was defined as an atmospheric boundary condition. The potential evapotranspiration (ET_p) equals to ET_0 as recommended by Šimůnek et al. [27], and the potential evaporation (E_p) and potential transpiration (T_p) were estimated from ET_p using Beer's law that partitions the solar radiation component of the energy budget via interception by the canopy [28,29] as follows in Equations (2) and (3):

$$E_p = ET_p \times e^{-k*LAI} \tag{2}$$

$$T_p = ET_p \times (1 - e^{-k*LAI}) \tag{3}$$

where k is a constant governing the radiation extinction by the canopy (-), the value is about 0.6 according to Belmans et al. [30]. LAI is the leaf area index and the values in grass site and tree site are 1.9 and 2.6, respectively. The lower boundary condition was variable pressure head that was monitored in the study site. The period of the inverse model was from 20 May 2019 to 5 June 2019 for K1 and from 24 May 2019 to 6 September 2019 for K14, when obvious diurnal fluctuations of water table and soil water moisture continuously occurred.

In the study site, the root depth was about 150 cm. An exponential root distribution model developed by Zeng [31] was used in our model, as follow in Equation (4):

$$Y = 1 - 0.5\,(e^{-ad} + e^{-bd}) \tag{4}$$

where Y is the cumulative root fraction from the surface to depth d; a and b are the empirical coefficients. For short grass, a and b are 10.74 and 2.61, respectively. The S-shaped water stress function of root water uptake was adopted [32].

2.3.2. Simulated Soil Moisture

Soil moisture was monitored at K1 and K14 and simulated by Hydrus-1D at other observation wells, because of the lack of monitoring equipment. The simulation domain is a 150 cm homogeneous soil profile and was divided uniformly into 150 elements with 151 nodes. The simulated period was from 1 April 2019 to 31 October 2019 and the soil hydraulic parameters were from the above-mentioned inverse modelling. The upper boundary condition was also atmospheric boundary condition. E_p and T_p were also estimated by Equations (2) and (3). The lower boundary condition was also defined as a variable pressure head condition. Root depth was 150 cm in the grass site and 200 cm in the areas with arboreous vegetation. The root water uptake model and root distribution model are same to the models used in the previous section. While the parameters a and b in the root distribution model are 7.07 and 1.95, respectively, for deciduous needleleaf tree, i.e., *Salix psammophila*, and are 5.99 and 1.96, respectively, for deciduous broadleaf tree, i.e., *Populus alba*, according to Zeng [31]. The observation points were set from the land surface to water table with 10 cm intervals to obtain soil moisture along the soil profile. To test the accuracy of the simulated data, the soil moisture of K1 (60 cm and 80 cm depth) and K14 (50 cm and 70 cm depth) were also simulated and then compared to the measured data.

2.3.3. Statistical Analysis

The root mean square error (RMSE), Nash–Sutcliffe efficiency (NSE) and percent bias (PBIAS) were used to compare the observed and simulated soil moisture. They are defined as Equations (5)–(7), respectively, as followed:

$$\mathrm{RMSE} = \sqrt{[(\Sigma_i{}^n(\theta_{obs,\,i} - \theta_{sim,\,i})^2)/n]} \tag{5}$$

$$\mathrm{NSE} = 1 - \Sigma_i{}^n(\theta_{obs,\,I} - \theta_{sim,\,i})^2/\Sigma_i{}^n(\theta_{obs,\,I} - \mathrm{ave}\theta_{obs,\,i})^2 \tag{6}$$

$$\mathrm{PBIAS} = 100 \times [\Sigma_i{}^n(\theta_{sim,\,I} - \theta_{obs,\,i})/\Sigma_i{}^n\theta_{obs,\,i}] \tag{7}$$

where θ_{obs} is the observed soil moisture in field, θ_{sim} is the simulated soil moisture, n is the number of observations, aveθ_{obs} is the average soil moisture of all the observed events.

RMSE indicates the degree of dispersion of the simulated soil moisture relative to the observed moisture [33], and a higher value indicates a higher error in the simulated moisture. NSE represents the extent to which the simulated moisture and the observed moisture follow a 1:1 slope line [34]. It ranges from $-\infty$ to 1 and higher values indicate a better agreement with the observed moisture [35]. PBIAS measures the average tendency of the simulated values to be larger or smaller than their observed ones. The optimal value of PBIAS is 0.0, with small values indicating accurate model simulation. Positive values indicate overestimation bias, and negative values indicate model underestimation bias.

2.4. *ET_G Estimation*

White [13] initially developed a method to quantify daily ET_G using diurnal water table fluctuations (hereafter referred to as the White method [36]), as follows in Equation (8):

$$\mathrm{ET_G} = S_y \times (24r \pm s) \tag{8}$$

where ET_G is the daily groundwater evapotranspiration ($L\cdot T^{-1}$), S_y is the specific yield of sediments (−), r is the rate of water table recovery during nighttime ($L\cdot T^{-1}$) and s is the daily change in storage ($L\cdot T^{-1}$)

In this study, a modified White method was adopted to estimate the daily ET_G. For the method, daily hydrostatic equilibrium recovery (R_e) is proposed to correct the groundwater inflow estimated by the White method, taking the disequilibrium of soil water into consideration. The hydrostatic equilibrium at the capillary zone is disturbed at the moment when root water uptake commences in the morning, then the degree of deviation from hydrostatic equilibrium increases in daytime due to the higher ET rate than the hydrostatic equilibrium recovery rate. So, capillary water continues to be replenished by groundwater to lower the degree of deviation from hydrostatic equilibrium at night, resulting in groundwater inflow being underestimated by water table recovery in the White method. Thus, daily R_e is estimated by extending the capillary water recovery rate at nighttime to the whole day, representing the error of daily groundwater inflow estimated by the White method. The procedure for calculating R_e is quite similar to that of the recovery rate (*r* in Equation (8)) in the White method. The first step is the calculation of the deviation of total soil water (TSW) from hydrostatic equilibrium, ($TSW_{eq} - TSW_s$), at two specified times (i.e., 22:00 and 6:00 in this study) during nighttime, respectively, then the difference between ($TSW_{eq} - TSW_s$) of both points of time is determined, and finally the daily R_e is calculated by multiplying by 24 h. Equations (9) and (10) are the mathematical expressions for R_e:

$$R_e = 24 \times [(TSW_{eq} - TSW_s)_t - (TSW_{eq} - TSW_s)_{t+\Delta t}]/\Delta t \quad (9)$$

and

$$TSW = \int_d \theta dz = \Sigma_j^d\, 0.5 \times (Z_{j+1} - Z_j) \times (\theta_{j+1} + \theta_j) \quad (10)$$

where TSW_{eq} is the TSW in the column for the corresponding DWT under hydrostatic equilibrium conditions (L); TSW_s is the TSW computed from the field measured water content values (L); θ is water content ($L^3 \cdot L^{-3}$); j is the soil moisture monitoring point of the soil column. Then the daily ET_G was estimated as follows in Equation (11):

$$ET_G = S_y\,(24r \pm s) + R_e \quad (11)$$

The soil moisture under a hydrostatic equilibrium condition was calculated using the van Genuchten soil water retention properties (Equations (12) and (13)).

$$\theta(h) = \theta_r + (\theta_s - \theta_r)/[1 + (\varphi\,|\,h\,|)^n]^m,\ h < 0 \quad (12)$$

$$\theta(h) = \theta_s,\ h \geq 0 \quad (13)$$

where h is the water pressure head (L), θ_s and θ_r are the saturated and residual soil moisture ($L^3 \cdot L^{-3}$), respectively, φ is the air entry value also referred as bubble pressure (L^{-1}), n is the pore size distribution parameter (−), m =1 − 1/n (−), and φ, n, and m are the empirical shape factors in the water retention function.

It should be noted that specific yield (S_y) is the most crucial factor for ET_G estimation by water table fluctuations [37–39]. S_y is defined as the volume of water released under gravity from storage per unit cross sectional area per unit decline in water table [40]. It is affected by soil texture, depth to water table and drainage time [38]. So, the depth-dependent and time-dependent S_y are widely used to estimate ET_G based on water table fluctuations [21,37].

In this study, the drainage time is not taken into account because the change of capillary water has been calculated. Therefore, the equation for depth-dependent S_y developed by Duke [41] with van Genuchten parameters was used (Equation (14))

$$S_y = \theta_s - [\theta_r + (\theta_s - \theta_r)/[1 + (\varphi\,|\,d\,|)^n]^m] \quad (14)$$

where d is the depth to water table (L).

3. Results and Discussion

3.1. Soil Hydraulic Parameters

The measured and simulated soil moisture from inverse modelling at two depths for each observation well (K1 and K14) are shown in Figure 3. The RMSE, NSE, PBIAS, and R-square (R^2) of measured and simulated data were 0.003 cm^3/cm^3, 0.99, −0.04%, and 0.99, respectively, for K1, while for K14, the RMSE, NSE, PBIAS, and R^2 of measured and simulated data were 0.015 cm^3/cm^3, 0.86, 1.22%, and 0.81, respectively. According to the statistical analysis, the simulated soil hydraulic parameters of K1 and K14, as showed in Table 3, were a good representation of the actual field parameters.

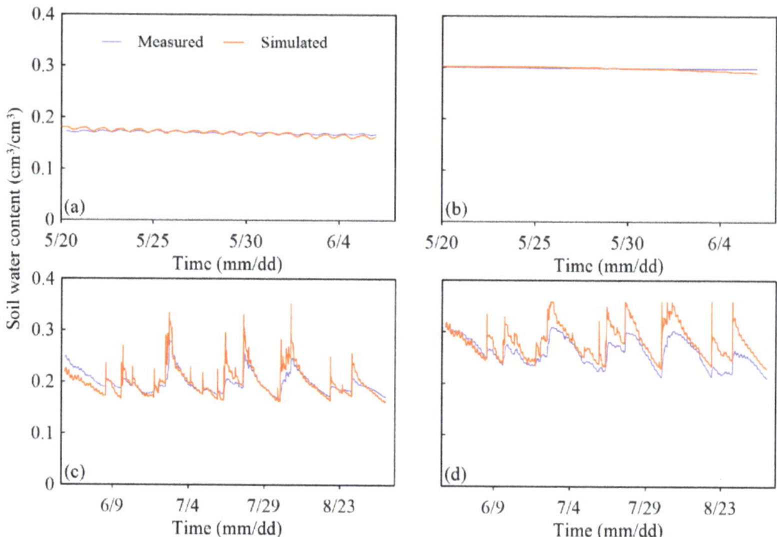

Figure 3. The measured and simulated data of soil moisture at (**a**) 20 cm depth and (**b**) 100 cm depth for K1, and (**c**) 20 cm depth and (**d**) 40 cm depth for K14.

Table 3. Simulated soil hydraulic parameters for K1 and K14.

Well	θ_r (cm^3/cm^3)	θ_s (cm^3/cm^3)	φ (1/cm)	n	K_s (cm/h)
K1	0.0453	0.302	0.0253	1.78	14.2
K14	0.0532	0.387	0.0282	2.68	29.4

3.2. Validation of the Simulated Soil Moisture

Soil moisture was also simulated at K1 and K14 to assess the accuracy of the modelled soil moisture, using the measured and simulated soil moisture at 60 cm and 80 cm for K1 and 50 cm and 70 cm for K14. As shown in Figure 4 and Table 4, the simulated soil moisture has a good agreement with the measured data, with the mean RMSE of 0.012 cm^3/cm^3 at K1 and 0.019 cm^3/cm^3 at K14. While the NSE and R^2 at K14 are 0.25 and 0.21, respectively. The possible reason is that the sensitivity of the monitoring instrument decreases due to the observation depth being close to water table.

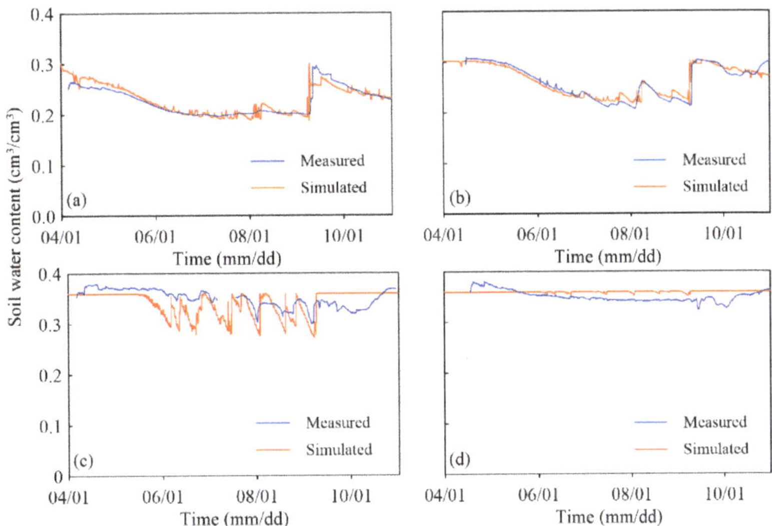

Figure 4. The measured and simulated data of soil moisture at (**a**) 60 cm depth and (**b**) 80 cm depth for K1, and (**c**) 50 cm depth and (**d**) 70 cm depth for K14.

Table 4. The root mean square error (RMSE), Nash-Sutcliffe efficiency (NSE), percent bias (PBIAS) and R-square (R^2) of the measured and simulated soil moisture for K1 and K14.

	RMSE (cm^3/cm^3)	NSE	PBIAS (%)	R^2
K1	0.012	0.88	0.18	0.89
K14	0.019	0.25	−1.82	0.21

3.3. Temporal and Spatial Variations of Water Table

Obvious diurnal water table fluctuation occurred during the whole growing season in the field. As shown in Figure 5a and b, groundwater table fell during daytime and recovered during nighttime. The variation pattern of water table at sub-daily scale is mainly caused by the evapotranspiration of vegetation. While solar radiation is a dominant factor controlling plant water use in sub-daily scale [42,43], and further influences water table variations (Figure 5c).

The changes in water table at seasonal scales are reflected in daily amplitudes. For the *Carex stenophylla* site, the average daily amplitudes of water table at K10 are 1.7 cm, 2.3 cm, and 1.4 cm in the early, middle, and late of the growing season, respectively (Figure 5a). While they are 2.3 cm, 4.1 cm, and 1.3 cm, respectively, at K12 in the *Populus alba* site (Figure 5b). Obviously, the daily amplitude of water table in the middle of the growing season is larger than that in the early and late of the growing season. It is mainly caused by the vegetation vitality that depends on air temperature (Figure 5d).

The patterns of water table fluctuation in the field are also different for various DWT. In the *Achnatherum splendens* site, the mean daily amplitude of water table is 0.2 cm with an average DWT of 1.85 m at K2 from 21 May to 25 May (Figure 6a). At K3, the mean daily amplitude of water table is 0.6 cm with an average DWT of 0.71 m in the same period. In addition, the mean daily amplitude of water table at K6 and K7 are 5.8 cm and 1.8 cm with average DWT of 0.99 m and 1.67 m, respectively, from 1 July to 5 July in the *Salix psammophila* site (Figure 6b). Obviously, the daily amplitude of water table with shallow water table is larger than that with deep water table in the same vegetation condition, mainly resulting from the decrease of root distribution and increase of specific yield with DWT.

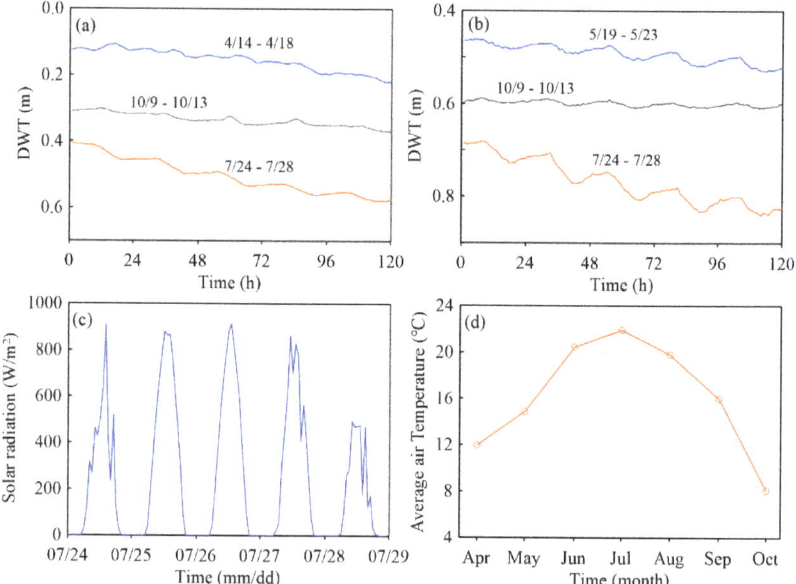

Figure 5. The depth of water table (DWT) of K10 (**a**) and K12 (**b**) at the different periods of the growing season; (**c**) the variation of hourly solar radiation during 24 to 28 July and (**d**) monthly average air temperature.

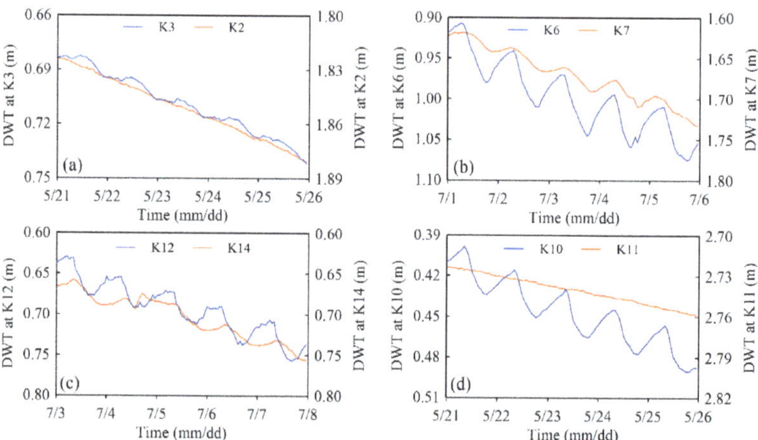

Figure 6. The water table fluctuations vary with (**a**) the DWT in *Achnatherum splendens* site and (**b**) the DWT in *Salix psammophila* site, (**c**) the vegetation conditions in the same DWT, and (**d**) the microrelief.

Vegetation condition is also a vital influencing factor of water level fluctuation pattern. For example, the daily amplitude of water table (3.7 cm) at K12 in the *Populus alba* site is larger than that (2.0 cm) at K14 in the *Carex stenophylla* site with the same average DWT of 0.70 m (Figure 6c). Furthermore, K10 is in the *Carex stenophylla* site and a sand dune at a distance of 20 m from K10 is free of vegetation. There is obviously water table fluctuation at K10 but no water table fluctuation at K11 (Figure 6d). The water table shows no fluctuation at K5 and K13 either, indicating that water table fluctuation is not influenced by other factors, such as tidal [44,45] and barometric pressure [46].

3.4. *Groundwater Evapotranspiration*

3.4.1. Temporal Variations of ET_G

ET_G was estimated except for the wells with no diurnal water table fluctuations, i.e., K5, K11, and K13, indicating that *Caragana korshinskii* is a plant independent of groundwater. There were no ET_G data on rainy days due to the inapplicability of the estimated method. Average daily ET_G and the standard deviation of ET_G at all the observation wells are showed in Figure 7. Temporal variations of ET_G are obvious in the growing season. The average ET_G is small (1.5 mm/d) in the early April and gradually increases to peak values (10.6 mm/d) at the end of July, then gradually decreases to 3.5 mm/d at the end of October. The variation trend is consistent with daily average air temperature (Figure 7), indicating that seasonal ET_G variation mainly depends on air temperature. However, the higher air temperature corresponds to the lower ET_G in April. It may result from the less vitality of vegetation due to the less leaf area in the early growing season [42].

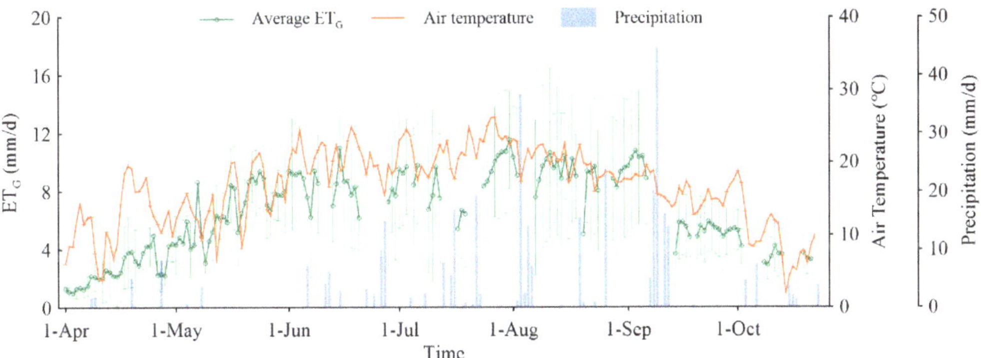

Figure 7. Average groundwater evapotranspiration (ET_G) and the standard deviation of ET_G at all the observation wells with air temperature and precipitation over the growing season.

3.4.2. Spatial Variations of ET_G

Spatial variation is significant indicated by the high standard deviation of the ET_G. The standard deviation shows a similar variation pattern to the average ET_G, which is small in the early and late growing season and large in the middle growing season. The large deviation (4.2 mm/d on average) in the middle of the growing season mainly depends on vegetations types and DWT of the observation wells. For example, the mean ET_G in June was 10.9 mm/d at K12 site with *Populus alba*, which was 1.7 times that of K14 (6.4 mm/d) in the *Carex stenophylla* site with the same average DWT of 0.65 m. On the other hand, the mean ET_G was 2.9 mm/d at K2 site with a mean DWT of 2.03 m and 4.1 mm/d at K3 site with a mean DWT of 0.87 m in June for the *Achnatherum splendens* site. The spatial variations of ET_G were also significant although the distance of two observations was small, for instance, the mean ET_G of K10 was 5.8 mm/d over the growing season in the *Carex stenophylla* site while there was no ET_G at K11 in a sand dune even though it is only 20 m away from K10.

The higher standard deviation of ET_G is also observed from the wells at the same vegetation site (Table 5). The mean standard deviation of average daily ET_G at monthly scale are 1.76 mm/d, 1.89 mm/d, and 3.43 mm/d with the average daily ET_G of 4.01 mm/d, 6.03 mm/d, and 8.96 mm/d in the *Achnatherum splendens* site, *Carex stenophylla* site, and *Salix psammophila* site, respectively. It reflects that the average standard deviation increases with the average ET_G in various vegetation sites.

Table 5. The average groundwater evapotranspiration (ET_G) and standard deviation at monthly scale for various vegetation sites.

	Achnatherum splendens Site		*Carex stenophylla* Site		*Salix psammophila* Site	
	Average ET_G (mm/d)	Standard Deviation (mm/d)	Average ET_G (mm/d)	Standard Deviation (mm/d)	Average ET_G (mm/d)	Standard Deviation (mm/d)
Apr-2019	3.11	2.40	2.05	0.86	6.58	0
May-2019	4.07	1.82	6.71	1.48	8.76	1.76
Jun-2019	3.69	2.20	7.75	2.18	10.51	5.60
Jul-2019	4.31	1.32	7.96	2.38	10.70	3.80
Aug-2019	4.54	1.45	7.93	2.58	12.60	5.45
Sep-2019	4.18	1.66	6.07	2.21	9.25	2.31
Oct-2019	3.19	1.49	3.71	1.53	4.29	1.67

The large spatial difference of ET_G at different wells reflects that ET_G estimated by one observation well can only represent point-scale ET_G. However, Butler et al. [42] used a point-scale ET_G to quantify the ET_G of an area of 150 m^2. According to our analysis, it may cause a large error to study regional ET_G by a point value and the mean ET_G of multiple wells in an area will be more representative, even though in the same vegetation site. Therefore, more observation wells are needed in an area to obtain a region-scale ET_G.

3.4.3. Average ET_G in Different Vegetation Sites

For various vegetations, the average ET_G over the whole growing season (April to October 2019) are 4.01 mm/d, 6.03 mm/d, 8.96 mm/d, and 12.26 mm/d at the *Achnatherum splendens* site, *Carex stenophylla* site, *Salix psammophila* site, and *Populus alba* site with the average DWT of 1.19 m, 0.49 m, 1.30 m, and 0.74 m, respectively. It is obvious that the ET_G in the tree sites is about twice as much as that of the grass sites. The same relationship has been found in previous studies [21,22]. For example, Satchithanantham et al. [22] found that the average ET_G at trees sites and grass sites were 4.1 mm/d and 2.1 mm/d, respectively, in the Assiniboine River drainage basin in southwestern Manitoba, Canada. Although the average ET_G of trees and grass also shows two to one ratio, both of them are much less than that of our study. There are several possible reasons. Firstly, the average DWT in the Assiniboine River drainage basin (1.4 m for trees sites and 1.0 m for grass sites) was larger than that in our study site (1.1 m and 0.5 m for trees and grass sites, respectively). While groundwater absorbed by vegetation decreases with DWT [43]. Secondly, the annual precipitation in the Assiniboine River drainage basin was 474 mm, being 44% larger than that of our study site (330 mm) due to the climate difference. Vegetation will use more soil water from rainfall infiltration than groundwater because the soil water is rich in nutrients and oxygen [47,48]. Finally, the ET_G estimated method in our study considers the daily hydrostatic equilibrium recovery in capillary zones that is considered to be a part of ET_G.

3.4.4. Sensitivity of ET_G to Influencing Factors

ET_0 reflects the combined effect of individual meteorological variables because it is determined by solar radiation, air temperature, relative humidity, and wind speed (Equation (1)). As shown in Figure 8, ET_G has a linear positive correlation with ET_0 in each observation well. The slope of the linear expression of them are considered to be the parameter that reflects the relationship between ET_G and ET_0, and they are different among the wells. The slope in K6, K7, and K12 are bigger than one (ranging from 1.03 to 1.78), indicating that the daily ET_G in these wells are generally larger than daily ET_0. The possible explanation is that the vegetation in K6, K7, and K12 are trees that have larger ET_G than grass. While K4 is close to the lake with dense grass, resulting in the larger ET_G. The slope in the other wells ranges from 0.46 to 0.91. It reflects that the proportion of ET_G and ET_0 shows obvious spatial difference. However, the coefficient of determination (R^2) of the linear expressions is not suitable for all wells, ranging from 0.06 to 0.67, resulting from other factors influencing ET_G. For example, R^2 are 0.06 and 0.08 in K2 and K7, respectively, where DWT is larger.

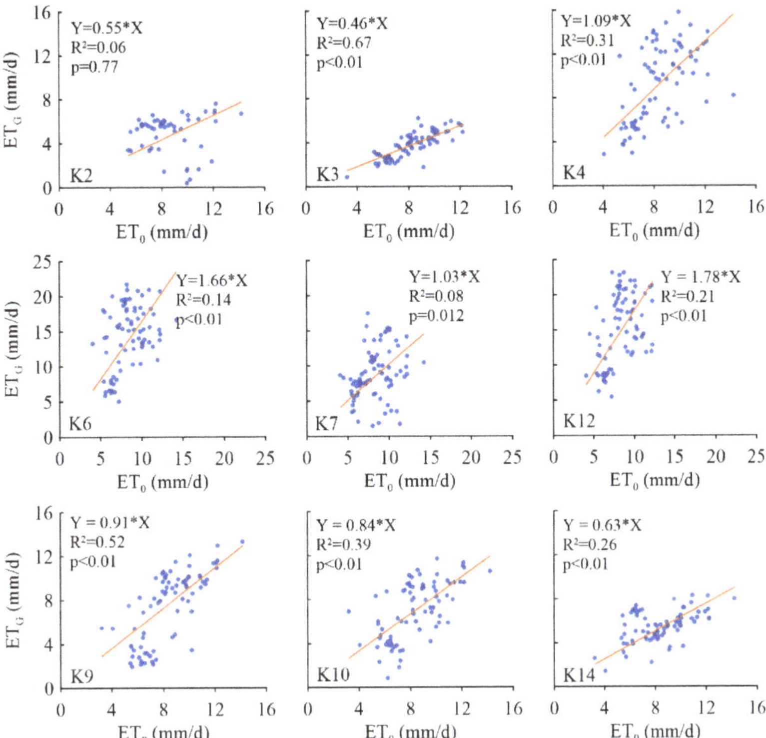

Figure 8. The relationship between ET_0 and ET_G for different observation wells.

In order to obtain the relationship between ET_G and DWT, a ratio of ET_G to ET_0 (ET_G/ET_0) are used to exclude the influence of climatic conditions. For various vegetation types, the exponential decay function between ET_G/ET_0 and DWT are obtained as shown in Figure 9, but the parameters of these functions are different. The maximum ET_G/ET_0 is 1.1 for the grass sites (the *Carex stenophylla* site and the *Achnatherum splendens* site), while the maximum ET_G/ET_0 is 3.6 for the tree sites (the *Populus alba* site and the *Salix psammophila* site), resulted from the larger ET_G in the tree sites. The coefficient of the independent variable (CIV) in the exponential function can reflect the decay rate of the $ET_G/ET_0$with DWT. For grass, ET_G/ET_0 decreases more rapidly with DWT in the *Carex stenophylla* site (CIV = −2.35) than in the *Achnatherum splendens* site (CIV = −1.55), indicating that ET_G is more sensitive to DWT in the *Carex stenophylla* site. While ET_G/ET_0 decreases more rapidly with DWT in the *Populus alba* site (CIV = −2.66) than in the *Salix psammophila* site (CIV = −0.98) for the tree sites, suggesting that ET_G in the *Populus alba* site is more sensitive to DWT.

The groundwater extinction depth is assumed to be reached when ET_G/ET_0 decreases to 0.5% [49]. According to the relationship of ET_G/ET_0 and DWT in our study site, the groundwater extinction depth are 4.1 m, 2.4 m, 7.1 m, and 2.9 m in the *Achnatherum splendens* site, *Carex stenophylla* site, *Salix psammophila* site and *Populus alba* site, respectively. It reflects that roots of *Achnatherum splendens* and *Salix psammophila* are deeper and better able to adapt to lower water levels.

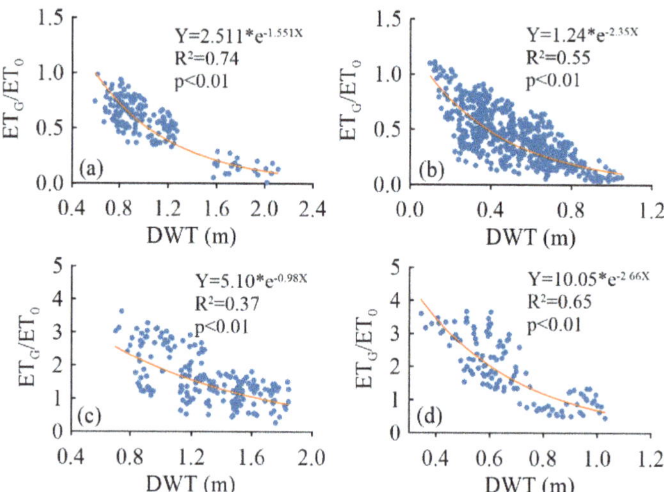

Figure 9. The relationship between the ratio of ET_G to ET_0 (ET_G/ET_0) and DWT for (**a**) the *Achnatherum splendens* site, (**b**) the *Carex stenophylla* site, (**c**) the *Salix psammophila* site, and (**d**) the *Populus alba* site.

4. Conclusions

In this study, ET_G of four vegetation types was estimated based on the diurnal water table and soil moisture fluctuations in the northeastern Mu Us sandy region. According to the variation of water table fluctuations at multiple wells, the influence of tidal and barometric pressure could be excluded. ET_G in all the observation wells shows a significant temporal and spatial variation in each vegetation site, indicating that the representativeness of ET_G from one observation well is poor and the mean ET_G of multiple wells in an area will be more representative. The average ET_G of multiple observation wells are 4.01 mm/d, 6.03 mm/d, 8.96 mm/d and 12.26 mm/d at the *Achnatherum splendens* site, *Carex stenophylla* site, *Salix psammophila* site, and *Populus alba* site with the average DWT of 1.19 m, 0.49 m, 1.30 m, and 0.74 m, respectively, in the whole growing season.

The main influencing factors of ET_G are ET_0 and DWT. The ET_G is more sensitive to DWT in the *Carex stenophylla* site than that in the *Achnatherum splendens* site for grass sites, and for tree sites, ET_G in the *Populus alba* site is more sensitive to DWT than that in the *Salix psammophila* site, according to the relationship between ET_G/ET_0 and DWT. In addition, the groundwater extinction depths are also predicted to be 4.1 m, 2.4 m, 7.1 m, and 2.9 m in the *Achnatherum splendens* site, *Carex stenophylla* site, *Salix psammophila* site, and *Populus alba* site, respectively. This typical geomorphic unit in the study area is thought to be able to represent the evapotranspiration characteristics of the groundwater-dependent vegetation covered area in the Mu Us Sandy region, northern China.

Author Contributions: Conceptualization, W.J. and L.Y.; methodology, W.J.; software, W.J.; validation, M.Z., F.H. and L.Y.; formal analysis, W.J.; investigation, W.J.; resources, L.Y.; data curation, W.J.; writing—original draft preparation, W.J.; writing—review and editing, L.Y.; visualization, K.Y.; supervision, L.W.; project administration, K.Y.; funding acquisition, L.Y. All authors have read and agreed to the published version of the manuscript.

Funding: This research was funded by China National Natural Science Foundation (41472228 and 41877199), Innovation Capability Support Program of Shaanxi (Program No. 2019TD-040), China Geological Survey (DD20160293), and Key Laboratory of Groundwater and Ecology in Arid Regions of China Geological Survey.

Institutional Review Board Statement: Not applicable.

Informed Consent Statement: Not applicable.

Data Availability Statement: The data presented in this study are available on request from the corresponding author. The data are not publicly available due to privacy.

Acknowledgments: We sincerely acknowledge the anonymous reviewers and the Editor for their useful comments.

Conflicts of Interest: The authors declare no conflict of interest.

References

1. Bullock, J.M.; Aronson, J.; Newton, A.C.; Pywell, R.F.; Rey-Benayas, J.M. Restoration of ecosystem services and biodiversity: Conflicts and opportunities. *Trends Ecol. Evol.* **2011**, *26*, 541–549. [CrossRef]
2. Mvungi, A.; Mashauri, D.; Madulu, N. Management of water for irrigation agriculture in semi-arid areas: Problems and prospects. *Phys. Chem. Earth Parts A/B/C* **2005**, *30*, 809–817. [CrossRef]
3. Alvarez, M.D.P.; Trovatto, M.M.; Hernández, M.A.; Gonzalez, N. Groundwater flow model, recharge estimation and sustainability in an arid region of Patagonia, Argentina. *Environ. Earth Sci.* **2012**, *66*, 2097–2108. [CrossRef]
4. Wang, L.; Good, S.P.; Caylor, K. Global synthesis of vegetation control on evapotranspiration partitioning. *Geophys. Res. Lett.* **2014**, *41*, 6753–6757. [CrossRef]
5. Li, X.; Gentine, P.; Lin, C.; Zhou, S.; Sun, Z.; Zheng, Y.; Liu, J.; Zheng, C. A simple and objective method to partition evapotranspiration into transpiration and evaporation at eddy-covariance sites. *Agric. For. Meteorol.* **2019**, *265*, 171–182. [CrossRef]
6. Xiu, L.; Yan, C.; Li, X.; Qian, D.; Feng, K. Monitoring the response of vegetation dynamics to ecological engineering in the Mu Us Sandy Land of China from 1982 to 2014. *Environ. Monit. Assess.* **2018**, *190*, 543. [CrossRef]
7. Lu, C.; Zhao, T.; Shi, X.; Cao, S. Ecological restoration by afforestation may increase groundwater depth and create potentially large ecological and water opportunity costs in arid and semiarid China. *J. Clean. Prod.* **2018**, *176*, 1213–1222. [CrossRef]
8. Gribovszki, Z.; Kalicz, P.; Balog, K.; Szabó, A.; Tóth, T.; Csáfordi, P.; Metwaly, M.; Szalai, S. Groundwater uptake of different surface cover and its consequences in great Hungarian plain. *Ecol. Process.* **2017**, *6*, 39. [CrossRef]
9. Yin, L.; Hu, G.; Huang, J.; Wen, D.; Dong, J.; Wang, X.; Li, H. Groundwater-recharge estimation in the Ordos Plateau, China: Comparison of methods. *Hydrogeol. J.* **2011**, *19*, 1563–1575. [CrossRef]
10. Doody, T.M.; Benyon, R.G.; Theiveyanathan, S.; Koul, V.; Stewart, L. Development of pan coefficients for estimating evapotranspiration from riparian woody vegetation. *Hydrol. Process.* **2013**, *28*, 2129–2149. [CrossRef]
11. Liou, Y.-A.; Kar, S.K. Evapotranspiration Estimation with Remote Sensing and Various Surface Energy Balance Algorithms—A Review. *Energies* **2014**, *7*, 2821–2849. [CrossRef]
12. Anapalli, S.S.; Fisher, D.K.; Reddy, K.N.; Wagle, P.; Gowda, P.H.; Sui, R. Quantifying soybean evapotranspiration using an eddy covariance approach. *Agric. Water Manag.* **2018**, *209*, 228–239. [CrossRef]
13. White, W. *A Method of Estimating Ground-Water Supplies Based on Discharge by Plants and Evaporation from Soil: Results of Investigations in Escalante Valley*; US Geological Survey: Salt Lake, UT, USA, 1932.
14. Cheng, D.-H.; Li, Y.; Chen, X.; Wang, W.-K.; Hou, G.-C.; Wang, C.-L. Estimation of groundwater evaportranspiration using diurnal water table fluctuations in the Mu Us Desert, northern China. *J. Hydrol.* **2013**, *490*, 106–113. [CrossRef]
15. Yin, L.; Zhou, Y.; Ge, S.; Wen, D.; Zhang, E.; Dong, J. Comparison and modification of methods for estimating evapotranspiration using diurnal groundwater level fluctuations in arid and semiarid regions. *J. Hydrol.* **2013**, *496*, 9–16. [CrossRef]
16. Gribovszki, Z.; Kalicz, P.; Szilágyi, J.; Kucsara, M. Riparian zone evapotranspiration estimation from diurnal groundwater level fluctuations. *J. Hydrol.* **2008**, *349*, 6–17. [CrossRef]
17. Wang, T.-Y.; Wang, P.; Yu, J.-J.; Pozdniakov, S.P.; Du, C.-Y.; Zhang, Y.-C. Revisiting the White method for estimating groundwater evapotranspiration: A consideration of sunset and sunrise timings. *Environ. Earth Sci.* **2019**, *78*, 78. [CrossRef]
18. Ii, S.P.L. A method for estimating subdaily evapotranspiration of shallow groundwater using diurnal water table fluctuations. *Ecohydrology* **2008**, *1*, 59–66. [CrossRef]
19. Li, N.; Yan, C.; Xie, J. Remote sensing monitoring recent rapid increase of coal mining activity of an important energy base in northern China, a case study of Mu Us Sandy Land. *Resour. Conserv. Recycl.* **2015**, *94*, 129–135. [CrossRef]
20. Liang, P.; Yang, X. Landscape spatial patterns in the Maowusu (Mu Us) Sandy Land, northern China and their impact factors. *Catena* **2016**, *145*, 321–333. [CrossRef]
21. Cheng, D.-H.; Duan, J.; Qian, K.; Qi, L.; Yang, H.; Chen, X. Groundwater evapotranspiration under psammophilous vegetation covers in the Mu Us Sandy Land, northern China. *J. Arid Land* **2016**, *9*, 98–108. [CrossRef]
22. Satchithanantham, S.; Wilson, H.; Glenn, A.J. Contrasting patterns of groundwater evapotranspiration in grass and tree dominated riparian zones of a temperate agricultural catchment. *J. Hydrol.* **2017**, *549*, 654–666. [CrossRef]
23. Allen, R.G.; Pereira, L.S.; Raes, D.; Smith, M. *Crop. Evapotranspiration-Guidelines for Computing Crop. Water Requirements*; FAO: Rome, Italy, 1998.
24. Schelle, H.; Durner, W.; Iden, S.C.; Fank, J. Simultaneous Estimation of Soil Hydraulic and Root Distribution Parameters from Lysimeter Data by Inverse Modeling. *Procedia Environ. Sci.* **2013**, *19*, 564–573. [CrossRef]

25. Nascimento, Í.V.; de Assis Júnior, R.N.; de Araújo, J.C.; de Alencar, T.L.; Freire, A.G.; Lobato, M.G.R.; da Silva, C.P.; Mota, J.C.A.; Nascimento, C.D.V. Estimation of van Genuchten Equation Parameters in Laboratory and through Inverse Modeling with Hydrus-1D. *J. Agric. Sci.* **2018**, *10*, 102. [CrossRef]
26. Mavimbela, S.S.W.; Van Rensburg, L.D. Estimating Soil Hydraulic Parameters Characterizing Rainwater Infiltration and Runoff Properties of Dryland Floodplains. *Comput. Water Energy Environ. Eng.* **2019**, *8*, 11–40. [CrossRef]
27. Šimůnek, J.; Šejna, M.; Saito, H.; Sakai, M.; Genuchten, M.T.V. *The HYDRUS-1D Software Package for Simulating the One-Dimensional Movement of Water, Heat, and Multiple Solutes in Variably-Saturated Media*; Version 4.17, HYDRUS Software Series 3; Department of Environmental Sciences, University of California Riverside: Riverside, CA, USA, 2013.
28. Li, Y.; Šimůnek, J.; Jing, L.; Zhang, Z.; Ni, L. Evaluation of water movement and water losses in a direct-seeded-rice field experiment using Hydrus-1D. *Agric. Water Manag.* **2014**, *142*, 38–46. [CrossRef]
29. Ritchie, J.T. Model for predicting evaporation from a row crop with incomplete cover. *Water Resour. Res.* **1972**, *8*, 1204–1213. [CrossRef]
30. Belmans, C.; Wesseling, J.; Feddes, R. Simulation model of the water balance of a cropped soil: SWATRE. *J. Hydrol.* **1983**, *63*, 271–286. [CrossRef]
31. Zeng, X. Global Vegetation Root Distribution for Land Modeling. *J. Hydrometeorol.* **2001**, *2*, 525–530. [CrossRef]
32. van Genuchten, M.T. *A Numerical Model for Water and Solute Movement in and below the Root Zone*; Unpublished Report; U.S. Salinity Laboratory, USDA, ARS: Riverside, CA, USA, 1987.
33. Willmott, C.J.; Ackleson, S.G.; Davis, R.E.; Feddema, J.J.; Klink, K.M.; LeGates, D.R.; O'Donnell, J.; Rowe, C.M. Statistics for the evaluation and comparison of models. *J. Geophys. Res. Space Phys.* **1985**, *90*, 8995–9005. [CrossRef]
34. Nash, J.E.; Sutcliffe, J.V. River flow forecasting through conceptual models part I-A discussion of principles. *J. Hydrol.* **1970**, *10*, 282–290. [CrossRef]
35. LeGates, D.R.; McCabe, G.J., Jr. Evaluating the use of "goodness-of-fit" Measures in hydrologic and hydroclimatic model validation. *Water Resour. Res.* **1999**, *35*, 233–241. [CrossRef]
36. Laczniak, R.J.; Smith, J.L.; Elliott, P.E.; DeMeo, G.A.; Chatigny, M.A.; Roemer, G.J. *Ground-Water Discharge Determined from Estimates of Evapotranspiration, Death Valley Regional Flow System, Nevada and California*; Report 2001-4195; U.S. Geological Survey: Reston, VA, USA, 2001.
37. Nachabe, M.H. Analytical expressions for transient specific yield and shallow water table drainage. *Water Resour. Res.* **2002**, *38*, 11-1–11-7. [CrossRef]
38. Ii, S.P.L.; Butler, J.J., Jr.; Gorelick, S.M. Estimation of groundwater consumption by phreatophytes using diurnal water table fluctuations: A saturated-unsaturated flow assessment. *Water Resour. Res.* **2005**, *41*, 1–14. [CrossRef]
39. Gribovszki, Z. Comparison of specific-yield estimates for calculating evapotranspiration from diurnal groundwater-level fluctuations. *Hydrogeol. J.* **2017**, *26*, 869–880. [CrossRef]
40. Freeze, R.A.; Cherry, J.A. *Groundwater*; Prentice-Hall, Inc.: Englewood Cliffs, NJ, USA, 1979.
41. Duke, H.R. Capillary Properties of Soils-Influence upon Specific Yield. *Trans. ASAE* **1972**, *15*, 0688–0691. [CrossRef]
42. Butler, J.J.; Kluitenberg, G.J.; Whittemore, D.O.; Ii, S.P.L.; Jin, W.; Billinger, M.A.; Zhan, X. A field investigation of phreatophyte-induced fluctuations in the water table. *Water Resour. Res.* **2007**, *43*, 299–309. [CrossRef]
43. Yue, W.; Wang, T.; Franz, T.E.; Chen, X. Spatiotemporal patterns of water table fluctuations and evapotranspiration induced by riparian vegetation in a semiarid area. *Water Resour. Res.* **2016**, *52*, 1948–1960. [CrossRef]
44. Hughes, C.; Kalma, J.D.; Binning, P.J.; Willgoose, G.R.; Vertzonis, M. Estimating evapotranspiration for a temperate salt marsh, Newcastle, Australia. *Hydrol. Process.* **2001**, *15*, 957–975. [CrossRef]
45. Senitz, S. Untersuchung und Anwendung kurzperiodischer Schwankungen des Grundwasserspiegels in Thüringen. *Grundwasser* **2001**, *6*, 163–173. [CrossRef]
46. Jiang, X.-W.; Sun, Z.-C.; Zhao, K.-Y.; Shi, F.-S.; Wan, L.; Wang, X.; Shi, Z. A method for simultaneous estimation of groundwater evapotranspiration and inflow rates in the discharge area using seasonal water table fluctuations. *J. Hydrol.* **2017**, *548*, 498–507. [CrossRef]
47. Scott, M.L.; Lines, G.C.; Auble, G.T. Channel incision and patterns of cottonwood stress and mortality along the Mojave River, California. *J. Arid Environ.* **2000**, *44*, 399–414. [CrossRef]
48. Martín, D.; Chambers, J. Restoration of riparian meadows degraded by livestock grazing: Above- and belowground responses. *Plant Ecol.* **2002**, *163*, 77–91. [CrossRef]
49. Shah, N.; Nachabe, M.; Ross, M. Extinction Depth and Evapotranspiration from Ground Water under Selected Land Covers. *Groundwater* **2007**, *45*, 329–338. [CrossRef] [PubMed]

Article

Technological Spaces in the Semi-Arid High Plains: Examining Well Ownership and Investment in Water-Saving Appliances

Brock Ternes

Department of Sociology and Criminology, University of North Carolina Wilmington, Wilmington, NC 28403, USA; ternesb@uncw.edu

Abstract: Groundwater depletion has been a consequential problem in Kansas, a drought-prone state widely reliant on the High Plains aquifer. This manuscript explores well ownership's moderating effects on the relationships between awareness of water supplies and the use of water-saving devices. It assesses one of the only quantitative datasets of private water well owners used in social scientific research (n = 864) and discusses the intricate results of multi-group structural equation models with respondents organized by their water supplies. Well ownership and water literacy are significantly correlated to owning water-conservation technologies, and well ownership combined with access to municipal water weakens the correlations between awareness and owning water-saving appliances.

Keywords: sociology of water use; well owners; groundwater; water supplies; infrastructure; water-saving appliances

1. Introduction

Around the planet, groundwater supplies are facing precipitous declines, and one third of Earth's largest aquifers are extremely stressed [1]. As anthropogenic warming intensifies droughts, groundwater losses present a challenging hydroclimatic hazard, and the diminishment of aquifers (underground reservoirs of freshwater) will continue to create global water shortages [2,3]. These vital supplies of freshwater underlie many of the planet's arid and semi-arid regions, and they remain humanity's greatest defense against droughts, though researchers anticipate further dependency on aquifers into the future [4].

Brutal droughts, extreme heat waves, and colossal extractions from tens of thousands of irrigation wells have occasioned terrible declines for one of the largest aquifer systems in the world, the High Plains aquifer. Located in the central United States, this expansive groundwater formation underlies eight states (see Figure 1). The High Plains aquifer contains multiple segments that react differently to overdrafting (the removal of water from of an aquifer faster than it can be regained) and recharge (the natural percolation of surface water into groundwater sources). The disparity between recharge and withdrawal has been so acute that the central and southern portions of the aquifer are undergoing fateful declines [5]. In the state of Kansas, low recharge rates and rapid extractions render segments of the High Plains aquifer virtually unrenewable [6]. The aquifer is now 30% depleted, and if current rates of extraction continue, it will be 69% depleted by 2060 [7].

Broadly, the High Plains aquifer is being exploited for its prized irrigation water; this is particularly true in Kansas, which over-allocated irrigation permits for large-capacity wells in the middle twentieth century. Without sweeping curtailments, Kansas will remain "extremely vulnerable to the occurrence of drought" [8] (p. 255). However, exploiting groundwater in the High Plains has sustained global food production for decades. The area overlying the aquifer annually produces more than $20 billion worth of crops, and it covers the largest irrigation-sustained cropland on Earth [10,11]. Climate forecasts indicate the High Plains will experience warmer droughts, meaning that the region will require even more groundwater to meet its future needs as water tables continue to shrink [12].

Citation: Ternes, B. Technological Spaces in the Semi-Arid High Plains: Examining Well Ownership and Investment in Water-Saving Appliances. *Water* **2021**, *13*, 365. https://doi.org/10.3390/w13030365

Academic Editor: Mirko Castellini
Received: 28 December 2020
Accepted: 26 January 2021
Published: 31 January 2021

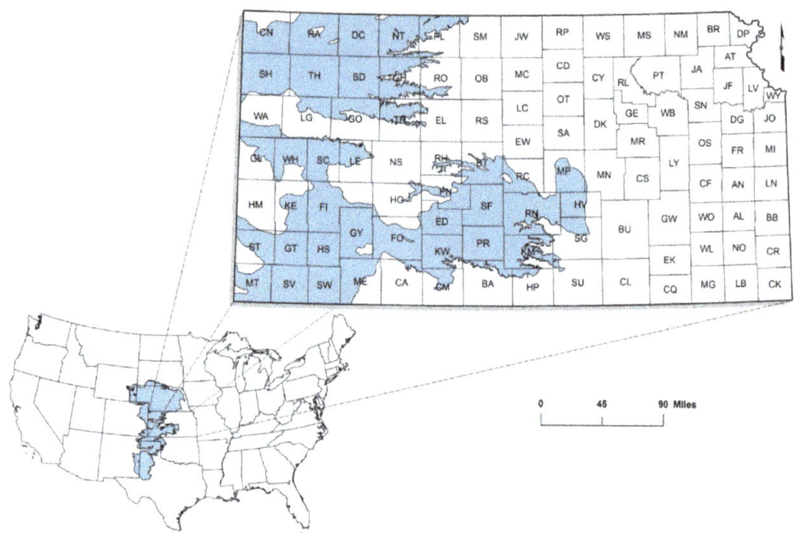

Figure 1. The High Plains aquifer [9].

Drought vulnerability has been a longstanding concern of Kansas policymakers. State law requires water rights owners submit an annual report of their usage to the Kansas Water Office [13], and while Kansas's groundwater regulation mostly focuses on high-capacity wells, the groundwater withdrawals of domestic wells do not have to follow water rights restrictions and remain unmonitored. Since low-capacity wells do not require permits or meters in many parts of the state, if private well owners relied on domestic wells, they could conceivably tap aquifers without acquiring additional permits. Groundwater consumption for domestic usage is not monitored in Kansas, which further complicates well ownership in the Midwest.

Analyzing well ownership in Kansas facilitates an assessment of the relationships between infrastructural conditions and water-conservation. Kansas only monitors high-capacity wells, but evidence suggests that low-capacity wells can also contribute to groundwater losses [14]. Any extraction that exceeds recharge ultimately threatens aquifers and the communities overlying them. Furthermore, if water-conservation technologies are not widely used for both domestic and agricultural purposes, then domestic usage could eventually compete with the water needed for food production. As climate change has a "growing impact on agriculture due to changing rainfall patterns ... warming temperatures, aridity, and greater uncertainty" [15] (p. 109), conserving groundwater will remain paramount for weathering the Anthropocene. Examining how water-saving technologies are utilized within specific groundwater formations and infrastructural systems can provide insight into well owners' management of their water supplies. Aquifers and well owners form a unified socio-ecological system, and looking into the technologies well owners use to conserve water is beneficial for preserving aquifers.

This study of water conservation demonstrates how household water supplies shape Kansans' investment in water-saving technologies. Investigating the individuals who depend on the High Plains aquifer, this research scrutinized Kansans' water literacy (the understanding of water supplies and how water is used) and how water supply infrastructure is associated with owning water efficient appliances. Generally, the paper unpacks the technologies affected by well ownership. What are well owners' and non-well owners' investments in water-saving appliances? Are relationships between water supply awareness and water conservation arbitrated by systems of water provision? Ultimately, this project demonstrates that water supplies contour the associations between the awareness of water supplies and using water-saving appliances. The present study assessed the ownership of water-conservation technologies among Kansans with different water supply systems. It found that well ownership and water literacy are significantly correlated to owning water-conservation technologies—but those relationships change across different infrastructural contexts. As the climate of the High Plains becomes less predictable, communities prone to drought must acquire a precise understanding of the public's deployment of water-saving technologies to prepare for a new hydrologic reality defined by more frequent and intense water shortages.

With the prospect of warmer droughts looming, analyzing water supply infrastructures will be critical. The resilience of rural communities rests on environmental awareness and conservation efforts, since groundwater is the drinking water source for 90% of the rural population in the United States and 80% of rural Canadians [16,17]. Private well owners are susceptible to groundwater contamination and reduced well yields, and they are disproportionately burdened by groundwater loss compared to citizens with municipally provided water. The next section outlines the literature on water supply infrastructure and consumption, and it discusses how water conservation has been studied.

Literature Review and Framing

In many ways, water infrastructure distinguishes urban communities (which provide water to residents via centralized supplies) from rural communities (where agricultural water usage and groundwater extractions provided by private wells are common). Contrasting public and private water supplies reveals how water consumption is defined across urban and rural contexts; additionally, modern drinking sources fundamentally revolutionized standards of water usage. As water became easily accessible within cities, new interpretations of cleanliness took over everyday life. Rates of consumption increased as plumbing became omnipresent in cities and urbanites affixed status to improved hygiene [18,19]. Since technological backdrops shape water-use routines, individual water usage is a consequence of material settings [19]. While the literature on water supply infrastructure spotlights urban centers with consolidated utilities [20–22], this project probed how private wells change water-conservation efforts.

Infrastructure arranges individuals' interactions with—and knowledge of—natural resources; municipal water outreach campaigns show that most citizens in the United States do not know where their water comes from [23]. Grasping environmental issues and awareness of personal involvement in natural resource decline motivates pro-environmental behavior (PEB) [24–26], and proximity to environmental hazards can impel PEBs [27,28]. This article connects environmental awareness to water saving while assessing differences in water supply infrastructure. By framing well owners within a distinct hydrologic setting, the present study shows how shifts in infrastructural context relate to investments in water-saving technologies.

Well owners have legitimate uneasiness about water security, including the regional impacts of climate change (specifically floods and droughts), groundwater contamination, hydraulic fracturing, and aging septic systems [29]. Studying environmental perceptions within specific infrastructural systems can improve how practice theorists [30,31] view consumption. Water conservation is an emergent property of peoples' involvement with their technical surroundings, " ... Infrastructure does not dictate habits, but it does emit cues that are responsible for physical and symbolic channeling of ... attractive opportunities [for behavior] ... " [32] (p. 118). Theories of practice have yet to explore the considerable variation in water supplies, although individuals embedded within private infrastructures likely make distinctive investments in water-efficient technologies. Practice scholarship [32] analyzes the structural conditions that guide routines, and this project assessed how spatiality influences water literacy and the propensity to use water-saving appliances.

Well owners monitor agricultural runoff, well yields, pumping costs, depth to the water tables, and well water quality; therefore, they constitute a "community of practice" [33,34], a group defined by similar routines and boundaries of performance. Studying well ownership can expose new ways to investigate the propensity to save water via efficient technologies. Analyzing well owners and non-well owners empowers a closer look at the relationships between environmentalism and environmental awareness. Different supplies provoke different demands; municipal water systems can deliver seemingly endless supplies of safe, affordable water and therefore changed the evolution of domestic water usage [35–37]. On the other hand, wells are more modest water provision instruments and draw from nearby, finite groundwater sources. Private wells, due to their limited access and reliance on groundwater supplies, are likely the sites of greater water-conservation adaptation strategies, and it seems appropriate to expect structurally based patterns of conservation efforts.

The contexts of private wells and centralized public infrastructures are different "systems of provision" [38]—the activities, technologies, and institutional arrangements that unite to provide a good or service. Systems of provision include the material constraints that guide performances related to water consumption; therefore, it should be expected that the infrastructure delivering water shapes water literacy and the use of appliances that save water. This exploration of different systems of provision outlines patterns of consumption and resource awareness, and it applies the concept of spatiality to water supplies. Interview data indicated that farmers attribute conservation practices to an upbringing without running water, and households deprived of municipal supplies develop unique watering schedules to keep their usage to a minimum [39,40]. Accessing water without public infrastructure requires the exhaustive chores of digging a private well or hauling water; therefore, differences of spatiality are likely associated with the presence of water-saving appliances.

The sociology of water usage [20,21] has not investigated private wells, even though sociologists have called for examinations of "relationships and interactions between processes of supply and the dynamics of consumption" [20] (p. 37). Extant water provision research generally defines water supply infrastructure as a centrally controlled utility within urban contexts, a definition that overlooks the infrastructural differences associated with rural sources. While water experts have explored water consumption's "technological, infrastructural, or behavior-based changes" [21] (p. 1020), they have not sufficiently analyzed how wells, as systems of water provision, contribute to environmental decisions. This research asks how non-centralized water supplies effect water conservation in Kansas. An online survey of well-owning and non-well-owning Kansans' investments in water-saving appliances was conducted to test whether owning efficient watering technologies was influenced by infrastructure.

2. Materials and Methods

The researcher mailed 7021 notification postcards inviting Kansas well owners to participate in an online survey that measured their household water supply, reactions to droughts, and awareness of water policies and supplies. The well owners' home addresses were obtained from the Kansas Geological Survey's database of well completion records. Additionally, a sample of 420 Kansans was obtained from Qualtrics, which had a high percentage of non-well owners. Qualtrics uses a sampling frame from the Survey Sampling International's (SSI) multi-sourcing panel recruitment model, which has a large number of diverse frames that generate representative random samples. This strategy enabled research generalizable to Kansas and comparisons of well owners to non-well owners.

This survey generated one of the only quantitative datasets on well owners used for social science research. Survey responses were collected in 2015; the overall response rate for the postcards sent to well owners was 6.3%, which produced 444 respondents. The timing of this data collection coincided with a number of important events, as 2015 was an important year for the High Plains aquifer. Kansas was clawing out of one of its worst droughts in several decades, and 2015 was the first full year of the "Long-Term Vision for the Future of Water Supply in Kansas," a plan for securing the state's water for the next fifty years. In response to irrigation's demand for tremendous amounts of water, the Kansas Water Office and US Army Corps of Engineers studied the feasibility of diverting water from the Missouri River to western Kansas. Known as the Kansas Aqueduct, this proposal was first drafted in 1982 and revisited in January 2015, when it was established that the 360-mile (580 km) aqueduct would take 20 years to construct and cost $18 billion. Given the statewide concerns surrounding drought vulnerability and groundwater losses, 2015 was an opportune time to collect data on well owners in Kansas.

The entire dataset was comprised of 864 respondents, 452 non-well owners (52%), and 412 well owners. Of those well owners, 20 were former well owners, 143 were without municipal water supplies, and 249 had both wells and municipal water. This research made strides towards understanding the generalizability of Kansans' relationships with water within different infrastructural contexts.

To analyze these data, the researcher performed multi-group structural equation modeling (SEM) with respondents organized by household water supplies, well owners who were without municipal water connections ("off the grid"), well owners with municipal water connections, and well owners who had domestic, lawn and garden, feedlot, and irrigation wells. SEM is a practical multivariate technique that estimates constructs, which are collections of related survey items. Constructs are latent concepts that are not measured directly; their qualities are inferred by relying on a collection of variables selected as proxies. The indicators measuring respondents' investment in water conservation and awareness of water supplies were organized so that they accurately represented the pertinent constructs by occasionally applying an organizational technique known as parceling, whereby modelers take two or more items and average them and then use the average as a manifest indicator rather than relying on individual items. This project's analytical software was M*plus*, a computer program designed for SEM. SEM estimates causal relations between constructs. A study of causality between these associated constructs, SEM is a complex series of hypotheses that consist of a measurement model (a set of multiple variables that form latent constructs) and a path model that describes the relations of dependency between the constructs.

Two constructs were at the heart of this analysis: one measured investment in indoor water-saving appliances (the information shared by indicators measuring if respondents own a low-flow showerhead, a low-flow toilet, and a water-saving washing machine); the other measured investment in outdoor water-saving appliances (the indicators of which were the ownership of timed sprinkler systems and drip irrigation).

The outcome variables were regressed on a construct that measured water supply awareness (which included three parceled indicators measuring familiarity with xeriscaping and greywater systems; familiarity with the High Plains aquifer and groundwater management districts; and familiarity with the Kansas Water Office, the Governor's Long-Term Vision, and the Kansas Aqueduct; as well as an indicator measuring if respondents could correctly identify irrigation as the biggest water user in Kansas). Since agricultural water usage is the biggest groundwater consumer in Kansas, it was important to include drip irrigation as an indicator for the construct measuring outdoor investments.

Factorial invariance was tested for using a goodness-of-fit index (GFI) that examined the change in the comparative fit index (CFI; a comparison of the fit of a target model to the fit of a null model with a cut-off for acceptable fit of greater than 0.90) from each model. Also provided are RMSEAs, or root mean square errors of approximation, which is a parsimony adjusted index to determine fit with a threshold of less than 0.08 for adequate fit, and chi-square values, which gauge discrepancies between observed and expected values. These are among the most commonly reported fit indices for SEM. Following the standards laid out by previous SEM researchers who used the GFI [41,42], these constructs were comparable across groups of well owners. The models examined how the construct measuring water literacy directly influenced the ownership of indoor and outdoor water-saving technologies.

Multi-Group Data and Moderation

How do water-conservation efforts change across communities when organizing Kansans along the lines of well ownership? Multi-group SEM was employed to measure how constructs performed in certain groups, and these respondents were organized based on their household water supply. Watering practices are nested within distinct contexts. Community resilience to aquifer decline is a complex set of interactions that occur within larger hydrologic and agronomic systems that reveals the nested qualities of groundwater communities: they exist within precise infrastructural settings, experience different levels of overdrafting and recharge, and reside above specific supplies of groundwater. Analyzing Kansas well owners without considering the nested effects of water supplies ignores the glaring infrastructural differences between well-owning and non-well-owning Kansans; clarifying how water usage is embedded in unique systems of provision was a key goal for this research.

Moderation is an important causal effect that had implications for this research agenda, which hypothesized that the relationship between water literacy and water conservation would be moderated by the well's function. The type of well in use (domestic, lawn and garden, irrigation, etc.) may influence the association between watering routines and water supply infrastructure. Low-capacity domestic wells may not have the pumping capabilities, depth, or groundwater supply to deliver an abundance of water to their owners' households. This would obviously limit domestic usage and lead to water-conservation routines out of necessity. While agricultural (e.g., irrigation and livestock) wells often do not provide households with water, the owners of these types of wells may be motivated to conserve water domestically out of stewardship for their groundwater supplies. Water is a metabolic piece of urban and rural life in Kansas, and it remains differentiated in its use. The theme of moderation is therefore central to water management.

Discovering that the relationships between water awareness and using water-saving devices are moderated by the type of well in use would reveal an understudied association between resource conservation and water supply infrastructure. Social scientists studying water management hold that separate groups have distinct interests in conserving their water supply [43]. Moderation should therefore be expected and empirically demonstrated among well owners, specifically regarding their well's function. The next brief section examines well ownership's role as a predictor with a simple SEM in which owning indoor and outdoor water-saving technologies were outcome constructs. These results, and the following section covering multi-group modeling, describe the regression coefficients between the dependent and independent constructs; consult the Appendix A for the complete results of the factor loadings, residual variance parameters, and standard errors of the indicators.

3. Results

3.1. Well Ownership and Investment in Water-Saving Appliances

Figure 2 shows the model regressing the outcome constructs of investing in water-saving devices on the independent variable measuring well ownership. Well ownership was found to have a positive association with indoor water-saving investments ($b = 0.305$; $p < 0.001$) and a stronger positive correlation with outdoor water-saving devices ($b = 0.452$; $p < 0.001$). The model supported the claim that well ownership is an important predictor for many of the constructs of interest, along with geography, sex, income, and age [44]. How do these constructs influence each other in a structural equation model, and does well ownership change the associations between these constructs? How does a well's specific function change the relationship between water usage and well reliance? Are the relationships different for "off the grid" well owners compared to well owners who have access to municipal water? The next section answers these questions by reviewing the results of multi-group structural equation models.

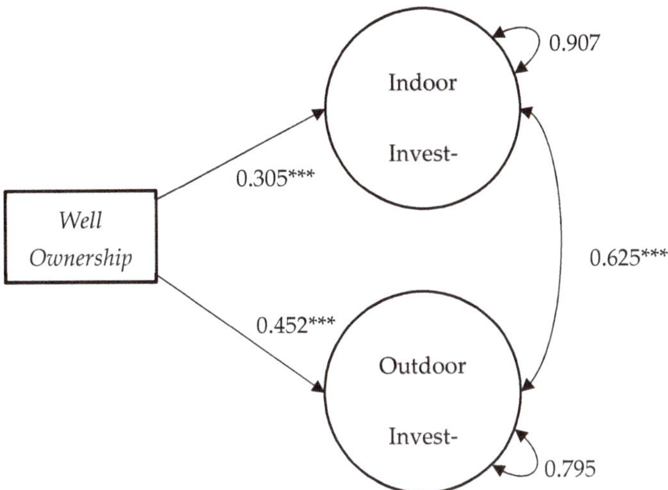

Model Fit: $\chi2$ (8) = 18.739; $p < 0.001$; CFI = 0.979; RMSEA = 0.040
* $p < 0.05$; ** $p < 0.01$; *** $p < 0.001$.

Figure 2. Model of owning water-saving appliances for indoor and outdoor usage regressed on well ownership (n = 847). CFI: comparative fit index; RMSEA: root mean square error of approximation.

3.2. *Moderation: Well Ownership, Municipal Connections, and Specific Well Function*

To examine how water supplies moderate the associations between water supply awareness and investing in water-saving technologies, a series of multi-group structural equation models organized respondents along the boundaries of water supply infrastructure. The results are sequenced as follows: first, they demonstrate how levels of water-related knowledge correlate to the ownership of water-saving technologies for well owners and non-well owners, then for well owners who do not have municipal water connections and those who have public utilities, and finally for owners of different types of wells (domestic, lawn and garden, feedlot, and irrigation).

3.3. *Water-Saving Appliances and Water Literacy*

3.3.1. Non-well Owners and Well Owners

Figures 3 and 4 depict the models of the multi-group analysis for non-well owners and well owners. For non-well owners, the correlations between awareness and indoor investments ($b = 0.471$; $p < 0.001$) and outdoor investments ($b = 0.579$; $p < 0.001$) were found to be positive and significant. The same could be said about well owners' slope for indoor appliances regressed on awareness levels ($b = 0.338$; $p < 0.001$) and their outdoor appliances regressed on awareness levels ($b = 0.438$; $p < 0.01$). The standardized beta values—which represented the associations between the constructs—for awareness levels and investments among well owners were weaker than they were for non-well owners.

3.3.2. Well Owners with and without Municipal Water Connections

Examining the multi-group SEMs that organized well-owning respondents by well owners without a connection to public utilities and well owners with connections to public utilities showed how water usage and well ownership is moderated by the presence or absence of publicly provided water (Figures 5 and 6). For well owners with no public water, there was a positive association between awareness and investing in indoor water-saving appliances ($b = 0.406$; $p < 0.01$) and a stronger positive relationship between awareness and outdoor investments ($b = 0.777$; $p < 0.01$). For well owners with a public water connection, the associations between awareness levels and indoor and outdoor water-saving investments were not significant, a key difference compared to the off the grid well owners. The combination of public and private water supplies was found to moderate the association between knowledge of water supplies and owning water-saving appliances.

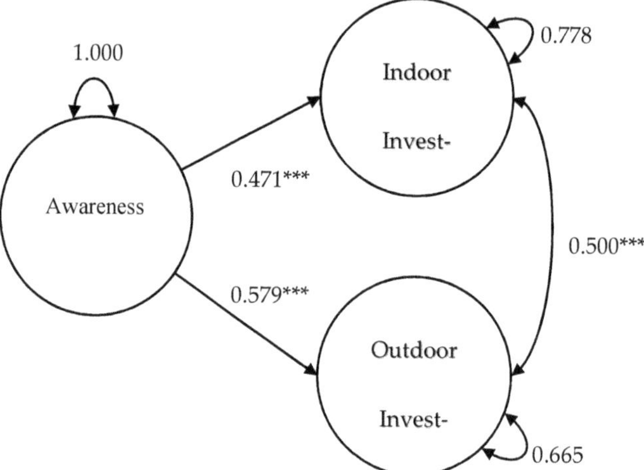

Model Fit: χ2 (62) = 101.142; $p < 0.01$; CFI = 0.943; RMSEA = 0.038
* $p < 0.05$; ** $p < 0.01$; *** $p < 0.001$.

Figure 3. Model of owning water-saving appliances for indoor and outdoor usage regressed on awareness of water supplies for non-well owners (n = 448).

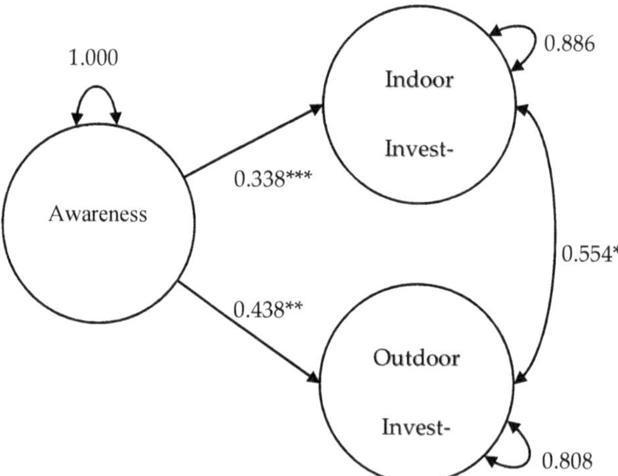

Model Fit: χ2 (62) = 101.142; $p < 0.01$; CFI = 0.943; RMSEA = 0.038
* $p < 0.05$; ** $p < 0.01$; *** $p < 0.001$.

Figure 4. Model of owning water-saving appliances for indoor and outdoor usage regressed on awareness of water supplies for well owners (n = 407).

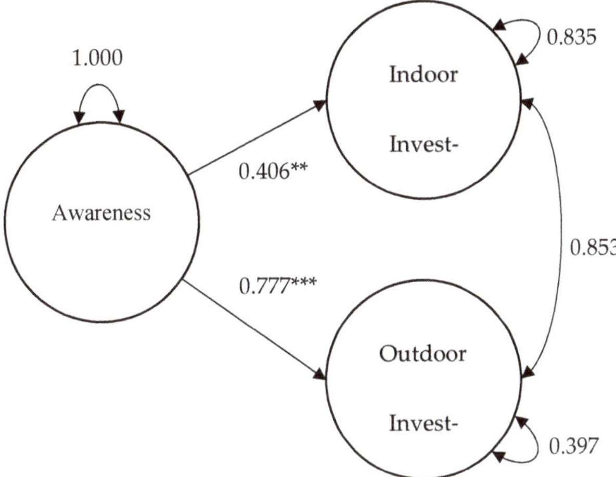

Model Fit: $\chi 2$ (62) = 86.019; $p < 0.05$; CFI = 0.902; RMSEA = 0.045
* $p < 0.05$;** $p < 0.01$;*** $p < 0.001$

Figure 5. Model of owning water-saving appliances for indoor and outdoor usage regressed on awareness of water supplies for well owners without municipal connections (n = 141).

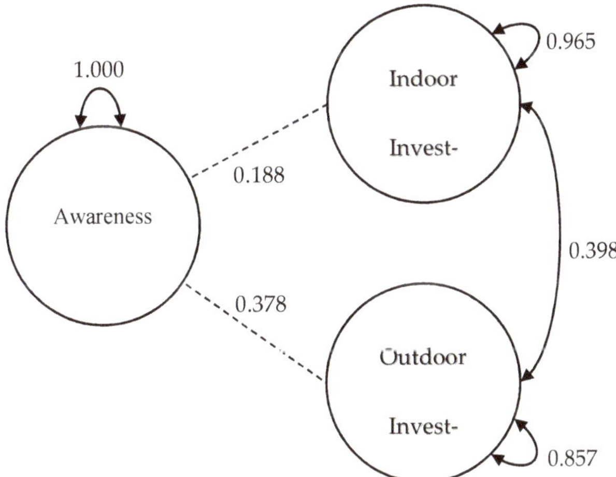

Model Fit: $\chi 2$ (62) = 86.019; $p < 0.05$; CFI = 0.902; RMSEA = 0.045
* $p < 0.05$;** $p < 0.01$;*** $p < 0.001$

Figure 6. Model of owning water-saving appliances for indoor and outdoor usage regressed on awareness of water supplies for well owners with municipal connections (n = 246).

3.3.3. Domestic, Lawn and Garden, Feedlot, and Irrigation Well Owners

Figures 7–10 show the models for each specific type of well owner. The positive slopes between water supply awareness levels and owning indoor water-saving appliances and between awareness and owning outdoor water-saving appliances were only significant for domestic well owners (b = 0.514 and $p < 0.01$; b = 0.578, and $p < 0.01$, respectively). Respondents who owned lawn and garden wells, feedlot wells, and irrigation wells were found to have no significant correlations between these constructs. Therefore, the function of the well changes the association between water supply awareness and investments; it was a moderated association. Increasing awareness levels could theoretically yield a significantly higher investment in water-saving technologies for domestic well owners than for well owners without domestic wells. Overall, familiarity with water supplies was found to be positively associated with owning such appliances, and these results suggested that owning a domestic well changes the association between awareness of water supplies and investing in water-saving appliances.

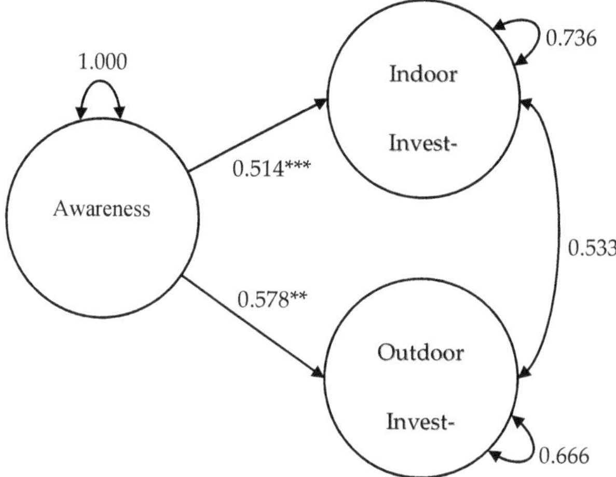

Figure 7. Model of owning water-saving appliances for indoor and outdoor usage regressed on awareness of water supplies for domestic well owners ($n = 145$).

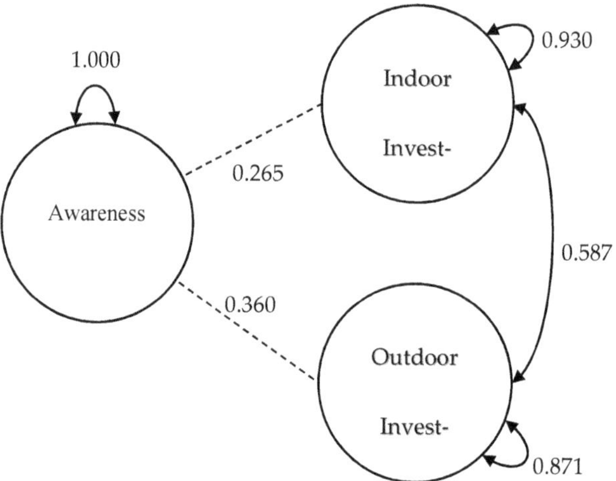

Figure 8. Model of owning water-saving appliances for indoor and outdoor usage regressed on awareness of water supplies for lawn and garden well owners ($n = 135$).

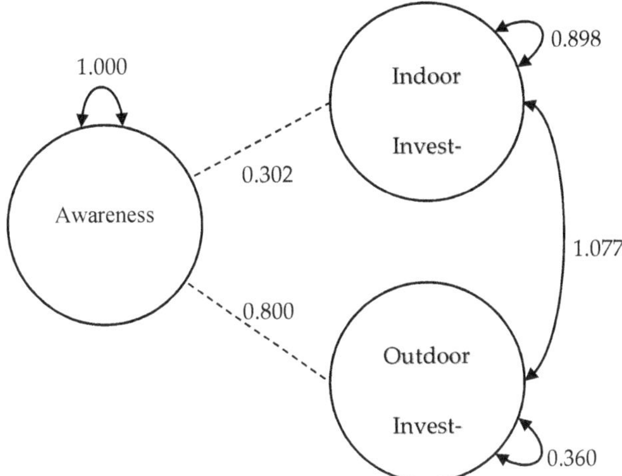

Figure 9. Model of owning water-saving appliances for indoor and outdoor usage regressed on awareness of water supplies for feedlot well owners ($n = 66$).

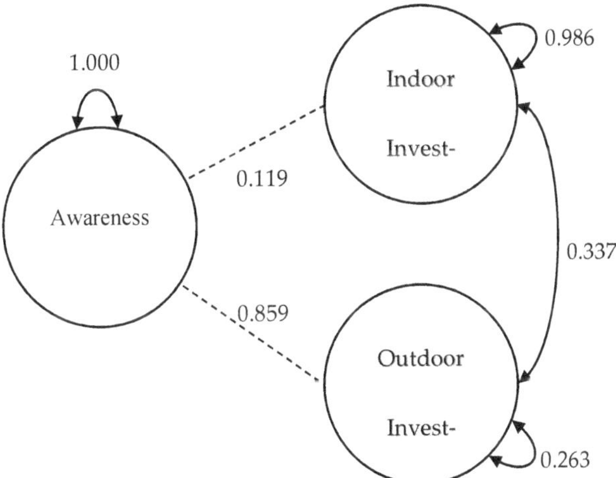

Figure 10. Model of owning water-saving appliances for indoor and outdoor usage regressed on awareness of water supplies for irrigation well owners ($n = 61$).

Therefore, water supply infrastructure should be analyzed as an important moderating variable for researchers who investigate resource conservation via technological efficiencies and cognizance of natural resources.

Four valuable findings arose. First, well ownership was found to be significantly correlated to owning water-conservation technologies. Second, combining public and private water supplies was found to slightly weaken the relationships between awareness and owning water-saving appliances. Water supply awareness was found to have a significant positive correlation with owning efficient watering devices for off the grid well owners but not for well owners who also have city water. Third, the well's function was found to change the association between water supply awareness and investments; it was a moderated association. Increasing awareness levels could theoretically yield a significantly higher investment in water-saving technologies for domestic well owners than for well owners without domestic wells. Fourth, awareness levels were positively associated with the ownership of water-saving appliances.

4. Discussion

The positive associations between well ownership and awareness levels with investments in efficient watering technologies is a promising finding that implies a commitment to conservation. Well owners require a series of "environmental heuristics" [45] to establish how to live sustainably. Such heuristics can harness the social rationale for conserving resources in a way that fits practically into daily life and would require well owners to make informed decisions about cautiously extracting groundwater. A communal identity can influence individuals' behaviors and coping reactions to disasters; residents living in drought-prone areas tend to integrate adaptation strategies into their daily lives. Well owners, while responsible for groundwater extractions, have an important responsibility as aquifer caretakers. Paradoxically, the very people who are most active in the depletion of aquifers are also their guardians; well owners' decisions play a large role in the severity of overdrafts. Groundwater citizenship [46] above the Ogallala Aquifer entails a reliance on groundwater during times of aridity, but that does not necessarily imply irresponsible resource usage. This recasting of stewardship follows Lister's [47] and Curtin's [48] arguments that citizenship should be described in terms of the nested sites in which practices of citizenship are performed and individuals establish their identity as ecological citizens in relation to their natural ecosystems. The book *New Farmers* [49], which chronicled the changing face of Kansas farmers, supports the claim that farming—when done correctly—translates into acts of preservation.

These results expose another possibility. Conceivably, well ownership can enable what Szasz [50] characterized as an "inverted quarantine," half-measures that consumers take to modestly adapt to climate change or resource scarcity. Instead of politically mobilizing, letting their lawns fend for themselves, or redefining their watering routines, people install a lawn and garden well and invest in efficient watering devices so that they can continue to use water "normally" without having to pay higher water bills or consciously change their behaviors during droughts. Instead of aggressively switching to a low-water lifestyle, small capacity lawn and garden wells can enable water consumption for some Kansans (as opposed to reducing it).

The research questions probing whether well function changes Kansans' relationships with water were designed to explore the influence of moderation and nested effects via multi-group analysis. The "liquid dynamics" [51]—the interactions between the social, technological, and hydrological dimensions of water systems—vary across Kansas due to the reliance on surface or groundwater supplies. Non-well owners and well owners were both found to have positive associations between the constructs measuring awareness and water-saving devices, but the correlations were found to be slightly weaker for well owners. Isolating the well owners specifically and conducting a multi-group model along the lines of the type of well they use suggested that domestic wells are the only type of well that yield a significant positive relationship between levels of water supply awareness and investments in water-saving technology. The significant correlations between water supply awareness and investing in efficient watering technologies appeared to be stronger for well owners without city water than for well owners with municipal connections. While owning water-saving appliances was not significantly predicted by awareness levels for lawn and garden, feedlot, and irrigation well owners, it was for domestic well owners. These distinctions were evidence of nested effects, findings similar to scholarship probing well owners' reactions to droughts [52].

When measuring the outcomes of investing in water-saving appliances, familiarity with water supplies was found to be positively correlated with the outcome constructs.

This implies that increasing Kansans' levels of awareness would theoretically increase the ownership of water saving devices, and increasing public awareness of water usage is a precondition for better water management. These results suggest that water supply infrastructure is an important component of citizens' resource awareness and investments in efficient technologies, which is similar to the echoes of previous research providing evidence that environmental considerations are influential precursors of PEBs [53].

Owning a well apparently inspires water conservation that is enabled through the use of water-saving devices. Furthermore, well function changes the relationship between knowledge of water supplies and investing in water-saving appliances. Overall, Kansans with private wells could be electing to save water through technological fixes. While it is challenging to establish a sense of well owners' specific daily routines with these findings, their awareness levels and commitment to technological adjustments are encouraging. Largely, infrastructural differences remain an important component for researchers interested in nested effects. Society's embeddedness within ecosystems requires what Catton and Dunlap [54,55] called a new ecological paradigm to ensure that sociologists consider the natural limits of human activities. This research contributes to this paradigm by studying the behaviors, awareness, and attitudes of water consumption with separate technological backdrops via the implementation of multi-group work that required a perspective that acknowledges the importance of nested effects and private, systemic nuances that may not have been detected if all of the data were to be analyzed at the state level.

As stated on numerous occasions, this study frames well owners as a distinct social group defined by their practices and their exposure to groundwater depletion and drought. According to the project's findings, this claim appears to be accurate but heavily nuanced. Well ownership increases technological conservation efforts, but the positive association between water literacy and using efficient appliances is not always significant in certain hydrologic contexts. Household water supplies are important predictors—and moderators—of watering routines, and they should be included in future research on water conservation.

5. Conclusions

The High Plains aquifer is being tapped beyond its natural rates of replenishment, which has profound implications for sustaining drought-prone communities in Midwestern states like Kansas. This study examined how private water wells influence conservation by comparing the watering technologies of well owners to those of non-well owners across the state. The results indicated that controlling for water supplies uncovers important differences across many associations. Investments in indoor and outdoor water-saving devices were found to be frequently correlated with levels of water literacy for these respondents, who had been delineated along boundaries of water supplies. Well ownership was found to change some associations surrounding water supply awareness and investments in efficient watering appliances.

Seeing modernity through the prism and distribution of infrastructures affords a new understanding of how modern interactions with nature have been shaped by the specific technological spaces of drinking water supplies. Water consumption is firmly rooted in the organization of infrastructural and social systems, and meaningful conservation efforts require fundamental changes to these systems. This research explored how watering practices and environmental awareness are formed around larger socio-ecological structures—aquifers, aridity, and systems of water provision. By establishing how hydrologic infrastructure organizes citizens' relationships with water, this project demonstrates how material assemblages have the potential to transform environmental dispositions. Individuals with private water wells can be framed as a distinct social group that is disproportionately burdened by drought, and well owners represent a unique community of practice that can improve how sociologists understand water supply management. Researching well owners' technological investments, awareness of water supplies, and acts of environmental stewardship will be key to aquifer preservation in the Anthropocene.

Funding: This research was partially supported by the University of Kansas under the Doctoral Student Research Grant, and the Institute for Policy & Social Research at the University of Kansas under the Doctoral Research Fellows Grant. The Article Processing Charge was funded by the Randall Library Open Access Publication Fund at the University of North Carolina Wilmington.

Institutional Review Board Statement: The study was conducted according to the guidelines of the Human Subjects Committee Lawrence Campus, and approved by the Institutional Review Board of the University of Kansas (protocol code #00001050; date of approval October 28, 2014).

Informed Consent Statement: Informed consent was obtained from all subjects involved in the study.

Data Availability Statement: The data presented in this study are available on request from the corresponding author. The dataset is not publicly available due to its sensitive information.

Acknowledgments: The author effusively thanks Paul Stock, Bob Antonio, David Smith, Ebenezer Obadare, and Terry Loecke at the University of Kansas, Todd Little at Texas Tech University, and Stephanie Crowe at the University of North Carolina Wilmington. Thanks to Valerie Peterson, Cassie Butts, Sofiia Filatova, Yulduz Kuchkarova, Liz Blackburn, Halle McCourt, Katelyn Whitt, Chelsea Martell, and Murphy Maiden for their research assistance. The editorial staff and reviewers at *Water* offered perceptive feedback on this manuscript and were extraordinarily punctual.

Conflicts of Interest: The author declares no conflict of interest.

Appendix A. SEM Measures

Table A1. Standardized model results of owning water-saving appliances for appliances for indoor and outdoor usage regressed on well ownership.

	Factor Loadings	Residual Variances	S.E.
Indoor water-saving appliances			
Low-flow showerhead	0.685	0.462	0.046
Low-flow toilet	0.790	0.137	0.043
Water-efficient washing machine	0.600	0.212	0.047
Outdoor water-saving appliances			
Timed sprinklers	0.615	0.795	0.049
Irrigate more	0.615	1.471	0.049

Table A2. Standardized model results of owning water-saving appliances for appliances for indoor and outdoor usage regressed on awareness levels for non-well owners (n = 448).

	Factor Loadings	Residual Variance	S.E.
Awareness			
Xeriscaping and Grey Water	0.561	0.685	0.039
High Plains aquifer and GMDs	0.804	0.353	0.050
KWO, Vision, and Kansas Aqueduct	0.617	0.619	0.036
Agriculture as biggest water user in Kansas	0.349	0.581	0.057
Indoor water-saving appliances			
Low-flow showerhead	0.732	0.458	0.052
Low-flow toilet	0.728	0.184	0.050
Water-efficient washing machine	0.591	0.178	0.052
Outdoor water-saving appliances			
Timed sprinklers	0.736	0.835	0.064
Drip irrigation	0.736	1.516	0.064

Table A3. Standardized model results of owning water-saving appliances for appliances for indoor and outdoor usage regressed on awareness levels for well owners (n = 407).

	Factor Loadings	Residual Variance	S.E.
Awareness			
Xeriscaping and Grey Water	0.440	0.806	0.047
High Plains aquifer and GMDs	0.746	0.444	0.066
KWO, Vision, and Kansas Aqueduct	0.543	0.705	0.050
Agriculture as biggest water user in Kansas	0.349	0.581	0.057
Indoor water-saving appliances			
Low-flow showerhead	0.709	0.458	0.053
Low-flow toilet	0.714	0.184	0.051
Water-efficient washing machine	0.579	0.178	0.052
Outdoor water-saving appliances			
Timed sprinklers	0.437	0.835	0.103
Drip irrigation	0.437	1.516	0.103

Table A4. Standardized model results of owning water-saving appliances for appliances for indoor and outdoor usage regressed on awareness levels for well owners without municipal utility connections (n = 141).

	Factor Loadings	Residual Variance	S.E.
Awareness			
Xeriscaping and Grey Water	0.340	0.885	0.068
High Plains aquifer and GMDs	0.829	0.313	0.099
KWO, Vision, and Kansas Aqueduct	0.558	0.689	0.074
Agriculture as biggest water user in Kansas	0.495	0.118	0.088
Indoor water-saving appliances			
Low-flow showerhead	0.750	−0.019	0.086
Low-flow toilet	0.803	−0.217	0.087
Water-efficient washing machine	0.575	−0.237	0.084
Outdoor water-saving appliances			
Timed sprinklers	0.562	0.709	0.100
Drip irrigation	0.562	0.967	0.100

Table A5. Standardized model results of owning water-saving appliances for appliances for indoor and outdoor usage regressed on awareness levels for well owners with municipal utility connections (n = 246).

	Factor Loadings	Residual Variance	S.E.
Awareness			
Xeriscaping and Grey Water	0.336	0.887	0.065
High Plains aquifer and GMDs	0.829	0.313	0.118
KWO, Vision, and Kansas Aqueduct	0.542	0.707	0.065
Agriculture as biggest water user in Kansas	0.495	0.117	0.088
Indoor water-saving appliances			
Low-flow showerhead	0.697	−0.019	0.077
Low-flow toilet	0.747	−0.217	0.076
Water-efficient washing machine	0.535	−0.237	0.077
Outdoor water-saving appliances			
Timed sprinklers	0.382	0.079	0.120
Drip irrigation	0.382	0.967	0.120

Table A6. Standardized model results of owning water-saving appliances for appliances for indoor and outdoor usage regressed on awareness levels for domestic well owners (n = 145).

	Factor Loadings	Residual Variance	S.E.
Awareness			
Xeriscaping and Grey Water	0.332	0.896	0.060
High Plains aquifer and GMDs	0.687	0.527	0.095
KWO, Vision, and Kansas Aqueduct	0.586	0.657	0.069
Agriculture as biggest water user in Kansas	0.499	0.302	0.081
Indoor water-saving appliances			
Low-flow showerhead	0.711	0.036	0.078
Low-flow toilet	0.795	−0.173	0.079
Water-efficient washing machine	0.595	−0.195	0.087
Outdoor water-saving appliances			
Timed sprinklers	0.627	0.803	0.127
Drip irrigation	0.627	1.054	0.127

Table A7. Standardized model results of owning water-saving appliances for appliances for indoor and outdoor usage regressed on awareness levels for lawn and garden well owners (n = 135).

	Factor Loadings	Residual Variance	S.E.
Awareness			
Xeriscaping and Grey Water	0.320	0.897	0.067
High Plains aquifer and GMDs	0.718	0.485	0.107
KWO, Vision, and Kansas Aqueduct	0.605	0.634	0.073
Agriculture as biggest water user in Kansas	0.499	0.302	0.081
Indoor water-saving appliances			
Low-flow showerhead	0.696	0.036	0.079
Low-flow toilet	0.777	−0.173	0.091
Water-efficient washing machine	0.582	−0.195	0.085
Outdoor water-saving appliances			
Timed sprinklers	0.379	−0.518	0.246
Drip irrigation	0.379	1.054	0.246

Table A8. Standardized model results of owning water-saving appliances for appliances for indoor and outdoor usage regressed on awareness levels for feedlot well owners (n = 66).

	Factor Loadings	Residual Variance	S.E.
Awareness			
Xeriscaping and Grey Water	0.307	0.906	0.081
High Plains aquifer and GMDs	0.834	0.305	0.144
KWO, Vision, and Kansas Aqueduct	0.620	0.616	0.097
Agriculture as biggest water user in Kansas	0.499	0.302	0.081
Indoor water-saving appliances			
Low-flow showerhead	0.591	0.036	0.108
Low-flow toilet	0.660	−0.173	0.120
Water-efficient washing machine	0.494	−0.195	0.101
Outdoor water-saving appliances			
Timed sprinklers	0.370	0.803	0.378
Drip irrigation	0.370	1.054	0.378

Table A9. Standardized model results of owning water-saving appliances for appliances for indoor and outdoor usage regressed on awareness levels for irrigation well owners (n = 61).

	Factor Loadings	Residual Variance	S.E.
Awareness			
Xeriscaping and Grey Water	0.316	0.900	0.083
High Plains aquifer and GMDs	0.838	0.298	0.150
KWO, Vision, and Kansas Aqueduct	0.603	0.636	0.103
Agriculture as biggest water user in Kansas	0.499	0.302	0.081
Indoor water-saving appliances			
Low-flow showerhead	0.776	0.036	0.112
Low-flow toilet	0.868	−0.173	0.111
Water-efficient washing machine	0.650	−0.195	0.085
Outdoor water-saving appliances			
Timed sprinklers	0.252	0.803	0.403
Drip irrigation	0.252	1.054	0.403

References

1. Richey, A.S.; Thomas, B.F.; Lo, M.-H.; Reager, J.T.; Famiglietti, J.S.; Voss, K.; Swenson, S.; Rodell, M. Quantifying renewable groundwater stress with grace. *Water Resour. Res.* **2015**, *51*, 5217–5238. [CrossRef] [PubMed]
2. Shah, T.; Molden, D.J.; Sakthivadivel, R.; Seckler, D. The global groundwater situation: Overview of opportunities and challenges. *Econ. Politic. Wkly.* **2000**, *36*, 4142–4150. [CrossRef]
3. Kallis, G. Droughts. *Annu. Rev. Environ. Resour.* **2008**, *33*, 85–188. [CrossRef]
4. Famiglietti, J.S. The global groundwater crisis. *Nat. Clim. Chang.* **2014**, *4*, 945–948. [CrossRef]
5. Haacker, E.M.K.; Kendall, A.D.; Hyndman, D.W. Water level declines in the high plains aquifer: Predevelopment to resource Senescence. *Ground Water* **2015**, *54*, 231–242. [CrossRef]
6. Padget, S. The water/energy/carbon nexus and triple bottom line solutions. Presented at the Global Water: Drought, Conservation and Security in the 21st Century Conference, Lawrence, KS, USA, 12 April 2013.
7. Steward, D.R.; Bruss, P.J.; Yang, X.; Staggenborg, S.A.; Welch, S.M.; Apley, M.D. Tapping unsustainable groundwater stores for agricultural production in the High Plains Aquifer of Kansas, projections to 2110. *Proc. Natl. Acad. Sci. USA* **2013**, *110*, E3477–E3486. [CrossRef]
8. Logan, K.E.; Brunsell, N.A.; Jones, A.R.; Feddema, J.J. Assessing spatiotemporal variability of drought in the U.S. central plains. *J. Arid Environ.* **2010**, *74*, 247–255. [CrossRef]
9. Buchanan, R.C.; Wilson, B.B.; Buddemeier, R.R.; Butler, J.J., Jr. *The High Plains Aquifer. Public Information Circular 18*; Kansas Geological Survey: Lawrence, KS, USA, 2015.
10. Peterson, J.; Daniel, B. High Plains regional aquifer study revisited: A 20-year retrospective for Western Kansas. *Great Plains Res.* **2003**, *13*, 179–197.
11. Ashworth, W. *Ogallala Blue: Water and Life on the High Plains*; W.W. Norton: New York, NY, USA, 2006.
12. Wilhite, D. Drought: Past and future. Presented at the Drought in the Life, Cultures, and Landscapes of the Great Plains, Lincoln, NE, USA, 2 April 2014.
13. Stover, S. Kansas Ogallala aquifer conservation policies. Presented at the Global Water: Drought, Conservation and Security in the 21st Century Conference, Lawrence, KS, USA, 12 April 2013.
14. Wilson, B.; Gaisheng, L.; Whittemore, D.; Butler, J.J., Jr. *Smoky Hill River Valley Groundwater Model*; Kansas Geological Survey: Lawrence, KS, USA, 2008. Available online: http://hercules.kgs.ku.edu/geohydro/ofr/ofr2008_20/index.htm (accessed on 1 October 2014).
15. White, W.A. *Biosequestration and Ecological Diversity: Mitigating and Adapting to Climate Change and Environmental Degradation*; CRC Press: Boca Raton, FL, USA, 2013.
16. Lemley, A.; Wagenet, L. Rural water quality database: Educational program to collect information. *J. Ext.* **1993**, *31*, 11–13.
17. Expert Panel on Groundwater. *The Sustainable Management of Groundwater in Canada*; Council of Canadian Academies: Ottawa, ON, Canada, 2009.
18. Shove, E. *Comfort, Cleanliness and Convenience: The Organization of Normality*; Berg Press: New York, NY, USA, 2003.
19. Southerton, D.; van Vliet, B.; Chappells, H. Introduction: Consumption, infrastructures and environmental sustainability. In *Sustainable Consumption: The Implications of Changing Infrastructure Provision*; Southerton, D., Chappells, H., van Vliet, B., Eds.; Edward Elgar: Northampton, MA, USA, 2004; pp. 1–11.

20. Medd, W.; Elizabeth, S. *The Sociology of Water Use*; Water Industry Research Limited: London, UK, 2007.
21. Browne, A.L.; Medd, W.; Anderson, B. Developing novel approaches to tracking domestic water demand under uncertainty—a reflection on the "up scaling" of social science approaches in the United Kingdom. *Water Resour. Manag.* **2013**, *27*, 1013–1035. [CrossRef]
22. Smith, C.S. *City Water, City Life: Water and the Infrastructure of Ideas in Urbanizing Philadelphia, Boston, and Chicago*; University of Chicago Press: Chicago, IL, USA, 2013.
23. Hickey, D. Regional service through unity: Meeting our region's needs today and Tomorrow. Presented at the Governor's Water Conference, Manhattan, KS, USA, 14 November 2016.
24. Oskamp, S.; Harrington, M.J.; Edwards, T.C.; Sherwood, D.L.; Okuda, S.M.; Swanson, D.C. Factors influencing household recycling behavior. *Environ. Behav.* **1991**, *23*, 494–519. [CrossRef]
25. Gilg, A.W.; Stewar, B. Behavioral attitudes towards water saving? evidence from a study of environmental Actions. *Ecol. Econ.* **2006**, *57*, 400–414.
26. Barr, S.; Gilg, A.W. A conceptual framework for understanding and analyzing attitudes towards environmental behavior. *Hum. Geogr.* **2007**, *89*, 361–379.
27. Baldassare, M.; Katz, C. The personal threat of environmental problems as predictor of environmental practices. *Environ. Behav.* **1992**, *24*, 602–616. [CrossRef]
28. De Young, R. Some psychological aspects of reduced consumption behavior: The role of intrinsic motivation and competence motivation. *Environ. Behav.* **1996**, *28*, 358–409. [CrossRef]
29. Fox, M.A.; Nachman, K.E.; Anderson, B.; Lam, J.; Resnick, B. Meeting the public health challenge of protecting private wells: Proceedings and recommendations from an expert panel workshop. *Sci. Total. Environ.* **2016**, *555*, 113–118. [CrossRef]
30. Warde, A. Consumption and theories of practice. *J. Consum. Cult.* **2005**, *5*, 131–153. [CrossRef]
31. Stillerman, J. *The Sociology of Consumption: A Global Approach*; Polity Press: Malden, MA, USA, 2015.
32. Warde, A. *The Practice of Eating*; Polity Press: Malden, MA, USA, 2016.
33. Lave, J.; Wenger, E. *Situated Learning: Legitimate Peripheral Participation*; Cambridge University Press: Cambridge, UK, 1991.
34. Wenger, E. *Communities of Practice: Learning, Meaning, and Identity*; Cambridge University Press: Cambridge, UK, 1998.
35. Wright, L. *Clean and Decent: The Fascinating History of the Bathroom and the Water Closet*; Routledge and Kegan Paul: London, UK, 1960.
36. Glennon, R. *Unquenchable: America's Water Crisis and What to do About It*; Island Press: Washington, DC, USA, 2009.
37. Fishman, C. *The Big Thirst: The Secret Life and Turbulent Future of Water*; Free Press: New York, NY, USA, 2011.
38. Fine, B. *The World of Consumption: The Material and Cultural Revisited*, 2nd ed.; Routledge: London, UK, 2002.
39. Foth, V. Buckets, barrels, and backaches: Carrying water for the farm home. In *Water and the Making of Kansas*; Kansas Natural Resource Council: Topeka, KS, USA, 2010.
40. Parfit, M. Sharing the Wealth of Water. *National Geographic*, November 1993; 20–37.
41. Little, T.D. Mean and covariance structures (macs) analyses of cross-cultural data: Practical and theoretical issues. *Multivar. Behav. Res.* **1997**, *32*, 53–76. [CrossRef]
42. Cheung, G.W.; Rensvold, B.R. Evaluating goodness-of-fit indexes for testing measurement invariance. *Struct. Equ. Modeling* **2002**, *9*, 233–255. [CrossRef]
43. Mollinga, P. Water, Politics and development: Framing a political sociology of water resources management. *Water Altern.* **2008**, *1*, 7–23.
44. Ternes, B.; Donovan, B. Hydrologic habitus. *Nat. Cult.* **2020**, *15*, 32–53. [CrossRef]
45. Spaargaren, G. Sustainable Consumption: A theoretical and environmental policy perspective. In *Sustainable Consumption: The Implications of Changing Infrastructure Provision*; Southerton, D., Chappells, H., van Vliet, B., Eds.; Edward Elgar: Northampton, MA, USA, 2004; pp. 15–31.
46. Ternes, B. Groundwater citizenship and water supply awareness: Investigating water-related infrastructure and well ownership. *Rural. Sociol.* **2017**, *83*, 347–375. [CrossRef]
47. Lister, R. *Citizenship: Feminist Perspectives*; New York University Press: New York, NY, USA, 1997.
48. Curtin, D. *Chinnagounder's Challenge: The Question of Ecological Citizenship*; Indiana University Press: Indianapolis, IN, USA, 1999.
49. Darby, D.; Hossler, T.; Stock, P. *New Farmers 2014/2018*; P. & T. Committee: Lawrence, KS, USA, 2019.
50. Szasz, A. *Shopping Our Way to Safety: How We Changed from Protecting the Environment to Protecting Ourselves*; University of Minnesota Press: Minneapolis, MN, USA, 2007.
51. Mehta, L.; Synne, M. Liquid Dynamics: Challenges for Sustainability in the Water Domain. In *Water and Development: Good Governance after Neoliberalism*; Munck, R., Asignwire, M., Fagan, H., Kabonesa, C., Eds.; Zed Books: London, UK, 2015; pp. 30–59.
52. Ternes, B. Saving for a dry day: Investigating well ownership and watering practices during droughts. *Environ. Sociol.* **2018**, *5*, 93–107. [CrossRef]
53. Yeboah, F.K.; Michael, D.K. Explaining energy conservation and environmental citizenship behaviors using the value-belief-norm framework. *Hum. Ecol. Rev.* **2016**, *22*, 137–159.
54. Catton, W.R.; Dunlap, R.E. Environmental sociology: A new paradigm. *Am. Behav. Sociol.* **1978**, *13*, 41–49.
55. Catton, W.R.; Dunlap, R.E. A New ecological paradigm for post-exuberant sociology. *Am. Behav. Sci.* **1980**, *24*, 15–47. [CrossRef]

Article

Silicon Fractionation of Soluble Silicon in Volcanic Ash Soils That May Affect Groundwater Silicon Content on Jeju Island, Korea

Won-Pyo Park [1], Hae-Nam Hyun [1] and Bon-Jun Koo [2,*]

1 Major of Plant Resources and Environment, Jeju National University, Jeju 63243, Korea; oneticket@jejunu.ac.kr (W.-P.P.); hnhyun@jejunu.ac.kr (H.-N.H.)

2 Department of Biological Sciences, California Baptist University, Riverside, CA 92504-3297, USA

* Correspondence: bonjunkoo@calbaptist.edu; Tel.: +1-951-343-4621

Received: 16 August 2020; Accepted: 21 September 2020; Published: 25 September 2020

Abstract: Silicon (Si) is found in various fractions of soil, depending on the pedogenic processes of the environment. Dissolved Si (DSi) is adsorbed in soil particles or leaches through the soil profile into the groundwater. The objective of this study is to quantify, using the sequential extraction method, the different Si fractions in volcanic ash soils on Jeju Island that may affect groundwater Si content, and to compare them with those in forest soils on mainland Korea. Most of the Si in these soils was bound in unavailable forms as primary and secondary silicates. The second largest proportion of Si in the non-Andisols of Jeju Island and Korean mainland soils was accumulated as amorphous Si, while in the Andisols of Jeju Island, the second most significant Si fraction was in pedogenic oxides and hydroxides. The products of these soil formations were short-range-order minerals such as allophane (4–40%). The adsorbed Si concentration tended to increase at lower depths in Andisols (100–1400 mg kg^{-1}) and was approximately five times higher than that in non-Andisols. The results indicate that Si is more soluble in the Andisols of high precipitation regions and that Andisols on Jeju Island potentially affect groundwater Si concentration.

Keywords: allophane; Andisol; readily soluble silicon; sequential silicon extraction

1. Introduction

On oceanic islands, groundwater aquifers replenished by infiltrating rainfall are a major source of freshwater. About 92% of the drinking water on Jeju Island, the largest volcanic island in Korea, is supplied from groundwater [1]. In South Korea, bottled water is very popular and is produced from groundwater sources on Jeju Island [2]. As the consumption of bottled water increases, interest in the quality of the drinking water also grows. A high concentration of silicon (Si) along with Ca and K in drinking water improves water taste and benefits human health [3]. Many researchers have reported that Si in drinking water inhibits the absorption of Al in the body, which is effective in preventing Alzheimer's disease and has effects on atherosclerosis, bone formation, and cholesterol levels [4–8].

Si dissolves into mono-silicic acid (H_4SiO_4) by the weathering of silicate minerals [9]. Dissolved Si (DSi) in the soil solution is absorbed by plants, and leached Si is moved from the river to the sea. Jeju Island has no permeant stream or river due to the high permeability of its volcanic ash soils and lava and pyroclastic flows [10,11]. Hence, much of the precipitation permeates the soil and enters groundwater, which makes it easy to leach DSi into groundwater. The median Si concentration in Jeju Island's groundwater is 33.6 mg SiO_2 L^{-1} (up to 63 mg SiO_2 L^{-1}), which is 1.8 times higher than that in bottled water from mainland Korea [12,13]. Si accounts for more than 80% of the ingredients that make Jeju Island's groundwater taste good [12]. Previous studies on the hydro-chemical composition of Jeju Island's groundwater were mostly conducted for preservation and control against contamination. In particular, their major focus included increasing Na and Cl concentrations by mixing with seawater, nitrogenous fertilizers used on farmland, and nitrate nitrogen originating from livestock wastewater [14,15].

Since it is easy on Jeju Island for DSi to move below the soil and flow into groundwater, the Si concentration in groundwater has a close correlation with the solubility of Si compounds in the soil. Si in the soil consists of primary crystalline silicates (e.g., quartz, feldspars), secondary silicates (e.g., clay minerals), poorly crystalline or microcrystalline silicates (e.g., allophane, imogolite, opal-CT), and amorphous silica (biogenic and pedogenic silica) [9,16].

Si released by weathering is either dissolved into the soil solution or absorbed into soil particles and exists in various forms, such as mono-silicic acid, poly-silicic acid, or organic-inorganic complexes [9,16,17]. As pyroclastic materials in volcanic ash soil are weathered quickly, Si, Al, and Fe are dissolved easily, and because Al preferentially binds with organic matter and produces Al-humus complexes in warm, humid climate conditions, a large amount of DSi can exist in the soil solution [18,19]. This DSi concentration decreases under high precipitation conditions, and DSi exists as monomeric Si or binds with polymerized Al to produce allophane [19,20]. If allophane undergoes extreme desilification over time, it sometimes produces gibbsite [19,21,22]. Eighty percent of Jeju Island's soils are Andisols, which consist of allophane and Al-humus complexes and are mainly distributed in the island's center and southeast regions where precipitation is high. In the coastal and middle mountainous areas of Jeju Island's west and north where precipitation is relatively low and evapotranspiration is high, non-Andisol soils (e.g., Alfisols, Inceptisols, Mollisols, and Ultisols) mainly consist of layered silicates [23,24]. DSi that fails to form layered silicates and allophane during weathering is likely to exist either as weakly bonded forms within the soil or in groundwater due to downward movement by rainwater.

Georgiadis et al. [25] reported that water-soluble and adsorbed Si, among other Si fractions from Si sequential extraction, is easy to mobilize. In addition, amorphous Si and Si occluded in pedogenic oxides and hydroxides can become potential sources of easy-to-mobilize Si because amorphous Si is dissolved more easily than crystalline minerals [26]. Miretzky et al. [27] reported that the dissolution of amorphous Si from volcanic pyroclastic or biogenic origin most likely determines the high DSi concentration in groundwater. Park et al. [23] compared the soluble Si content in soils from Jeju Island and mainland Korea to identify why Jeju Island's groundwater has a higher DSi content. However, since they used the 1N sodium acetate extraction method, the soluble Si content in the analyzed soils included Si in silicate minerals, amorphous silica, and allophane. Thus, it is necessary to estimate the readily dissolvable Si content by quantifying different Si fractions from compounds in the soils. Since precipitation and permeability are high on Jeju Island, Si can be easily dissolved in the soil solution and then continue to be leached, which may contribute to increasing the SiO_2 concentration in groundwater.

This study compared the SiO_2 concentration in groundwater from Jeju Island and mainland Korea. We compared the distribution characteristics of sequentially extracted Si forms in Andisols (Noro and Pyeongdae series), Ultisols (Jeju series), and Alfisols (Gangjeong series) on Jeju Island, the parent materials of which are pyroclastics, and Inceptisols (Samgag and Oesan series) derived from granite and granite gneiss on mainland Korea. This study will illustrate a correlation between the content of soluble Si in the soil and the DSi concentration in groundwater.

2. Materials and Methods

2.1. Sampling and Analyzing Groundwater

Sixty-five groundwater wells were selected for Jeju Island's groundwater samples. After opening the well valve and letting water flow for about 5 min, the groundwater samples were collected into polyethylene bottles (500 mL). For the samples, 2 mL of concentrated nitric acid solution at 68% was added into the samples to reach a 0.5% nitric acid concentration. The samples were transported to the laboratory, filtered by a 0.45 μm glass fiber filter, measured for Si concentration by inductively coupled plasma optical emission spectrometry (ICP-OES) (Integra XL dual GBC, Melbourne, Australia), and calculated as SiO_2. Data analyzed in Gangwon Province (58 samples), Gyeonggi Province (64 samples), Chungcheong Province (88 samples), Gyeongsang Province (126 samples), and Jeolla Province (103 samples) were provided by Geochemistry Laboratory, Korea University, and used for the groundwater SiO_2 concentrations in mainland Korea in comparison to those of Jeju Island's groundwater.

2.2. Soil Samples

We analyzed 28 soil samples collected from horizons with six different soil profiles in Jeju Island and mainland Korea (Figure 1). For the volcanic ash soil samples from Jeju Island, four soils with different properties (Noro, Pyeongdae, Jeju, and Gangjeong series) were selected according to the Taxonomical Classification of Korean Soils [28]. To compare with volcanic ash soils on Jeju Island, two soils (Samgag and Oesan series) from mainland Korea were selected, which were representative soils (typifying Pedon) collected for reclassification research on Korean soils [28] (Table 1). The parents of the four soil series on Jeju Island are pyroclastic materials. The Noro and Pyeongdae series are classified as Andisols. While the Noro series is a cinder cone soil distributed in high elevation mountainous areas (646 m asl) south of Hallasan on Jeju Island, the Pyeongdae series is distributed in lava plains (320 m asl) east of Jeju Island. The Jeju and Gangjeong series are classified as Ultisols and Alfisols, respectively, and have argillic horizons, or layers of accumulated clay. These two soils are distributed in 2–7% sloped lava plains north of Jeju Island. The Samgag and Oesan series on mainland Korea are developed from residuum parent materials in granite and gneiss, respectively, located in 30–60% sloped mountainous areas, and are classified as Inceptisols.

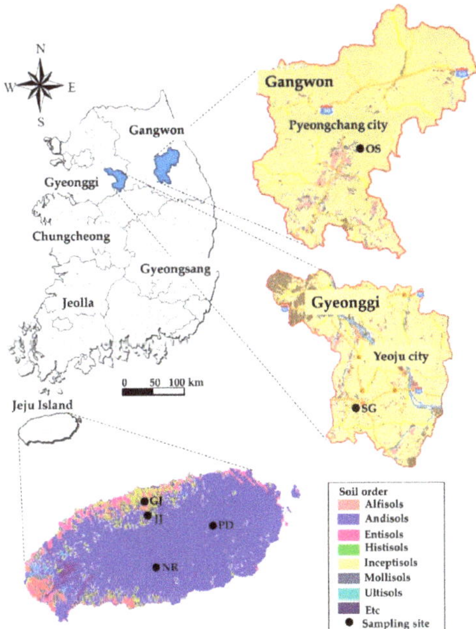

Figure 1. Map of the study area with soil sampling locations in Jeju Island [Andisols: Noro (NR) and Pyeongdae (PD), Ultisols: Jeju (JJ), Alfisols: Gangjeong (GJ)] and in Korean mainland forest soils [Inceptisols: Samgag (SG) and Oesan (OS)].

Table 1. Site characteristics of six soil profiles.

Region	Soil Series	MAP [(a)]	Land Use/Vegetation	Parent Material	Coordinates	Soil Taxonomy/WRB [(b)]
Jeju	Noro (NR)	3430	Forest/Latifoliate trees	Pyroclastic materials	33°18′17.4″ N 126°28′19.1″ E	Acrudoxic Fulvudands/Fulvic Silandic Andosols
Jeju	Pyeongdae (PD)	2660	Pasture/Wide grasses	Pyroclastic materials	33°26′01.5″ N 126°43′30.4″ E	Acrudoxic Melanudands/Melanic Silandic Andosols
Jeju	Jeju (JJ)	1860	Forest/Pinus thunbergii,	Pyroclastic materials	33°26′58.5″ N 126°31′09.7″ E	Andic Palehumults/Umbric Cutanic Alisols
Jeju	Gangjeong (GJ)	1740	Forest/Pinus thunbergii,	Pyroclastic materials	33°27′36.8″ N 126°29′07.7″ E	Mollic Paleudalfs/ Cutanic Luvisols
Gyeonggi	Samgag (SG)	1370	Forest/Pinus	Granite	37°11′17.9″ N 127°36′36.5″ E	Typic Dystrudepts/Haplic Cambisols
Gangwon	Oesan (OS)	1220	Forest/Coniferous and deciduous	Mica schist, mica gneiss	37°28′8.4″ N 128°33′9.6″ E	Typic Dystrudepts/ Leptic Cambisols

[(a)] Mean annual precipitation (mm year^{-1}); [(b)] Taxonomical Classification of Korean Soils [28].

2.2.1. Physicochemical Analyses of Soils

The soil samples were air-dried and passed through a 2-mm sieve for analysis (Table 2). Unless otherwise noted, the soils' physicochemical properties were analyzed according to the Soil Survey Laboratory Methods Manual [29]. Bulk density (Bd) was determined by the dry weight of 100 cm^3 undisturbed core samples. After removing organic matter, the sand, silt, and clay contents were measured by wet sieving (sand fractions) and pipette (silt and clay) methods. Soil pH was measured by the Orion Star A211 pH meter (Thermo Fisher Scientific, Waltham, MA, USA) at a 1:2 ratio of soil to 0.01 M $CaCl_2$ solution. Soil organic carbon (SOC) was analyzed by the Walkley and Black wet digestion method [30].

Table 2. Physical and chemical properties of the soils studied in Jeju Island (Noro, Pyeongdae, Jeju, and Gangjeong) and mainland Korea (Samgag and Oesan).

Horizon	Depth (cm)	Bd (g cm^{-3})	pH ($CaCl_2$)	SOC (g kg^{-1})	Sand (%)	Site	Clay	Fe_d	Al_o + 1/2Fe_o	Allo.	Si/Al (Molar Ratio)
Noro											
A	0–22	0.49	5.39	122	12.1	54.8	33.0	2.59	6.10	11.6	2.12
BA	22–44	0.53	5.32	70.1	30.0	64.8	5.19	3.65	7.64	20.9	1.76
Bw	44–87	0.50	5.76	31.2	28.7	62.0	9.34	3.82	10.3	30.8	1.61
C	87–160	0.51	5.90	13.9	48.7	43.2	8.17	1.88	11.8	39.0	1.51
Pyeongdae											
A	0–18	0.51	5.24	120	26.3	38.8	34.8	4.59	5.44	7.27	2.19
AB	18–44	0.52	5.19	87.3	13.3	65.5	21.2	4.50	7.47	15.9	2.41
Bw1	44–73	0.60	5.35	43.7	15.5	63.1	21.4	4.08	6.17	13.3	2.44
Bw2	73–100	0.71	5.46	23.0	10.1	61.9	28.0	4.13	7.53	15.2	2.12
BC	100–160	0.72	5.74	15.4	34.3	47.1	18.6	3.92	8.34	19.5	2.01
Jeju											
Ap	0–20	0.92	5.08	37.8	7.71	58.6	33.7	1.92	2.46	-	5.69
AB	20–41	1.03	5.02	19.9	3.29	59.9	36.8	2.10	2.13	-	6.23
Bt1	41–65	1.29	5.13	9.80	2.65	62.4	35.0	2.23	1.45	-	6.16
Bt2	65–92	1.48	5.19	7.12	3.97	61.0	35.1	2.42	1.26	-	6.07
Bt3	92–160	1.48	5.12	6.40	5.06	57.6	37.3	2.54	1.21	-	6.21
Gangjeong											
Ap	0–24	1.05	5.52	30.2	10.6	66.1	23.3	1.82	1.38	-	6.36
BAt	24–38	1.36	5.22	13.9	4.76	70.1	25.1	1.94	0.74	-	6.35
Bt1	38–53	1.48	5.29	5.15	5.09	63.5	31.4	2.12	0.59	-	6.75
Bt2	53–85	1.53	5.27	2.86	3.82	59.9	36.3	2.32	0.59	-	6.78
Bt3	85–160	1.53	5.23	1.92	3.87	58.2	38.0	1.93	0.89	-	6.74

Table 2. *Cont.*

Horizon	Depth (cm)	Bd (g cm^{-3})	pH ($CaCl_2$)	SOC (g kg^{-1})	Sand (%)	Site	Clay	Fe_d	Al_o + 1/2Fe_o	Allo.	Si/Al (Molar Ratio)
Samgag											
A	0–15	1.05	4.27	22.8	78.4	13.	7.81	0.24	0.11	-	4.25
BA	15–32	1.36	4.64	9.44	74.9	16.5	8.58	0.25	0.11	-	4.22
Bw	32–50	1.48	4.77	7.69	74.0	16.6	9.37	0.32	0.11	-	4.05
C1	50–78	1.53	5.19	3.94	70.9	15.2	13.9	0.45	0.08	-	4.03
C2	78–180	1.53	5.43	2.39	69.7	14.9	15.4	0.51	0.11	-	3.53
Oesan											
A	0–15	-	4.77	61.1	23.8	49.6	26.6	1.40	1.15	-	5.72
BA	15–34	-	4.90	24.3	21.0	54.4	24.6	1.28	1.09	-	6.41
Bw	34–63	-	4.95	11.6	24.1	54.1	21.8	1.14	0.87	-	6.31
C	63–98	-	5.22	5.13	48.3	38.7	13.0	0.87	0.62	-	6.92

Bd, bulk density; SOC, soil organic carbon; Fe_d, dithionite-citrate-extractable Fe; Fe_o, Al_o, acid-oxalate-extractable Al_o and Fe_o; Allo., Allophane; -, not determined.

Organically complexed Al (Al_p) was extracted for 16 hours using 0.1 M sodium pyrophosphate (pH 10). Free iron oxides (Fe_d), which consist of crystalline and non-crystalline hydrous oxides, were extracted for 16 h using 0.4 g sodium dithionite and 0.57 M sodium citrate. Al, Fe, and Si (Al_o, Fe_o, Si_o, respectively) included in organically complexed Al, non-crystalline hydrous oxides of Fe and Al, and allophane were protected from light and extracted for 12 hours in the dark with 0.2 M acid ammonium oxalate (pH 3.0) using a mechanical vacuum extractor (SampleTek, MAVCO, Lawrenceburg, KY, USA). The allophane content was calculated by the equation $100 \times Si_o/\{-5.1[(Al_o - Al_p)/Si_o \text{ atomic ratio}] + 23.4\}$ [31]. The total Si and Al in the soils were dissolved in HNO_3, HCl, HF, and H_3BO_3 in a microwave oven. The Al, Fe, and Si contents were measured by ICP-OES (JY 138 Ultrace, Jobin Yvon, Longjumeau, France).

X-ray diffraction (XRD) analyses were performed by a Geigerfiex 2301 diffractometer (Rigaku, Tokyo, Japan), using Cu Kα radiation at 30 kV and 15 mA to measure powder samples and clay fractions of less than 2 µm treated by heating and ethylene glycol at 550 °C. Semi-quantitative analysis (Table S1) was performed using a quantitative analysis program (SIEROQUANT™, version 3.0).

2.2.2. Sequential Si Extraction Procedures

The sequential Si extraction method developed by Georgiadis et al. [32] was used for different Si fractions in the soils, in combination with other previous methods (Table 3).

Table 3. Modified sequential extraction procedure for Si fractions in soil.

Step	Si Fraction	Extractant	Extraction Conditions
1	Mobile Si	0.01 M $CaCl_2$	SSR [(a)] 1:10 at 20–25 °C, 24 h
2	Adsorbed Si	0.01 M acetic acid	SSR 1:10 at 20–25 °C, 24 h
3	Si bound to soil organic matter	Hot concentrated H_2O_2	SSR 1:30 in a water bath at 85 °C until the reaction is completed
4	Si occluded in pedogenic oxides and hydroxides	0.2 M Ammonium oxalate (pH 3)	SSR 1:50 at 20–25 °C in the dark, 4 h
5	Si in total amorphous silica	0.2 M NaOH	SSR 1:400 at 20–25 °C in the steps (5 to 240 h)
6	Residual Si in crystalline silicates [(b)]		
	Total Si [(c)]		

[(a)] SSR = soil to solution ratio. [(b)] Residual Si = Total Si minus the sum of the Si fractions extracted in Steps 1 to 5 above. [(c)] Total Si = digestion with a mixture of HNO_3, HCl, HF, and H_3BO_3.

Hence, to extract Si in allophane produced from volcanic ash soils, this study used 0.2 M ammonium oxalate extraction in darkness [31,33–35]. As amorphous silica can be extracted as biogenic and minerogenic Si [32], both reservoirs are potential sources for mobile Si [25]. Accordingly, this study did not distinguish between the two forms of Si but extracted Si in total amorphous silica.

One gram of soil sample was placed in a 50-mL centrifuge tube with a polyethylene screw cap and extracted sequentially with the extractant for each step. Mobile Si was extracted by adding 10 mL 0.01 M $CaCl_2$ solution and shaking it at 100 rpm for 24 h. Adsorbed Si was extracted by adding 10 mL 0.01 M acetic acid solution and shaking it at 100 rpm for 24 h. After adding 20 mL 17.5% H_2O_2 solution, shaking it 4–5 times, leaving it for 1 hour at 20–25 °C adding 10 mL 35% H_2O_2 solution, and placing the sample in a water bath at 85 °C, Si bound to soil organic matter was reacted to completion. After adding 50 mL 0.2 M ammonium oxalate solution (pH 3) and protecting it from light, Si occluded in pedogenic oxides and hydroxides was shaken for 4 h in tubes wrapped with aluminum foil. After adding 400 mL 0.2 M sodium hydroxide solution, Si in total amorphous silica was shaken slowly for 10 days at 20–25 °C at 100 rpm. After 5, 24, 48, 72, 120, 144, 168, 192, 216, and 240 h of shaking, 5 mL of the solution was drawn by pipette, filtered by a 0.45-μm syringe filter, diluted, and then analyzed (Figure S1). The amount of Si in total amorphous silica was determined by SigmaPlot 10.0 software (Systat Software Inc., Chicago, IL, USA) using the first-order dissolution model [36]. Residual Si in crystalline silicates was calculated as the difference between total Si and the sum of the Si fractions extracted in steps 1–5. All extractions were executed on four replicates. After each extraction step, the extracts were centrifuged at 3000 rpm for 15 min to separate soils and supernatants. To remove the extract from the previous step after separating the supernatant, 10 mL distilled water was added, and the solution was centrifuged in each extraction step. Si in the extract was analyzed by ICP-OES (JY 138 Ultrace, Jobin Yvon).

Additionally, for the Bw horizon in the Noro series, XRD and scanning electron microscope/ energy-dispersive X-ray spectroscopy (SEM/EDS) analyses were performed on the soils before sequential extraction and on residual soils after steps 4 and 5 to observe allophane and crystalline minerals in the soils. SEM/EDS analysis was performed at an acceleration voltage of 15 kV using FE-SEM SUPRA25 (Zeiss, Jena, Germany) and ELPHY Quantum (Raith, Dortmund, Germany).

2.3. Statistical Analyses

SPSS 18.0 (SPSS Inc., Chicago, IL, USA) was used for statistical analyses, and correlation analysis was evaluated by Spearman's rank correlation coefficients due to the partially normal and non-normal distribution patterns of the data.

3. Results and Discussion

3.1. Comparison of the Silica Concentration in Groundwater between Jeju Island and Mainland Korea

The SiO_2 concentration in groundwater from Jeju Island and mainland Korea is illustrated in Figure 2. The SiO_2 concentration in Jeju Island's groundwater ranged from 29.5 to 66.7 mg L^{-1} (average of 39.2 mg L^{-1}). The SiO_2 concentration in mainland Korea's groundwater ranged from 0.79–29.2 mg L^{-1} (average of 10.3 mg L^{-1}), and the average SiO_2 concentration in Gangwon Province's groundwater was the lowest at 6.0 mg L^{-1}. These results were consistent with those previously reported, showing that the SiO_2 concentration in groundwater derived from volcanic rocks on Jeju Island was higher than that from granite, gneiss, and metamorphic rocks [2,13]. The silica concentration in groundwater in Argentina, which is similarly of volcanic pyroclastic origin, ranges from 19.6–72.9 mg L^{-1}, similar to the silica content in Jeju Island's groundwater [27]. Weathering in quartz, amorphous silica, and aluminosilicate minerals may have an effect on the Si concentration in groundwater [2,9,16,27]. In general, extrusive rocks (e.g., basalt, andesite) have a lower total Si content but are weathered more quickly than intrusive rocks (e.g., granite) [27,37]. Therefore, weathering in Jeju Island's basaltic pyroclastic materials releases a higher percentage of soluble Si than do granitic rocks.

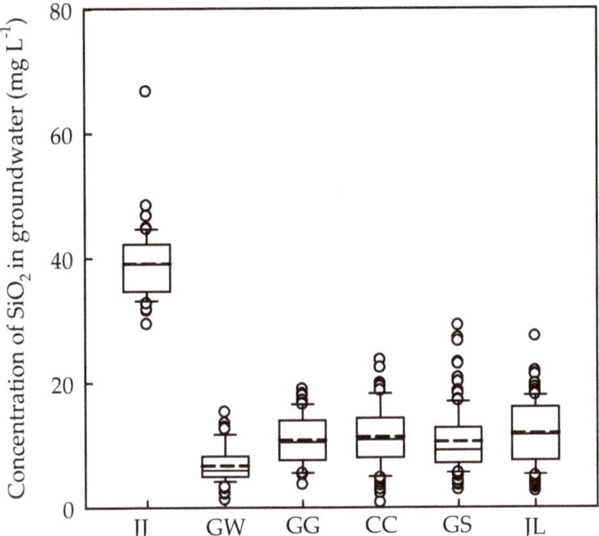

Figure 2. Box plots of SiO_2 concentration in groundwater of South Korea. JJ: Jeju Island, GW: Gangwon province, GG: Gyeonggi province, CC: Chungcheong province, GS: Gyeongsang province, JJ: Jeolla province. Boxes represent 25th, 50th (median), and 75th percentiles, and the whiskers indicate the minimum and maximum values. Mean values (dashed line); outliers (o).

As precipitation increases while pyroclastic materials are weathered, the leached amount of the base and Si increases while Al is preserved. Hence, the Si concentration in the soil solution in Andisols decreases, producing more noncrystalline materials (e.g., allophane, ferrihydrite, and noncrystalline hydrous oxides of Fe and Al) than crystalline layer silicates [18–20]. Under wet climate conditions, Si losses in volcanic ash soils have been reported by mass-balance calculations [38–40]. In this study's soils, the SiO_2/Al_2O_3 molar ratio was below 2 in Andisols, while other soils showed higher ratios from 3.5–7 (Table 3). The low SiO_2/Al_2O_3 molar ratio in Andisols suggests that a large amount of DSi in the soil solution has been leached during the pedogenic processes. A large amount of DSi that fails to form layered silicates or allophane during the soils' pedogenesis (depending on weathering and climate conditions) may exist as readily soluble Si in Si pools from Jeju Island's volcanic ash soils.

3.2. Sequential Si Extraction Procedures

3.2.1. Mobile and Adsorbed Si

In sequential extractions, we compared the Si fraction content among Andisols, Ultisols, and Alfisols from Jeju Island and Inceptisols from mainland Korea. The Si fraction content varied greatly depending on the depth and development of the soil (Figure 3, Table S2).

The mobile Si content in the soil profiles of Jeju Island and mainland Korea ranged from 6.9–57.1 $\mu g \cdot g^{-1}$ and increased with increasing soil depth (Figure 3a). The highest mobile Si was recorded as 57.1 and 52.9 $\mu g \cdot g^{-1}$ in the C horizon from the Noro series, Jeju Island's Andisols, and the Bt3 horizon from the Gangjeong series Alfisols. Compared with the mobile Si content, the adsorbed Si content increased more along with increasing soil depth in Andisols than in other soils. The adsorbed Si content was similar in topsoil horizons across the soils, ranging from 9.4–24 $\mu g \cdot g^{-1}$. In subsoil horizons, however, the adsorbed Si content in Andisols increased up to 400 $\mu g \cdot g^{-1}$, while it ranged from 16–35 $\mu g \cdot g^{-1}$ in subsoil horizons from other soils, a slightly greater increase than in topsoil horizons (Figure 3b). In soils other than Andisols, the mobile and adsorbed Si contents were similar to the mobile Si (0.3–28 $\mu g \cdot g^{-1}$) and adsorbed Si (1.2–39 $\mu g \cdot g^{-1}$) contents in six soil profiles from Southwest

Germany [25]. Nevertheless, the adsorbed Si content in Andisols from Jeju Island was 20 times higher in subsoil horizons than in other soils. This seems attributable to the fact that DSi may have been leached from topsoil horizons to subsoil horizons since the areas in Jeju Island where Andisols are developed have a high level of precipitation.

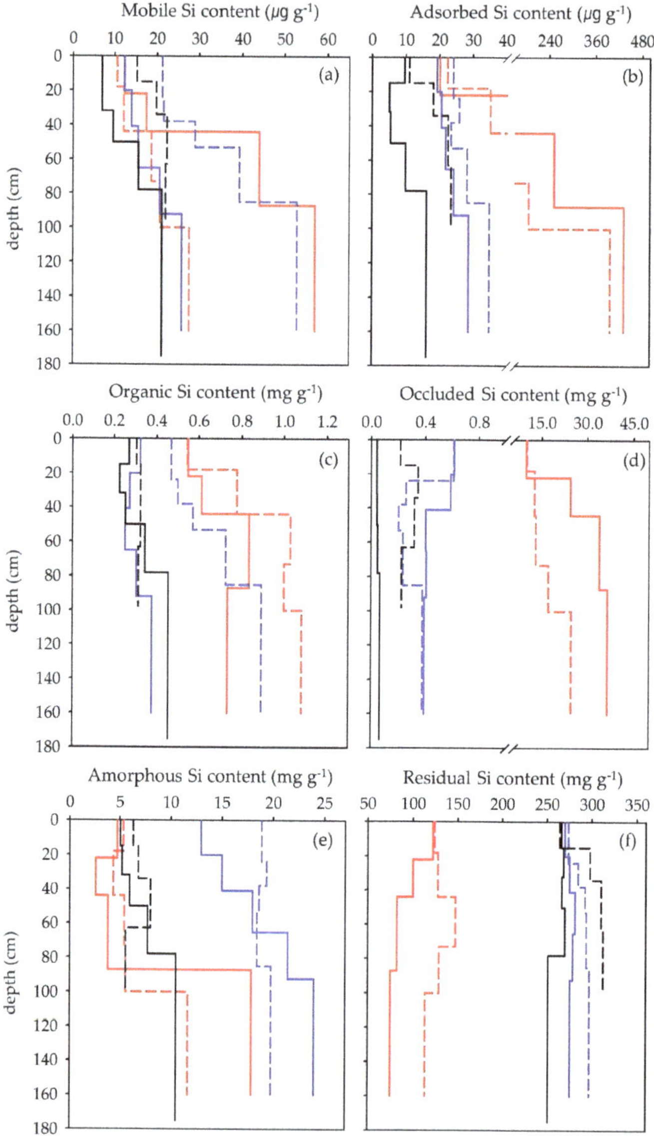

Figure 3. Depth fractionation of silicon in Jeju Island [Andisols; Noro (solid red), Pyeongdae (dashed red), Ultisols; Jeju (solid blue), and Alfisols; Gangjeong (dashed blue)] and Korean mainland forest soils [Inceptisols; Samgag (solid black) and Oesan (dashed black)] according to sequential extraction. (**a**) mobile Si, (**b**) adsorbed Si, (**c**) Si bound to soil organic matter, (**d**) Si occluded in pedogenic oxides and hydroxides, (**e**) Si in total amorphous silica, and (**f**) residual Si in crystalline silicates.

Si is chemically adsorbed by soil compounds such as oxides and hydroxides from Al and Fe, crystalline Fe oxides, allophane, and clay minerals [9,16,17,41]. In non-Andisols (Inceptisols from mainland Korea and Ultisols and Alfisols from Jeju Island), the mobile and adsorbed Si and the clay content had a positive correlation (r = 0.682 and 0.858, respectively, p < 0.01 for each) (Table 4). Supporting findings by Georgiadis et al. [25], the Samgag series, which had the lowest clay content versus high sand content, showed the lowest mobile and adsorbed Si contents. In Andisols, the allophane and noncrystalline hydrous oxides of Fe and Al (Al_o + $1/2Fe_o$) contents increased with increasing soil depth (Table 2), and abundant crystalline Fe oxides as hematite and goethite were found (Table S1). These soil components had a positive correlation with mobile Si and adsorbed Si (p < 0.01, Table 4). Hence, this suggests that most of the DSi has been absorbed into subsoil horizons and moved downward during weathering.

Table 4. Spearman's rank correlation between readily soluble Si fraction (mobile Si and adsorbed Si) and selected physicochemical properties of the soils.

	Total (n = 28)		Non-Andisols (n = 19)		Andisols (n = 9)	
	Mobile Si	Adsorbed Si	Mobile Si	Adsorbed Si	Mobile Si	Adsorbed Si
pH	0.418 *	0.756 **	0.608 **	0.505 *	0.850 **	0.950 **
Organic C	−0.444 *	0.125	−0.550 **	−0.318	−0.900	−0.983 **
Sand	−0.337	−0.317	−0.613 **	−0.800 **	0.517	0.617
Silt	0.103	0.451 *	0.270	0.547 *	−0.167	−0.183
Clay	0.272	0.129	0.682 **	0.858 **	−0.600	−0.617
Fe_d	0.161	0.711 **	0.544 *	0.779 **	−0.567	−0.333
Al_o + $1/2Fe_o$	0.164	0.720 **	0.057	0.426	0.883 **	0.883 **
Allophane	-	-	-	-	0.767 *	0.767 *
Fe_o/Fe_d	0.174	0.691 **	0.080	0.439	0.750 *	0.880 **

*,**: significant difference at 0.05 and 0.01 probability levels, respectively.

3.2.2. Si in Soil Organic Matter (SOM)

The SOM-bound Si content ranged from 0.2–1.0 $mg{\cdot}g^{-1}$ across the profiles (Figure 3c). The A horizon in Andisols with higher organic carbon content showed 0.55 $mg{\cdot}g^{-1}$, which was higher than the A horizon in other soils. While the organic carbon content in the soils decreased with increasing soil depth (Table 3), the Si content in the fractions rose with increasing soil depth or was similar to that in the A horizon. In particular, the Si content in the Jeju Island's Noro series and Pyeongdae series Andisols and the Gangjeong series Alfisols tended to increase noticeably with increasing soil depth. This may be because Si from amorphous and poorly crystalline silica, which are extracted from the latter fractions, was extricated by H_2O_2 treatment during the sequential extraction procedure [25].

Georgiadis et al. [32] reported that the Si content in SOM may be overestimated, since a large amount of Si is released from clay minerals, sesquioxides, and amorphous silica when treated by hot H_2O_2 solution. In this study, the Si content in SOM and the oxalate-extractable Si content from sequential extraction step 4 was correlated (r = 0.651 **, Table S3). Furthermore, the Si content in SOM from soils other than Andisols was found to correlate with amorphous silica and clay fractions (r = 0.603 ** and r = 0.606 **, respectively) from sequential extraction step 5. As subsoil horizons from Jeju Island's soils have more amorphous and poorly crystalline silica, it is possible that the SOM-bound Si content was higher, and Si in these fractions can serve as a potential source for readily soluble Si.

3.2.3. Si Occluded in Pedogenic Oxides and Hydroxides

In Andisols, the content of Si occluded in pedogenic oxides and hydroxides increased with increasing soil depth, ranging from 10–36 $mg{\cdot}g^{-1}$. In other soils, it ranged from 0.04–0.6 $mg{\cdot}g^{-1}$, and the lowest content was found in the Samgag series (0.04–0.06 $mg{\cdot}g^{-1}$) from mainland Korea (Figure 3d).

Acid ammonium oxalate reagents are known to be very effective in extracting short-range-order minerals with weak crystallinity, such as allophane, imogolite, and ferrihydrite [19,31,34]. Andisols in this study contained allophane in the range of 7.27–39.0% (Table 3). In Andisols, the content of Si occluded in pedogenic oxides and hydroxides showed a significant correlation with the allophane content (r = 0.917 **), which suggests that oxalate-extractable Si in these fractions constitutes allophane. Furthermore, the SEM/EDS analysis results of the Bw horizon from the Noro series Andisols before sequential extraction confirmed that allophane was exfoliated and formed on the surface of feldspars, and it was removed after acid ammonium oxalate extraction (step 4) (Figure 4).

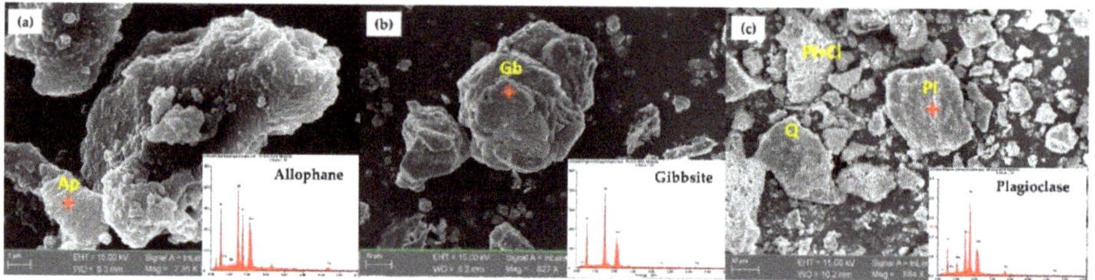

Figure 4. Scanning electron microscope/energy-dispersive X-ray spectroscopy (SEM/EDS) analysis after sequential extraction from Bw horizon of Noro in Jeju Island, showing (**a**) allophane formed on the surface of plagioclase before extracting, (**b**) removal of allophane after step 4, and (**c**) crystalline minerals after step 5. Ap: allophane, Gb: gibbsite, Q: Quartz, Pl: Plagioclase, Cl: Clay minerals.

Assuming that soil is lacking allophane if the acid ammonium oxalate extractable Si lower than 6 mg·g^{-1} [42,43], a silicon phase only exists in Andisols. In such soils, oxalate-extractable Si and oxalate-extractable Al (r = 0.555 *) and Fe (r = 0.594 **) were correlated. This is the result of Si extracted from the soils, which was occluded or co-precipitated with noncrystalline hydrous oxides of Fe and Al [25,44].

3.2.4. Si in Amorphous Silica

In soil Si pools, amorphous silica includes pedogenic and biogenic silica [32]. In amorphous silica, the Si content was the highest in Jeju Island's Ultisols and Alfisols, ranging from 13.0–24.0 mg g^{-1}. In contrast, Andisols had the lowest Si content with a range of 2.62–5.58 mg g^{-1} from the A horizon to the Bw horizon. However, the Si contents in the C horizon in the Noro series and the BC horizon in the Pyeongdae series were 17.8 and 11.7 mg g^{-1}, respectively (Figure 3e).

Amorphous silica is formed by precipitation in the soil solution where silica is over-saturated due to surface evaporation [9,19]. Amorphous silica accumulation may be related to repeated redox changes if Si bound to Fe oxides is present [45,46]. In soils other than Andisols, the Si content in amorphous silica had a close correlation with clay fractions (r = 0.93 **). This demonstrates the accumulation of amorphous silica in clayey horizons [25,45]. Drouza et al. [47] reported that, in the soils in Nisyros Island, Greece, where volcanic materials are developed, the SiO_2-rich parent material combined with low levels of rainfall restricted leaching, and an extended dry season resulted in the accumulation of Si in the soil solution and its deposition in amorphous forms.

As the two soils that develop into Ultisols and Alfisols on Jeju Island have pyroclastic materials as their parent material, weathering takes place at a rapid pace. However, these soils develop in areas where precipitation is relatively low and evapotranspiration is high rather than in areas with Andisols [23,48], which increases the Si concentration in the soil solution, inhibiting the pedogenic process of allophane and producing amorphous silica in large quantities. The Yongdang series developed in the western and norther regions of Jeju Island where non-Andisols, such as the two tested soils in this study, contain fragipan in the subsoil horizons [49]. The formation of fragipan in non-Andisols is associated with Si accumulation [9].

After NaOH extraction, XRD analysis for the Bw horizon from the Noro series Andisols showed that it is composed of primary minerals, such as plagioclase, orthoclase, pyroxene, olivine, amphibole, and magnetite; and secondary minerals, such as clay minerals, iron oxides, and quartz (Table 5, Figure S2). In addition, SEM/EDS analysis found that gibbsite was removed by NaOH extraction, while crystalline minerals remained (Figure 4). As with the results of Georgiadis et al. [25], 0.2M NaOH solution at 20–25 °C did not extract crystalline silicate in Andisols in this study while amorphous and poorly crystalline clay compounds seem to have been fully extracted.

Table 5. X-ray diffraction (XRD) semi-quantitative analysis after sequential extraction from the Bw horizon of Noro.

Mineral Compositions	Untreated	NH_4-Oxalate	NaOH
	wt.%		
Quartz	10.9	12.7	14.9
Plagioclase	16.6	10.5	24.9
K-Feldspars	4.0	4.2	4.2
Micas/Illite	5.4	5.1	3.3
Hornblende	2.0	3.1	3.2
Pyroxene	5.7	7.3	15.6
Olivine	9.5	9.5	7.7
Kaolinite	0.7	0.0	0.0
Chlorite	3.0	3.7	4.0
Gibbsite	19.5	26.6	0.0
Hematite/Goethite	16.8	12.7	17.5
Magnetite	5.9	4.5	4.8

3.2.5. Residual Si

The Si of the residual fraction comprises Si in primary and secondary minerals. The residual Si content in Ultisols and Alfisols from Jeju Island and Inceptisols from mainland Korea ranged from 251–312 $mg \cdot g^{-1}$, which was two times higher than that in Jeju Island's Andisols. The residual Si in Andisols decreased with an increasing soil depth, and the Si of the lowest residual fraction was 74.8 $mg \cdot g^{-1}$ in the C horizon from the Noro series (Figure 3f).

Not only amorphous silica but also a large amount of quartz and layer silicate minerals such as kaolinite, illite, chlorite, and vermiculite are produced in Ultisols and Alfisols from Jeju Island. As plagioclase and feldspars remained in Inceptisols from mainland Korea, the residual Si content was high (Table S1) [50]. In contrast, since Al formed Al-humus complexes in surface horizons from Jeju Island's Andisols during pedogenesis of volcanic ash soils, DSi moved downward. Since Al-rich allophane was mostly dissolved, as the Si concentration in the soil solution was low in subsoil horizons, DSi continued to be leached, leading to a very low residual Si content.

3.3. *Relationships between Readily Soluble Si and Other Si Fractions in Soils*

Table 6 describes the correlation between mobile and adsorbed Si and other Si fractions. In the study soils, the mobile Si and adsorbed Si contents had a close correlation ($r = 0.661$, $p < 0.01$). In subsoil horizons from the Samgag and Gangjeong series, the adsorbed Si content was lower than the mobile Si content. The Bw and C horizons from the Noro series Andisols showed a higher mobile Si content than did the other soils. These results are attributable to the fact that, since there is no boundary between mobile and loosely adsorbed Si in the soils, some loosely adsorbed Si was already desorbed from the soils at sequential extraction step 1 with $CaCl_2$ [25]. Consequently, the sum of the two fractions is a readily soluble, easy-to-mobilize Si fraction in Si pools from the soils [25,51]. As mentioned earlier, Jeju Island's Andisols, which have a high content of soil components that absorb monosilicic acid, have a larger amount of readily soluble Si in subsoil horizons compared with mainland Korea's Inceptisols.

Table 6. Spearman's rank correlation between readily soluble Si fraction (mobile Si and adsorbed Si) and potential sources in Si fraction (Si_{org}, Si_{occ}, Si_{ma}).

	Total (*n* = 28)		Non-Andisols (*n* = 19)		Andisols (*n* = 9)	
	Mobile Si	Adsorbed Si	Mobile Si	Adsorbed Si	Mobile Si	Adsorbed Si
Adsorbed Si	0.661 **		0.849 **		0.933 **	
Organic Si	0.539 **	0.814 **	0.819 **	0.672 **	0.536	0.644
Occluded Si	0.256	0.782 **	0.251	0.591 **	0.900 **	0.933 **
Amorphous Si	0.514 **	0.116	0.643 **	0.839 **	0.450	0.550
Residual Si	0.216	−0.286	0.691 **	0.655 **	−0.533	−0.450

*,**: significant difference at 0.05 and 0.01 probability levels, respectively.

In the study soils, adsorbed Si had a correlation with SOM-bound Si (r = 0.814 **) and occluded Si (r = 0.782 **), while mobile Si had a correlation with Si in amorphous silica (r = 0.514 **) (Table 6). In Si pools of the soils, Si in allophane and amorphous silica fractions is more soluble than crystalline minerals, so not only is Si dissolved more easily, but the desorption of monosilicic acid adsorbed to mineral surfaces may also become a potential source for readily soluble Si (mobile + adsorbed Si) [17,25]. The solubility of amorphous Si ranges between 1.8–2 mM compared with quartz's 0.10–0.25 mM Si [17,52] Miretzky et al. [27] reported that, under a wet climate in Buenos Aires Province, Argentina, amorphous Si of volcanic pyroclastic or biogenic origin is likely to determine a high concentration of DSi in groundwater.

Especially in Andisols, mobile and adsorbed Si had the highest correlation with occluded Si (r = 0.900 ** and r = 0.933 **, respectively). Occluded Si comprises allophane, and allophane with high adsorption capacity for monosilicic acid can make the biggest contributions to readily soluble Si [41,53]. During weathering under high precipitation conditions, allophane produces gibbsite by extreme desilication [19,21,22]. Subsoil horizons from Andisols with high allophane content contain more than 17% gibbsite content (Table S1, Figure 4). The formation of gibbsite in Andisols means that a large amount of Si has been leached and has moved down the soil, which may affect the Si content in groundwater.

Among extractants attempted by Georgiadis et al. [32], 0.01 M acetic acid showed considerable desorption efficiency with the smallest effect on clay minerals (except for smectite) and silica gel. Since the soils did not include smectite, adsorbed Si may not have been released from clay minerals. This concentration of acetic acid may have had an effect on allophane, a short-range-order mineral produced from Andisols in the study soils. However, since 0.01 M acetic acid had the smallest amount of Si released from the test materials compared with that of other extractants [32], Si released from allophane in this study's soils may also have been minimized.

3.4. Percentaget Distribution of the Different Si Fraction to Total Si

The distribution percentage of the total Si content in the soils by form is provided in Table 7. The largest amount of Si in the soils was residual Si in primary and secondary silicates, accounting for 57–98% of the total Si. Residual Si showed the largest distribution across the profiles of the Samgag and Oesan series derived from mainland Korea's granite and granite gneiss Inceptisols, while Si extracted from amorphous silica (including pedogenic and biogenic silica) is derived from basalt and has pyroclastic materials as its parent material. The Jeju and Gangjeong series on Jeju Island, where Ultisols and Alfisols are developed, and Inceptisols from mainland Korea had the second-largest amount of Si (1.74–8% of total Si). Meanwhile, Si occluded in pedogenic oxides and hydroxides showed the second-highest percentage (7.2–28% of total Si) in Jeju Island's Andisols. For SOM, Si was around 0.5% in Andisols, which was five times higher than that in other soils. Mobile Si and adsorbed Si showed the smallest distribution across the soil profiles. Mobile Si's distribution was 0.003–0.04% of the total Si. In soils other than Andisols, adsorbed Si's distribution was similar to that of mobile Si. However, adsorbed Si in the C horizon from the Noro series Andisols and the BC horizon from the

Pyeongdae series was approximately 0.3% of the total Si, which was 10 times higher than that in other soils. Furthermore, mobile Si was the highest in the Bw and C horizons from the Noro series at about 0.04% of the total Si.

Table 7. Distribution of different Si fractions by sequential extraction.

Horizon	Depth	Mobile Si	Adsorbed Si	Organic Si	Occluded Si	Amorphous Si	Residual Si
	(cm)	%					
Noro							
A	0–22	0.009	0.015	0.40	7.19	3.46	88.9
BA	22–44	0.014	0.037	0.48	19.2	2.05	78.2
Bw	44–87	0.036	0.209	0.69	28.0	3.19	67.9
C	87–160	0.044	0.332	0.56	27.9	13.7	57.5
Pyeongdae							
A	0–18	0.008	0.016	0.39	7.16	3.82	88.6
AB	18–44	0.008	0.024	0.53	8.61	3.00	87.8
Bw1	44–73	0.011	0.077	0.62	7.80	3.27	88.2
Bw2	73–100	0.013	0.121	0.65	11.2	3.65	84.3
BC	100–160	0.018	0.262	0.72	16.3	7.72	74.9
Jeju							
Ap	0–20	0.004	0.007	0.11	0.22	4.57	95.1
AB	20–41	0.005	0.007	0.09	0.20	5.15	94.5
Bt1	41–65	0.005	0.007	0.08	0.13	5.98	93.8
Bt2	65–92	0.007	0.008	0.10	0.14	7.12	92.6
Bt3	92–160	0.009	0.010	0.13	0.13	7.99	91.7
Gangjeong							
Ap	0–24	0.007	0.008	0.16	0.21	6.41	93.2
BAt	24–38	0.007	0.008	0.16	0.08	6.34	93.4
Bt1	38–53	0.009	0.007	0.18	0.06	5.95	93.8
Bt2	53–85	0.013	0.009	0.23	0.07	5.87	93.8
Bt3	85–160	0.017	0.011	0.28	0.12	6.22	93.4
Samgag							
A	0–15	0.003	0.003	0.10	0.01	1.85	98.0
BA	15–32	0.003	0.002	0.08	0.01	1.90	98.0
Bw	32–50	0.004	0.002	0.09	0.02	2.18	97.7
C1	50–78	0.006	0.003	0.12	0.02	2.77	97.1
C2	78–180	0.008	0.006	0.17	0.02	4.00	95.8
Oesan							
A	0–15	0.006	0.004	0.11	0.08	2.33	97.5
BA	15–34	0.006	0.006	0.11	0.11	2.22	97.5
Bw	34–63	0.007	0.007	0.10	0.10	2.51	97.3
C	63–98	0.007	0.007	0.10	0.07	1.74	98.1

The weathering of Jeju Island's Andisols is fast because the parent material is composed of pyroclastic deposits [23]. Under conditions where many organic matter sources exist due to high precipitation, Al preferentially forms humus complexes, which may lead to excessive Si [18–20]. However, since many silicic acids are leached from surface horizons to subsurface horizons during heavy rainfall, the readily soluble Si content is relatively low in the soil profile's upper part. In subsoil horizons, Si released by the weathering of primary minerals and soluble Si leached from the upper profile combines with polymerized Al to form allophane [19,20]. As the island has more than 2000 mm precipitation annually on average, significant leaching of Si takes place, which leads to a lower soluble Si concentration in the soil solution and the formation of Al-rich allophane with a lower Si content [24,48]. During Andisol formation, noncrystalline materials (e.g., allophane, ferrihydrite, and noncrystalline hydrous oxides of Fe and Al) are formed more abundantly than crystalline layer silicate. Noncrystalline materials present in Andisols can play an important role in adsorption of Si released during weathering, which may increase the readily soluble Si content. Therefore, in the volcanic ash soil of Jeju Island, where precipitation is high and water permeability is fast, the readily soluble Si fraction from the soil Si pools is released as DSi in the soil solution. This is likely to move into groundwater by infiltration. Non-Andisols in Jeju Island and the Korean mainland mainly produce quartz and layered silicates. As a result, Si is not leached much in non-Andisols.

4. Conclusions

This study was conducted to investigate the Si distribution and content in Andisols, Ultisols, and Alfisols in Jeju Island, and Inceptisols derived from granite and granite gneiss in the Korean mainland. The Si occurring as different soil compounds can be soluble, and could eventually migrate underground to increase the SiO_2 concentration. The average SiO_2 concentration in Jeju Island's groundwater was about four times higher than that in mainland Korea's groundwater. According to the results of Si sequential extraction, more than 90% of Jeju Island's Ultisols and Alfisols and mainland Korea's Inceptisols existed as crystalline Si and showed no difference by soil depth. While Jeju Island's Andisol soil had a smaller distribution percentage of crystalline Si compared with the non-Andisols of Jeju Island and mainland Korea, the percentage of Si constituting allophane, which is a noncrystalline material, absorbed Si. Soluble Si was noticeably higher. In the topsoil of Jeju Island's Andisols, Al-humus complexes are mostly formed, while allophane with a lower Si content is produced in the subsoil, which suggests that DSi is leached by high precipitation during weathering. At the same time, soluble or potentially soluble Si, a large amount of which is contained during the pedogenesis of Andisols, continues to be leached. This may be a factor in increasing the SiO_2 content in groundwater, which helps to improve the taste of drinking water and benefits human health.

Supplementary Materials: The following are available online at http://www.mdpi.com/2073-4441/12/10/2686/s1: Table S1: Mineral compositions of selected soil samples, Table S2: Results of the sequential silicon, aluminum, and iron extraction of analyzed soils, Table S3: Spearman's rank correlation between different Si fractions of analyzed soils, Figure S1: Silicon released by extraction with 0.2 M NaOH at a soil:solution ratio of 1:400 at 20–25 °C, Figure S2: X-ray diffraction pattern after sequential extraction from the Bw horizon of the Noro series.

Author Contributions: Conceptualization and methodology, W.-P.P. and H.-N.H.; formal analysis, W.-P.P.; data curation, W.-P.P. and H.-N.H.; writing—original draft, W.-P.P.; writing—review & editing, H.-N.H. and B.-J.K. All authors have read and agreed to the published version of the manuscript.

Funding: This research received no external funding.

Acknowledgments: We would like to express our gratitude to Seong-Taek Yun at Korea University who kindly provided SiO_2 concentration data in groundwater from the Korean mainland and also give thanks to Hyomin Lee for his technical assistance.

Conflicts of Interest: The authors declare no conflict of interest.

References

1. Mair, A.; Hagedorn, B.; Tillery, S.; El-Kadi, A.I.; Westenbroek, S.; Ha, K.; Koh, G.W. Temporal and spatial variability of groundwater recharge on Jeju Island, Korea. *J. Hydrol.* **2013**, *501*, 213–226. [CrossRef]
2. Lee, B.D.; Oh, Y.H.; Cho, B.W.; Yun, U.; Choo, C.O. Hydrochemical properties of groundwater used for Korea bottled waters in relation to geology. *Water* **2019**, *11*, 1043. [CrossRef]
3. Hashimoto, S.; Fujita, M.; Furukawa, K.; Minami, J.I. Indices of drinking water concerned with taste and health. *J. Ferment. Technol.* **1987**, *65*, 185–192. [CrossRef]
4. Gillette-Guyonnet, S.; Andrieu, S.; Vellas, B. The potential influence of silica present in drinking water on Alzheimer's disease and associated disorders. *J. Nutr. Health Aging* **2007**, *11*, 119–124. [PubMed]
5. Edwardson, J.A.; Moore, P.B.; Ferrier, I.N.; Lilley, J.S.; Newton, G.W.; Barker, J.; Templar, J.; Day, J.P. Effect of silicon on gastrointestinal absorption of aluminum. *Lancet* **1993**, *342*, 211–212. [CrossRef]
6. Calomme, M.R.; Vanden Berghe, D.A. Supplementation of calves with stabilized orthosilicic acid. *Biol. Trace Elem. Res.* **1997**, *56*, 153–165. [CrossRef] [PubMed]
7. Vasanthi, N.; Saleena, L.M.; Raj, S.A. Silicon in day to day life. *World Appl. Sci. J.* **2012**, *17*, 1425–1440.

8. Li, Z.; Karp, H.; Zerlin, A.; Lee, T.Y.A.; Carpenter, C.; Heber, D. Absorption of silicon from artesian aquifer water and its impact on bone health in postmenopausal women: A 12-week pilot study. *Nutr. J.* **2010**, *9*, 44. [CrossRef]
9. Sommer, M.; Kaczorek, D.; Kuzyakov, Y.; Breuer, J. Silicon pools and fluxes in soils and landscapes—A review. *J. Plant Nutr. Soil Sci.* **2006**, *169*, 310–329. [CrossRef]
10. Choi, H.M.; Lee, J.Y. Changes of groundwater conditions on Jeju volcanic island, Korea: Implications for sustainable agriculture. *Afr. J. Agric. Res.* **2012**, *7*, 647–661. [CrossRef]
11. Park, C.; Seo, J.; Lee, J.; Ha, K.; Koo, M.H. A distributed water balance approach to groundwater recharge estimation for Jeju volcanic island, Korea. *Geosci. J.* **2014**, *18*, 193–207. [CrossRef]
12. Song, Y.C.; Oh, S.S.; Hyun, I.H.; Oh, T.G.; Kim, S.M. Distribution of vital mineral groundwaters in Jeju. *Rep. JERI* **2009**, *2*, 254–267. (In Korean)
13. Kang, K.G. Studies on the Hydrogeochemical Processes and Characteristics of Groundwater in the Pyosun Watershed. Ph.D. Thesis, Jeju National University, Jeju, Korea, 2010. (In Korean).
14. Ko, H.S.; Kim, Y.; Koh, D.C.; Lee, K.S.; Lee, S.G.; Kang, C.H.; Seong, H.J.; Park, W.B. Hydrogeochemical characterization of groundwater in Jeju Island using principal component analysis and geostatistics. *Econ. Environ. Geol.* **2005**, *38*, 435–450. (In Korean)
15. Koh, D.C.; Cheon, S.H.; Park, K.H. Characterization of groundwater quality and recharge using periodic measurements of hydrogeochemical parameters and environmental tracers in basaltic aquifers of Jeju Island. *J. Soil Groundw. Environ.* **2007**, *12*, 57–68. (In Korean)
16. Cornelis, J.T.; Delvaux, B.; Georg, R.B.; Lucas, Y.; Ranger, J.; Opfergelt, S. Tracing the origin of dissolved silicon transferred from various soil-plant systems towards rivers: A review. *Biogeosciences* **2011**, *8*, 89–112. [CrossRef]
17. McKeague, J.A.; Cline, M.G. Silica in soil solutions II. The adsorption of monosilicic acid by soil and by other substances. *Can. J. Soil Sci.* **1963**, *43*, 83–96. [CrossRef]
18. Ugolini, F.C.; Dahlgren, R.A. Soil development in volcanic ash. *Glob. Environ. Res.* **2002**, *6*, 69–81.
19. Shoji, S.; Nanzyo, M.; Dahlgren, R.A. *Volcanic ash Soils-Genesis, Properties and Utilization*; Elsevier Publishers B.V.: Amsterdam, The Netherlands, 1993.
20. Ugolini, F.C.; Dahlgren, R.A. Weathering environments and occurrence of imogolite/allophane in selected Andisols and Spodosols. *Soil Sci. Soc. Am. J.* **1991**, *55*, 1166–1171. [CrossRef]
21. Ndayiragije, S.; Delvaux, B. Coexistence of allophane, gibbsite, kaolinite and hydroxy-Al-interlayered 2:1 clay minerals in a perudic Andosol. *Geoderma* **2003**, *117*, 203–214. [CrossRef]
22. Buol, S.W.; Southard, R.J.; Graham, R.C.; McDaniel, R.A. Andisols: Soils with andic soil properties. In *Soil Genesis and Classification*, 5th ed.; Iowa State Press: Ames, IA, USA, 1997; pp. 231–241.
23. Park, W.P.; Song, K.C.; Koo, B.J.; Hyun, H.N. Distribution of available silicon of volcanic ash soils in Jeju Island. *Appl. Environ. Soil Sci.* **2019**, *2019*. [CrossRef]
24. Shin, J.S.; Tavernier, R. Composition and genesis of volcanic ash soils in Jeju Island. II. Mineralogy of sand, silt, and clay fractions. *J. Min. Soc. Korea* **1988**, *1*, 40–47.
25. Georgiadis, A.; Sauer, D.; Herrmann, L.; Breuer, J.; Zarei, M.; Stahr, K. Testing a new method for sequential Si-extraction on soils of a temperate-humid climate. *Soil Res.* **2014**, *52*, 645–657. [CrossRef]
26. Makabe, S.; Kakuda, K.I.; Sasaki, Y.; Ando, T.; Fujii, H.; Ando, H. Relationship between mineral composition or soil texture and available silicon in alluvial paddy soils on the Shounai Plain, Japan. *Soil Sci. Plant Nutr.* **2009**, *55*, 300–308. [CrossRef]
27. Miretzky, P.; Conzonno, V.; Fernández Cirelli, A. Geochemical processes controlling silica concentrations in groundwaters of the Salado River drainage basin, Argentina. *J. Geochem. Explor.* **2001**, *73*, 155–166. [CrossRef]
28. National Academy of Agricultural Science (NAAS). *Taxonomical Classification* of Korean Soils; Rural Development Administration: Suwon-si, Korea, 2014.
29. Burt, R. *Soil Survey Laboratory Methods Manual, Soil Survey Investigations Report No. 42*; Version 4.0; Natural Resources Conservation Service, US Department of Agriculture: Washington, DC, USA, 2004.
30. Allison, L.E. Organic carbon. In *Methods of Soil Analysis: Part 2—Chemical and Microbiological Properties*; Black, C.A., Evans, D.D., White, J.L., Enisminger, L.E., Clark, F.E., Eds.; American Society of Agronomy: Madison, WI, USA, 1965; pp. 1372–1378.

31. Mizota, C.; van Reeuwijk, L.P. *Clay Mineralogy and Chemistry of Soils Formed in Volcanic Material in Diverse Climatic Regions, Soil Monograph 2*; International Soil Reference and Information Center (ISRIC): Wageningen, The Netherlands, 1989.
32. Georgiadis, A.; Sauer, D.; Herrmann, L.; Breuer, J.; Zarei, M.; Stahr, K. Development of a method for sequential Si extraction from soils. *Geoderma* **2013**, *209*, 251–261. [CrossRef]
33. Zhu, H.; Wang, C.; Wang, P.; Hou, J.; Qian, J.; Ao, Y.; Liu, C. Speciation of potentially mobile Si in Yangtze Estuary surface sediments: Estimates using a modified sequential extraction technique. *Environ. Sci. Pollut. Res.* **2016**, *23*, 18928–18941. [CrossRef]
34. Wada, K. Allophane and Imogolite. In *Minerals in Soil Environments*, 2nd ed.; Dixon, J.B., Weed, S.B., Eds.; Soil Science Society of America: Madison, WI, USA, 1989; pp. 1051–1088.
35. Tsai, C.C.; Chen, Z.S.; Kao, C.I.; Ottner, F.; Kao, S.J.; Zehetner, F. Pedogenic development of volcanic ash soils along a climosequence in Northern Taiwan. *Geoderma* **2010**, *156*, 48–59. [CrossRef]
36. Georgiadis, A.; Sauer, D.; Breuer, J.; Herrmann, L.; Rennert, T.; Stahr, K. Optimising the extraction of amorphous silica by NaOH from soils of temperate-humid climate. *Soil Res.* **2015**, *53*, 392–400. [CrossRef]
37. Klotzbücher, T.; Marxen, A.; Vetterlein, D.; Schneiker, J.; Türke, M.; Van Sinh, N.; Manh, N.H.; Van Chien, H.; Marquez, L.; Villareal, S.; et al. Plant-available silicon in paddy soils as a key factor for sustainable rice production in Southeast Asia. *Basic Appl. Ecol.* **2015**, *16*, 665–673. [CrossRef]
38. Chadwick, O.A.; Gavenda, R.T.; Kelly, E.F.; Ziegler, K.; Olson, C.G.; Crawford Elliott, W.; Hendricks, D.M. The impact of climate on the biogeochemical functioning of volcanic soils. *Chem. Geol.* **2003**, *202*, 195–223. [CrossRef]
39. Nieuwenhuyse, A.; van Breemen, N. Quantitative aspects of weathering and neoformation in selected Costa Rican volcanic soils. *Soil Sci. Soc. Am. J.* **1997**, *61*, 1450–1458. [CrossRef]
40. Di Figlia, M.G.; Bellanca, A.; Neri, R.; Stefansson, A. Chemical weathering of volcanic rocks at the island of Pantelleria, Italy: Information from soil profile and soil solution investigations. *Chem. Geol.* **2007**, *246*, 1–18. [CrossRef]
41. Wada, K.; Inoue, A. Adsorption of monomeric silica by volcanic ash soils. *Soil Sci. Plant Nutr.* **1974**, *20*, 5–15. [CrossRef]
42. Shoji, S.; Nanzyo, M.; Dahlgren, R.A.; Quantin, P. Evaluation and proposed revisions of criteria for Andosols in the world reference base for soil resources. *Soil Sci.* **1996**, *161*, 604–615. [CrossRef]
43. Imaya, A.; Inagaki, Y.; Tanaka, N.; Ohta, S. Free oxides and short-range ordered mineral properties of brown forest soils developed from different parent materials in the submontane zone of the Kanto and Chubu districts, Japan. *Soil Sci. Plant Nutr.* **2007**, *53*, 621–633. [CrossRef]
44. Cornelis, J.T.; Dumon, M.; Tolossa, A.R.; Delvaux, B.; Deckers, J.; Van Ranst, E. The effect of pedological conditions on the sources and sinks of silicon in the vertic planosols in south-western Ethiopia. *Catena* **2014**, *112*, 131–138. [CrossRef]
45. Saccone, L.; Conley, D.J.; Koning, E.; Sauer, D.; Sommer, M.; Kaczorek, D.; Blecker, S.W.; Kelly, E.F. Assessing the extraction and quantification of amorphous silica in soils of forest and grassland ecosystems. *Eur. J. Soil Sci.* **2007**, *58*, 1446–1459. [CrossRef]
46. Kendrick, K.J.; Graham, R.C. Pedogenic Silica accumulation in chronosequence soils, Southern California. *Soil Sci. Soc. Am. J.* **2004**, *68*, 1295–1303. [CrossRef]
47. Drouza, S.; Georgoulias, F.A.; Moustakas, N.K. Investigation of soils developed on volcanic materials in Nisyros Island, Greece. *Catena* **2007**, *70*, 340–349. [CrossRef]
48. Song, K.C.; Yoo, S.H. Andic properties of major soils in Cheju Island. III. Conditions for formation of allophane. *Korean J. Soil Sci. Fert.* **1994**, *27*, 149–157. (In Korean)
49. Song, K.C.; Hyun, B.G.; Moon, K.H.; Jeon, S.J.; Lim, H.C. Taxonomical classification of Yongdang series. *Korean J. Soil Sci. Fert.* **2009**, *42*, 393–398. (In Korean)
50. Hwang, J.; Lee, H. The applications and mineral compositions of residual soils distributed in South Korea. *Clay Sci.* **2009**, *14*, 253–262. [CrossRef]
51. Georgiadis, A.; Rinklebe, J.; Straubinger, M.; Rennert, T. Silicon fractionation in Mollic Fluvisols along the Central Elbe River, Germany. *Catena* **2017**, *153*, 100–105. [CrossRef]

52. Drees, L.R.; Wilding, L.P.; Smeck, N.E.; Senkayi, A.L. Silica in soils: Quartz and disordered silica polymorphs. In *Minerals in Soil Environments*, 2nd ed.; Dixon, J.B., Weed, S.B., Eds.; Soil Science Society of America: Madison, WI, USA, 1989; pp. 913–974.
53. Yanai, J.; Taniguchi, H.; Atsushi, N. Evaluation of available silicon content and its determining factors of agricultural soils in Japan. *Soil Sci. Plant Nutr.* **2016**, *62*, 511–518. [CrossRef]

Article

Sustainable Development of Arid Rangelands and Managing Rainwater in Gullies, Central Asia

Zheng Li, Wentai Zhang *, Yilahong Aikebaier, Tong Dong, Guoping Huang, Tao Qu and Hexin Zhang

Xinjiang Key Laboratory of Soil and Plant Ecological Processes, College of Grassland and Environmental Sciences, Xinjiang Agricultural University, Urumqi 830052, Xinjiang, China; Liz_94@163.com (Z.L.); akbarilahun@xjau.edu.cn (Y.A.); mr_dongtong@126.com (T.D.); xau_hgp@163.com (G.H.); qtao_xj@163.com (T.Q.); zhx19950826@163.com (H.Z.)

* Correspondence: zwt@xjau.edu.cn; Tel.: +86-0991-8763937

Received: 7 August 2020; Accepted: 9 September 2020; Published: 10 September 2020

Abstract: Along with the global climate change, gully erosion, flood and drought jointly restrict the sustainable development of arid rangeland in Central Asia. Rainwater harvesting (RWH) system in gully is a flexible practice that alleviate complex environmental problems. In the Kulusitai watershed of Xinjiang, China, our study presented a decision-making system using GIS combined with multi-criteria analysis and a field survey to identify suitability of gully for RWH. The results showed that nearly 40% of rangeland belonged to high runoff potential area, and gullies as the runoff collection channel became the potential site of RWH. The selection of RWH systems depended on catchment environment and gully characteristics. Therefore, based on the unique natural conditions of Xinjiang and successful RWH cases in other regions, we discussed some suitable low-cost RWH techniques to restore degraded grassland and promote community development. Our study will provide some suggestions for ecological restoration and pasture management in arid regions of Central Asia.

Keywords: gully erosion; runoff potential; rainwater harvesting; ecological restoration

---

1. Introduction

Land degradation caused by gully erosion is a long-standing and complex global environmental problem, especially in arid rangelands [1]. Climate change and overgrazing have exacerbated the rangeland water erosion and its consequences in recent years [2,3]. Severe gully erosion causes land fragmentation, destroys grassland resources and reduces livestock carrying capacity of rangeland [4,5]. In addition, gullies tend to enhance drainage and accelerate aridification processes [6]. Under heavy rainfall events, a large amount of surface runoff is rapidly lost through gullies, leading to the increase risk of flooding. Moreover, the waste of water resources will aggravate the drought in these areas. With global warming, extreme rainfall events and drought events occur frequently [7,8], which makes the management of gullies an urgent problem in the development of arid pastoral areas.

Rainwater harvesting (RWH) is defined as "the process of concentrating rainwater over catchments through runoff to be stored and beneficially used" [9]. RWH systems can control erosion and restore degraded land during the green revolution [10]. The gully is a temporary gathering channel of runoff in the process of rainstorm, so gullies are used as potential RWH site in some arid areas. Majalubas (Tanzania) and cisterns (Mediterranean region) collect and store runoff in the gullies for agricultural production, which solves the problem of water shortage in dry season, and then improves water efficiency and productivity in arid areas [11,12]. In addition, people have built dams to retain runoff in the channels, which control the soil erosion of arid rangeland, and maintain the natural function and service of watershed ecosystem. Jessours (Tunisia) and Limans (Israel) create water- and nutrient-enriched patches in gullies, which reduce the risk of flood and effectively improve grassland productivity [13–15]. In 2015, the United Nations mentioned, in "transforming our world: the 2030 agenda for sustainable development", that the implementation of flexible and future agricultural practices would be beneficial for improving land productivity, maintaining ecosystem functions and increasing resilience to climate change, extreme weather and natural disasters [16]. Therefore, gully RWH systems will be effective measure to promote the sustainable development of arid regions.

Previous reports on RWH were mostly concentrated in the Mediterranean, the Middle East, Central Africa and other arid areas. Less attention had been paid to arid rangelands in Central Asia, where research on RWH started late, resulting in few studies related to RWH [2,9,10,17,18]. Central Asia is a typical arid area and an important pasture in the world, so, based on the special natural conditions and the urgent demand for gully RWH technologies, this study evaluates the suitability of a gully for RWH.

Whereas successful RWH systems often depend on suitable sites and technical design, assessing the catchment environment is our priority [17,19]. The formation and development of gullies are closely related to the topography, soil texture and other environmental factors [20]. Environmental factors, such as vegetation and land use/cover also determine the resource potential of runoff [21]. Most scholars used GIS combined with multi-criteria analysis to identify suitable RWH sites [22–25]. This method could judge and weigh the importance of multiple variables and accurately evaluate the runoff potential of the catchment. Gully is an independent system, and each gully has its unique morphological characteristics, due to various environment factors (topography, soil, vegetation) of catchment. The characteristics of channel directly affect the suitability, implementation and cost of RWH systems [22,26]. However, in most scholarly evaluation systems, the characteristics of channels were often ignored. The economic conditions and scientific and technological level of pastoral areas in Central Asia often lag behind other areas, which requires us to consider the catchment environment, but also to integrate channel characteristics into the evaluation system. Only by fully understanding the relationship between the catchment environment and gully characteristics, can the selected RWH technique maximize its potential in the gully. Field surveys are often used to select suitable RWH techniques in smaller areas [17]. This study gathered the parameter information of gully through field survey, and also verified the catchment environment, so as to improve the accuracy of evaluation results.

As such, RWH system in gully is an effective practice to control gully erosion and improve water efficiency in arid rangelands, but how to select a suitable RWH system is the primary problem faced due to various gully characteristics. Our study presents a decision-making system using GIS combined with multi-criteria analysis and a field survey to identify suitability of gully for RWH. This method can help decision-makers to select more flexible RWH systems in gully. The purposes of this paper are: (1) get to know the development characteristics of gully and discuss its relationship with catchment environment in arid rangeland of Central Asia; and (2) to evaluate the suitability of gully for RWH accurately and discuss suitable techniques.

2. Materials and Methods

2.1. Study Area

Our study took place in the hills of Kulusitai watershed, which is located in a region of northwestern China's Yili River Valley, in Xinjiang. The area is a typical arid rangeland with a large number of gullies (Figure 1). Grassland in the study area has been seriously degraded, and the vegetation coverage is 10–50%. The main soil type is Calcisol, which has developed from the parent material of Quaternary loess [27]. The loess parent material contains more sand and silt, and therefore the Calcisols are porous and have a weak bonding strength [28]. During rainfall events, the Calcisols are prone to water erosion.

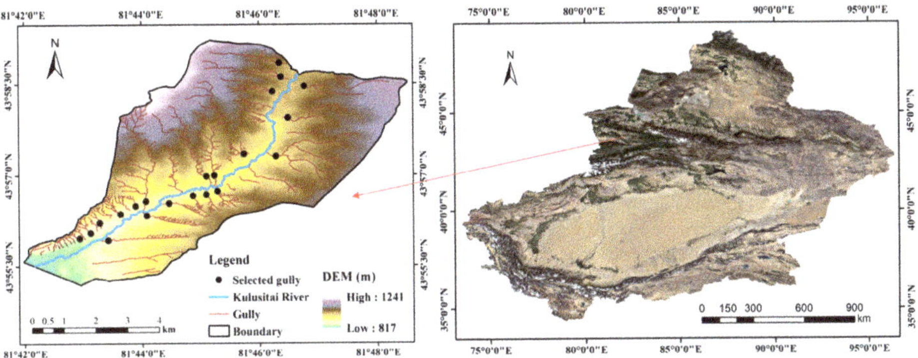

Figure 1. Location of study area and selected gullies.

The climate of study area is temperate continental climate. The rangeland has an average annual temperature of 10 °C and an average annual potential evapotranspiration of 1600 mm. However, the average annual precipitation is about 400 mm, and uneven in seasonal distribution (Figure 2). Precipitation is the only water source in natural grassland, and temperature is main driving factor of grassland water evapotranspiration. There was less evapotranspiration and a higher precipitation supplement from April to May. The evapotranspiration of grassland is so large in the dry season (July and August), but rainfall events are greatly reduced, so grass growth and development are often influenced by drought stress. Therefore, there is obvious seasonal water shortage in this rangeland. When daily precipitation reached 10 mm, we regard the rainfall as a heavy daily rainfall [7]. The contribution of precipitation on the heavy rainfall days to total precipitation is more than 30%, and there is at least one heavy rainfall day every month [29]. During severe rainfall events or high intensity precipitation, rainfall is easily converted into surface runoff, which often lead to floods. RWH techniques transform runoff into available water for grass growth, which are used for ecological restoration of degraded grassland. These special rainfall characteristics have a negative impact on the ecological development of arid rangelands in Xinjiang, but also create opportunities for RWH.

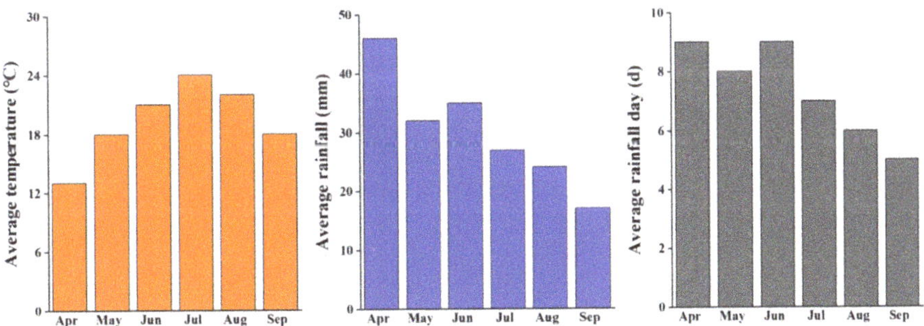

Figure 2. Monthly temperature, precipitation, daily rainfall of Kulusitai watershed.

2.2. Data Source

Digital evaluation models (DEM) with 30 m resolution and remote sensing data (Landsat 8 Operational Land Imager (OLI)) with 30 m resolution were accessible from the Geospatial Data Cloud website (http://www.gscloud.cn/). Land use/cover data was downloaded from the global land cover 10 m resolution map (http://data.ess.tsinghua.edu.cn) [30]. Monthly temperature and rainfall data (from 2010 to 2019) were obtained from the Weather China Website (http://www.weather.com.cn).

2.3. Measurement of Gully Characteristics

We selected 21 gullies in the study area for field survey and analyzed in GIS (Figure 1). The field survey mainly included the measurement of the width and depth of main gully and GPS positioning of some important geographical coordinates, such as gully head (Table 1). After the survey, ArcGIS 10.5 (ESRI, Redlands, CA, USA) was used to calculate the main gully length, area of catchment and other parameters. Principal component analysis (PCA) in SPSS 20.0 (IBM, Amonk, NY, USA) was used to evaluate and classify gullies.

Table 1. Details characteristics of gully and catchment.

Parameter	Measurement and Calculation of Parameters
Main gully length (MGL)	Field survey and calculated in GIS
Total gully length (TGL)	Field survey and calculated in GIS
Main gully width (MGW)	Field survey
Main gully depth (MGD)	Field survey
Circumference of catchment (C)	Field survey and calculated in GIS
Area of catchment (A)	Field survey and calculated in GIS
Relief ratio (Rr)	Rr = H/L, L is maximum catchment length (m); H is height (m) difference between highest point and lowest point in catchment [31].
Drainage density (Dd)	Dd = TGL/A [32]
Slope (S)	Analyzed DEM in GIS
Topographic wetness index (WI)	WI = ln ($Ac/\tan^S$), Ac is the potential contributing area [33]
Normalized difference vegetation index (NDVI)	NDVI = (NIR − R)/(NIR + R), NIR and R are the spectral reflectance in the near infrared band and red band [34]
Land use/cover	Data from Gong's classification system and field survey [30]

2.4. Evaluation of Runoff Potential in Catchment

2.4.1. Criteria Selection

The first step of quantifying runoff potential area was to select appropriate criteria. Based on the guidelines of the Food and Agriculture Organization of the United Nations (FAO) and a literature review, hydrology (rainfall-runoff relationship), topography (slope), land cover/use, agronomy (crop characteristics) and soil (depth and type) were the five important environmental factors. We first selected slope, land use/cover and vegetation characteristics. The accuracy of hydrological model highly depended on the spatial scale of study area and data availability [17]. In Central Asia, there were many pastoral areas similar to the Kulusitai Watershed, which had small spatial scale and little available information. Hydrological models were not applied in our study. We also did not consider soil, because there was only one soil type in our study area. Furthermore, topographic wetness index (WI) was selected to evaluate the runoff potential more accurately.

(1) Topographic wetness index (WI)

WI is a function of slope and catchment area, which accurately reflects the impact of topographic changes on runoff [33]. WI has been applied to predict the spatial distribution of soil water and runoff process widely [35]. The higher WI is, the easier runoff is generated in catchment, indicating the greater potential of runoff utilization.

(2) Slope

Slope is an important index to evaluate runoff potential. Slope affects runoff coefficient and kinetic energy. In the high slope area, the application of RWH system is difficult and requires a stronger structure [19,23].

(3) Land use/cover

Land use can change soil hydraulic characteristics of underlying surface, affect runoff infiltration, and then lead to different runoff coefficients [36,37]. It is generally believed that farmland and bare land have higher runoff coefficient compared with forest and grassland [38].

(4) Normalized difference vegetation index (NDVI)

Good vegetation conditions can enhance infiltration by improving soil structure, and increase surface roughness to delay runoff flow time [39,40]. In arid and semi-arid areas, the increase of vegetation coverage significantly reduces runoff coefficient [38,41].

2.4.2. Data Analysis

We merged 30 m DEM data and Landsat 8 OLI data into a 10 m-resolution spatial data set (WGS-84 geographic coordinate system) using the 'reclassify' tool in ArcGIS 10.5, and then calculated and classified the four criteria (Table 1). As a result of the complex terrain in study area, we divided the WI, and slope into five categories (Figure 3a,b). We used Gong's classification system to classify land use/cover into four categories (Figure 3c). The dominant vegetation type was grassland, and NDVI was divided into three categories (Figure 3d).

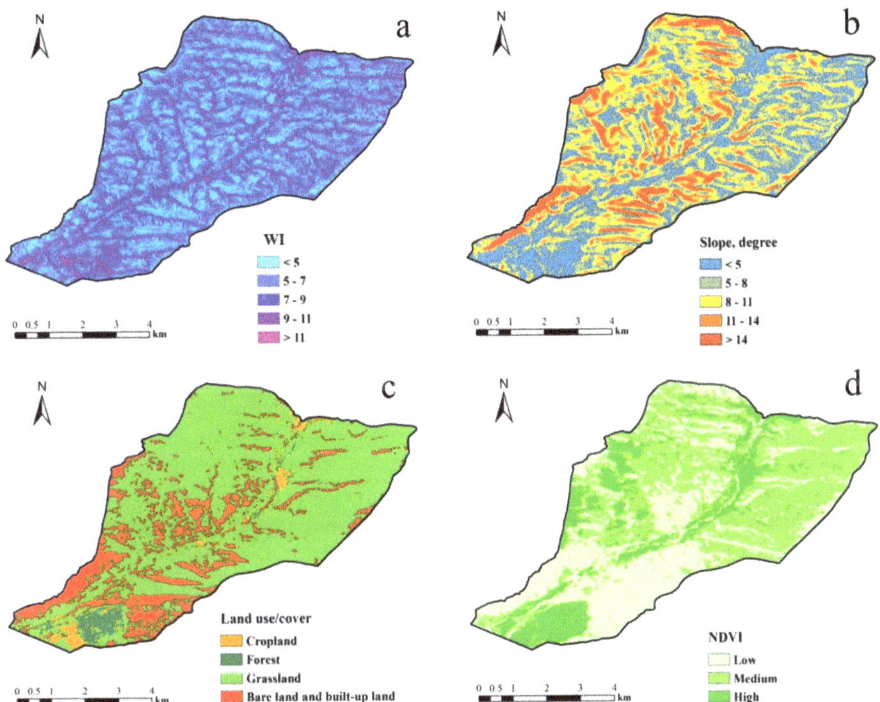

Figure 3. Environmental factors of catchment. (**a**) wetness index (WI); (**b**) slope; (**c**) land use/cover; (**d**) normalized difference vegetation index (NDVI).

The Analytic Hierarchy Process (AHP) is a multi-criteria decision-making method [42]. AHP allows us to understand the contribution of each factor to a given objective and is applied widely to identify runoff potential zones. According to the essential principle of AHP, we made the influence factors of runoff potential organized and hierarchical, and asked experts to score for relative importance of each criteria at each level to form a matrix of pairwise comparisons. We calculated the weight of each layer criteria to upper level criteria by applying a matrix, and then established a multi-level relationship model to quantify runoff potential. The weight of each criteria was calculated by AHP using the yaahp 12.4 software for Windows (Meta Decision Software Technology, Taiyuan, Shanxi, China) (Table 2). In addition, a consistency ratio was used to assess the accuracy of pairwise comparison in AHP [43]. Finally, we generated the runoff potential map by integrating four criteria maps using the 'weighted overlay' tool.

Table 2. Weight of each factor on runoff potential.

Characteristic	Theme Weight	Class	Assigned Class Weight	Consistency Ratio
WI	0.4918	<5	0.0422	0.0385
		5–7	0.0625	
		7–9	0.1367	
		9–11	0.2089	
		>11	0.5498	
Slope (degree)	0.3056	<5	0.0347	0.0523
		5–8	0.0696	
		8–11	0.1411	
		11–14	0.2136	
		>14	0.5437	
Land use/cover	0.1248	Bare land and built-up land	0.6106	0.0452
		Cropland	0.2160	
		Grassland	0.1297	
		Forest	0.0437	
NDVI	0.0778	Low	0.6370	0.0370
		Medium	0.2583	
		High	0.1047	

Theme weight' consistency ratio was 0.0181.

3. Results

3.1. Runoff Potential

By evaluating the runoff potential, we found that areas with a high density of gullies tended to have high runoff potential (Figure 4a). Especially from the WI map, it could be clearly seen that gullies were the area of runoff collection (Figure 3a). In the southwest of study area, topography fluctuates greatly, and vegetative cover was close to bare land, which made the area become the highest runoff potential area. When heavy rainfall occurred, a large amount of runoff would be generated in catchment and transported to river through channel, and then increased the risk of flood in the lower reaches of watershed. Therefore, it would become a key area for future RWH projects. On the contrary, in the northeast, the terrain was flat, and the vegetative cover had been greatly improved, so runoff potential was far lower than that in the southwest. Nearly 40% of the study area belonged to high runoff potential area (Figure 4b), which meant that there was a large potential for RWH in this rangeland. As a result of different environmental factors, catchments showed different RWH potential, which was of great reference value for understanding the formation and development of gullies and RWH projects in arid rangelands.

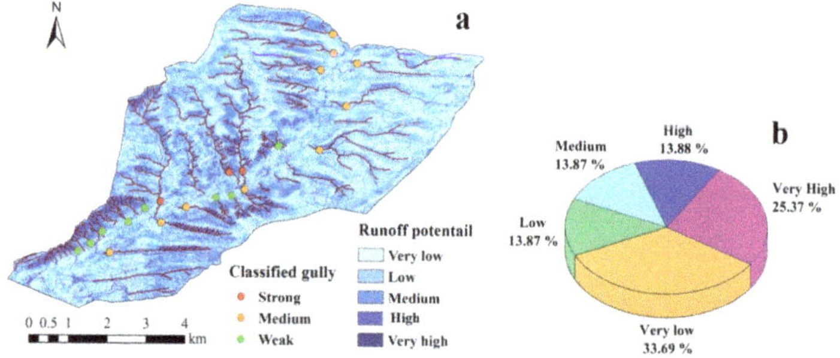

Figure 4. Runoff potential and gullies (strong, medium and weak) location. (**a**) runoff potential map; (**b**) percentage of area covered by different runoff potential.

3.2. Characteristics of Gullies

The results of a field investigation showed that there were severe gully erosion and diversified gullies in arid rangeland. The parameters of 21 gullies were analyzed by PCA (Figure 5). According to the development characteristics, gullies were divided into strong gully, medium gully and weak gully (Table 3). Although catchment of weak gully was small, the higher Dd and Rr enhanced the drainage ability of channels, and runoff was transported out rapidly from catchment. The strong gully had a big catchment, but its Dd was low, and the MGL accounted for a large proportion of the TGL, which led to a large quantity of runoff gathered in the main gully. The wide channel ensured the transport of runoff. The parameters of medium gully were between weak gully and strong gully. The deep main channel was the main feature of medium gully, especially the MGD of three gullies in the south was even more than 5 m.

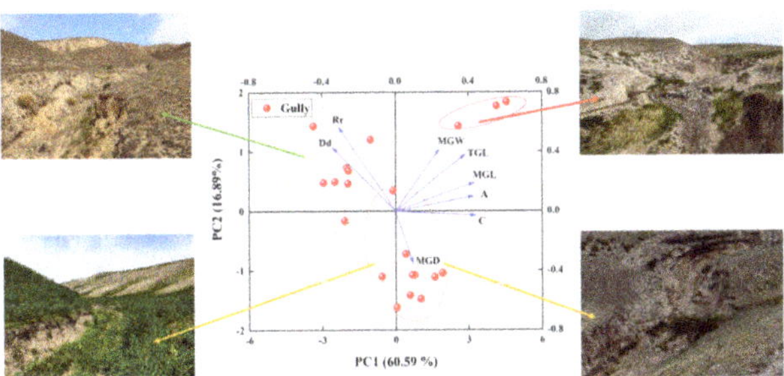

Figure 5. Principal component analysis (PCA) of gully developmental characteristics.

Table 3. Gully developmental characteristics.

Classified Gullies	MGL (km)	TGL (km)	MGW (m)	MGD (m)	C (km)	A (km^2)	Dd (km/km^2)	Rr
Weak gully	0.650c ± 0.05	1.459b ± 0.29	4.62b ± 1.04	1.63b ± 0.12	1.64c ± 0.15	0.13c ± 0.02	12.4a ± 1.75	0.133a ± 0.0049
Medium gully	1.483b ± 0.13	2.293b ± 0.26	5.41b ± 0.84	3.52a ± 0.55	4.60b ± 0.34	0.96b ± 0.14	3.05b ± 0.79	0.067b ± 0.0064
Strong gully	3.300a ± 0.27	6.366a ± 1.13	9.97a ± 0.70	1.71b ± 0.22	7.96a ± 0.71	2.43a ± 0.31	2.57b ± 0.17	0.091b ± 0.0045

Lowercase letters indicated a significant difference at the 0.05 level.

4. Discussion

4.1. Relationship between Gully Characteristics and Catchment

Weak gully catchment had the highest runoff potential. The steep slope provided high kinetic energy for runoff. The critical vegetative coverage for soil and water loss control was 40% in our study area [44]. The vegetative coverage in weak gully catchment was far less than the critical value, and the sparse vegetation would further enhance the sensitivity of gully erosion [18,45,46]. Therefore, gullies were very easy to form in high runoff potential areas, which was the reason why the Dd of weak gully was much higher than that of strong gully and medium gully. Moreover, due to small catchment, the runoff transported from the catchment to the channel was less. The erosion intensity of runoff on the main gully was weakened [47], and the width and depth of weak gully were smaller.

The catchment of strong gully was large, and elevation drop was even more than 300 m. Compared with weak gully catchment, the vegetation and terrain conditions in strong gully catchment were improved, and runoff potential was reduced. Therefore, large scale catchment would not necessarily result in higher Dd, and the same result also appeared in Northeast China [45]. Previous studies had demonstrated that channel width was strongly related to runoff flow [48]. Under heavy rainfall events, major runoff concentrated in gullies, which intensified the erosion of main gully, and then widened the main gully width.

The runoff potential of medium gully catchment was the lowest, but area of catchment was still much larger than that of weak gully. The huge volume of runoff would also be generated in catchment, which increased the erosion of main gully. However, different from strong gully, the main gully erosion of medium gully was mainly reflected in MGD. Headcut retreat, widening and deepening were the typical erosion processes for gullies, but the contribution of these processes might be different due to various catchment environment [6,49]. The depth of channel was closely related to soil erodibility and shear stress of runoff. The soil erodibility K of grassland in this area was about 0.08 t·hm^2·h/hm^2·MJ·mm, indicating that the soil was prone to water erosion [28]. During the field survey, we found that gully banks of medium gully formed steep steps. The initial step height significantly increased the shear stress of runoff. The overland concentrated flows with potential energy would cause severe plunge pool erosion, and then acted on the channel bed [50]. With the continuous erosion of runoff, headcut would gradually retreat and channel would deepen.

4.2. Selection of Suitable RWH Techniques in Gully

In the case of heavy rainfall, runoff was easily generated in weak gully catchment, and the volume of flow collected in channel was small. At the same time, runoff would be rapidly lost from the catchment due to strong drainage capacity of gullies. The earth check dam with biological control measures (similar to Jessour, Figure 6a) was a suitable RWH system for weak gully. These dams allowed on-site retaining of runoff in channel and plant roots increased runoff infiltration [13,51]. Runoff decreased in volume and speed gradually as it passed through each earth check dam. In addition, dams maintained high available water content in the gully soil, improved soil fertility and vegetation conditions, which was conducive to restoration of pastureland [14,15,52]. The width and depth of weak gully were small, so the construction cost of earth check dam was low and could be used widely in arid rangelands' gullies.

Figure 6. Suitable rainwater harvesting (RWH) techniques in arid rangeland. (**a**) earth check dam; (**b**) water cellar; (**c**) fence; (**d**) vines on the gully wall.

Strong gully had a large catchment with high runoff potential, which was easy to cause flood in the process of heavy rainfall. Although floods were destructive, appropriate RWH systems transformed floods into green water resources in arid rangeland. There were two problems in the construction of dams: one was technical difficulty; the other was high cost. We could build flood diversion channels and reservoirs to collect and store huge amounts of runoff. The flood diversion channel greatly reduced runoff volume in main channel and dispersed impact force of flood. The diverted flood was used to supply irrigation [53]. Storing runoff effectively solved the local seasonal water shortage problem. For example, storing runoff in a simple reservoir saved the runoff that was lost in three days for 40 days in Ethiopia [54]. The stored runoff would also be used for crop irrigation and vegetation restoration (Figure 6b) [10,55], and in desert areas of China, farmers even used the flood to develop aquaculture [56]. Water storage systems, such as water cellar and reservoir, transformed runoff into available resources and helped the green development of arid pastoral areas.

The runoff potential was low in medium gully catchment in the north, but grass was abundant in the lower part of the slope (Figure 5). In view of this special vegetation distribution pattern, we could refer to Tunisia's tabias. We regarded the entire slope as runoff area and developed artificial grassland in the gully and its adjacent areas. Artificial grassland collected runoff from runoff area and stored it in the soil [57]. Some herdsmen had fenced the grassland (Figure 6c). Moreover, we also needed to use simple RWH techniques to further improve grassland productivity. Loose rock dams and branch bundles, made of local materials, might be effective for lessening gully erosion and restoring degraded rangeland [58,59]. In the south, the medium gully had steep gully walls, which was the best place to build dams. Compared with weak gully, the dams retained more runoff because of deeper channels, and steep walls reduced water evaporation in the channels effectively. [23,60]. However, the dykes had to be more robust to resist runoff's forces. In addition, the erosion of gully wall could be controlled by planting lianas (Figure 6d). Some plants with medicinal value would also bring additional economic income to local herdsmen and farmers.

5. Conclusions

About 40% of the low mountain rangeland in Kulusitai watershed belonged to high runoff potential area, which was also the area of gullies. As the runoff gathering place, gullies became suitable sites to carry out RWH projects in arid pasture.

The environmental factors in catchment led to different runoff potential and diversification of gully morphology. According to the development characteristics, gullies were divided into strong gully, medium gully and weak gully. The runoff potential of weak gully catchment was the highest, which also led to the highest drainage density. The deep main channel was the main feature of medium gully, but runoff potential of catchment area was the lowest. Strong gully had a large catchment with high runoff potential, main gully length and width were much larger than those of the weak gully and medium gully.

Our study presents an approach combined runoff potential of catchment with gullies characteristics to further evaluate suitability of different gullies for RWH, and discusses some low-cost RWH techniques suitable for local area. Herdsmen can select flexible RWH techniques to restore degraded grassland and develop farming and aquaculture. The RWH systems in gullies will produce huge ecological benefits (improve ecosystem service value) and social benefits (increase herdsmen's economic income) in future, which has significance for the sustainable development of arid rangelands in Central Asia.

Author Contributions: Conceptualization, Z.L., W.Z. and Y.A.; field survey, Z.L., W.Z., G.H., T.Q. and H.Z., methodology and software, Z.L. and T.D.; formal analysis, Z.L. and Y.A.; writing—original draft preparation, Z.L.; writing—review and editing, W.Z., Y.A. All authors have read and agreed to the published version of the manuscript.

Funding: The research was supported by the National Natural Science Foundation of China (Grant No. 41761059) and the Joint Project of Nanjing Agricultural University and Xinjiang Agricultural University (Grant No. KYYJ201903).

Acknowledgments: The authors thank the editor and anonymous reviewers for their insightful and constructive comments.

Conflicts of Interest: The authors declare no conflict of interest.

References

1. Poesen, J.; Nachtergaele, J.; Verstraeten, G.; Valentin, C. Gully erosion and environmental change: Importance and research needs. *Catena* **2003**, *50*, 91–133. [CrossRef]
2. Zhang, W.T.; Zhou, J.Q.; Feng, G.L.; Weindorf, D.C.; Hu, G.Q.; Sheng, J.D. Characteristics of water erosion and conservation practice in arid regions of Central Asia: Xinjiang, China as an example. *Int. Soil Water Conserv.* **2015**, *3*, 97–111. [CrossRef]

3. Wilkinson, S.N.; Kinsey-Henderson, A.E.; Hawdon, A.A.; Hairsine, P.B.; Bartley, R.; Baker, B. Grazing impacts on gully dynamics indicate approaches for gully erosion control in northeast Australia. *Earth Surf. Process. Landf.* **2018**, *43*, 1711–1725. [CrossRef]
4. Stavi, I.; Perevolotsky, A.; Avni, Y. Effects of gully formation and headcut retreat on primary production in an arid rangeland: Natural desertification in action. *J. Arid. Environ.* **2010**, *74*, 221–228. [CrossRef]
5. Melliger, J.; Niemann, J.D. Effects of gullies on space-time patterns of soil moisture in a semiarid grassland. *J. Hydrol.* **2010**, *389*, 289–300. [CrossRef]
6. Avni, Y. Gully incision as a key factor in desertification in an arid environment, the Negev highlands, Israel. *Catena* **2005**, *63*, 185–220. [CrossRef]
7. Wang, B.; Zhang, M.J.; Wei, J.L.; Wang, S.J.; Li, S.S.; Ma, Q.; Li, X.F.; Pan, S.K. Changes in extreme events of temperature and precipitation over Xinjiang, northwest China, during 1960–2009. *Quat. Int.* **2013**, *298*, 141–151. [CrossRef]
8. Zhang, Q.; Singh, V.P.; Li, J.F.; Jiang, F.Q.; Bai, Y.G. Spatio-temporal variations of precipitation extremes in Xinjiang, China. *J. Hydrol.* **2012**, *434–435*, 7–18. [CrossRef]
9. Oweis, T.Y. Rainwater harvesting for restoring degraded dry agro-pastoral ecosystems; a conceptual review of opportunities and constraints in a changing climate. *Environ. Rev.* **2017**, *25*, 135–149. [CrossRef]
10. Biazin, B.; Sterk, G.; Temesgen, M.; Abdulkedir, A.; Stroosnijder, L. Rainwater harvesting and management in rainfed agricultural systems in sub-Saharan Africa-A review. *Phys. Chem. Earth* **2012**, 139–151. [CrossRef]
11. Senkondo, E.M.M.; Msangi, A.S.K.; Xavery, P.; Lazaro, E.A.; Hatibu, N. Profitability of rainwater harvesting for agricultural production in selected semi-arid areas of Tanzania. *J. Appl. Irrig. Sci.* **2004**, *39*, 65–81.
12. Larry, W.M. Use of cisterns during antiquity in the Mediterranean region for water resources sustainability. *Water Sci. Technol. Water Supply* **2014**, *14*, 38–47.
13. Castelli, G.; Oliveira, L.A.A.; Abdelli, F.; Dhaou, H.; Bresci, E.; Ouessar, M. Effect of traditional check dams (jessour) on soil and olive trees water status in Tunisia. *Sci. Total Environ.* **2019**, *690*, 226–236. [CrossRef] [PubMed]
14. Paz-Kagan, T.; Ohana-Levi, N.; Shachak, M.; Zaady, E.; Karnieli, A. Ecosystem effects of integrating human-made runoff-harvesting systems into natural dryland watersheds. *J. Arid. Environ.* **2017**, *147*, 133–143. [CrossRef]
15. Paz-Kagan, T.; Demalach, N.; Shachak, M.; Zaady, E. Resource redistribution effects on annual plant communities in a runoff harvesting system in dryland. *J. Arid. Environ.* **2019**, *171*, 1–14. [CrossRef]
16. United Nations. Transforming Our World: The 2030 Agenda for Sustainable Development, A/RES/70/1. 2015. Available online: https://sustainabledevelopment.un.org/content/documents/21252030%20Agenda%20for%20Sustainable%20Development%20web.pdf (accessed on 8 June 2020).
17. Ammar, A.; Riksen, M.; Ouessar, M.; Ritsema, C. Identification of suitable sites for rainwater harvesting structures in arid and semi-arid regions: A review. *Int. Soil Water Conserv.* **2016**, *4*, 108–120. [CrossRef]
18. Stavi, I.; Siad, S.M.; Kyriazopoulos, A.P.; Halbac-Cotoara-Zamfir, R. Water runoff harvesting systems for restoration of degraded rangelands: A review of challenges and opportunities. *J. Environ. Manag.* **2020**, *255*, 1–13. [CrossRef]
19. Mbilinyi, B.P.; Tumbo, S.D.; Mahoo, H.F.; Mkiramwinyi, F.O. GIS-based decision support system for identifying potential sites for rainwater harvesting. *Phys. Chem. Earth* **2007**, *32*, 1074–1081. [CrossRef]
20. Mukai, S. Gully erosion rates and analysis of determining factors: A case study from the semi-arid main Ethiopian rift valley. *Land Degrad. Dev.* **2017**, *28*, 602–615. [CrossRef]
21. Mahmoud, S.H.; Adamowski, J.; Alazba, A.A.; El-Gindy, A.M. Rainwater harvesting for the management of agricultural droughts in arid and semi-arid regions. *Paddy Water Environ.* **2016**, *14*, 231–246. [CrossRef]
22. Sushanth, K.; Bhardwaj, A. Assessing geomorphologic characteristics and demarcating runoff potential zones using RS and GIS in Patiala-Ki-Rao watershed of Shivalik foot-hills, Punjab. *J. Agric. Eng.* **2019**, *56*, 45–54.
23. Ammar, A.; Khamis, N.S.; Rasha, A.; Mohamed, A.A.; Jan, G.W.; Michel, R.; Luuk, F.; Usama, K.; Coen, J.R. A GIS-based approach for identifying potential sites for harvesting rainwater in the western desert of Iraq. *Int. Soil Water Conserv.* **2018**, *6*, 297–304.
24. Shadeed, S.; Judeh, T.; Riksen, M. Rainwater harvesting for sustainable agriculture in high water-poor areas in the west bank, Palestine. *Water* **2020**, *12*, 380. [CrossRef]
25. Ibrahim, G.R.F.; Rasul, A.; Hamid, A.A.; Ali, Z.F.; Dewana, A.A. Suitable site selection for rainwater harvesting and storage case study using Dohuk Governorate. *Water* **2019**, *11*, 864. [CrossRef]

26. Luo, H.; Hu, X.N.; Xie, Y.S.; Wang, J.J.; Guo, M.C. Constructions of an evaluation framework for soil and water conservation techniques. *Catena* **2020**, *186*, 1–10. [CrossRef]
27. IUSS Working Group WRB. *World Reference Base for Soil Resources 2014. International Soil Classification System for Naming Soils and Creating Legends for Soil Maps*; World Soil Resources Reports; FAO: Rome, Italy, 2015.
28. Zhu, C.G.; Li, W.H.; Li, D.L.; Liu, M.Z.; Fu, L. Feature analysis of soil physicochemical properties and erodibility of the Yili Valley. *Resour. Sci.* **2016**, *38*, 1212–1221. (In Chinese with English abstract)
29. Li, Z.; Zhang, W.T.; Xuan, J.W. Characteristics of heavy rainfall in Yining City from 1956 to 2015 and its impact on flood. *Water Resour. Protect.* **2018**, *34*, 49–55. (In Chinese with English abstract)
30. Gong, P.; Liu, H.; Zhang, M.N.; Li, C.C.; Wang, J.; Huang, H.B.; Clinton, N.; Ji, L.Y.; Li, W.Y.; Bai, Y.Q.; et al. Stable classification with limited sample: Transferring a 30-m resolution sample set collected in 2015 to mapping 10-m resolution global land cover in 2017. *Sci. Bull.* **2019**, *64*, 370–373. [CrossRef]
31. Schumm, S.A. The evolution of drainage systems and slopes in badlands at Perth Amboy, New Jersey. *Bull. Geol. Soc. Am.* **1956**, *67*, 597–646. [CrossRef]
32. Horton, R.E. Drainage basin characteristics. *Trans. Am. Geophys. Union* **1932**, *13*, 350–361. [CrossRef]
33. Beven, K.J.; Kirkby, M.J. A physically-based variable contributing area model of basin hydrology. *Hydrol. Sci. J.* **1979**, *24*, 43–69. [CrossRef]
34. Justice, C.O.; Townshend, J.R.G.; Holben, B.N.; Tucker, C.J. Analysis of the phenology of global vegetation using meteorological satellite data. *Int. J. Remote Sens.* **1985**, *6*, 1271–1318. [CrossRef]
35. Zhang, C.X.; Yang, Q.K.; Li, R. Advancement in topographic wetness index and its application. *Process. Geogr.* **2005**, *24*, 116–123, (In Chinese with English abstract).
36. Kadam, A.K.; Kale, S.S.; Pande, N.N.; Pawar, N.J.; Sankhua, R.N. Identifying potential rainwater harvesting sites of a semi-arid, basaltic region of western India, using SCS-CN method. *Water Resour. Manag.* **2012**, *26*, 2537–2554. [CrossRef]
37. Yu, M.Z.; Zhang, L.L.; Xu, X.X.; Feger, K.H.; Wang, Y.H.; Liu, W.Z.; Schwärzel, K. Impact of land-use changes on soil hydraulic properties of Calcaric Regosols on the Loess Plateau, NW China. *J. Plant. Nutr. Soil Sci.* **2015**, *178*, 486–498. [CrossRef]
38. Wang, D.D.; Yu, X.X.; Jia, G.D.; Wang, H.N. Sensitivity analysis of runoff to climate variability and land-use changes in the Haihe Basin mountainous area of north China. *Agric. Ecosyst. Environ.* **2019**, *269*, 193–203. [CrossRef]
39. Cao, L.X.; Zhang, Y.G.; Lu, H.Z.; Yuan, J.Q.; Zhu, Y.Y.; Liang, Y. Grass hedge effects on controlling soil loss from concentrated flow: A case study in the red soil region of China. *Soil Tillage Res.* **2015**, *148*, 97–105. [CrossRef]
40. Xiong, M.Q.; Sun, R.H.; Chen, L.D. Effects of soil conservation techniques on water erosion control: A global analysis. *Sci. Total Environ.* **2018**, *645*, 753–760. [CrossRef]
41. Zhang, W.T.; Hu, G.Q.; Dang, Y.; Weindorf, D.C.; Sheng, J.D. Afforestation and the impacts on soil and water conservation at decadal and regional scales in northwest China. *J. Arid. Environ.* **2016**, *130*, 98–104. [CrossRef]
42. Saaty, T.L. Decision making with the analytic hierarchy process. *Int. J. Serv. Sci.* **2008**, *1*, 83–98. [CrossRef]
43. Saaty, T.L. Rank generation, preservation, and reversal in the analytic hierarchy decision process. *Decis. Sci.* **1987**, *18*, 157–177. [CrossRef]
44. Zhang, S.B.; Li, J.G.; Huang, J.H.; Zhang, W.T.; Yang, W.Y.; Wang, F. Impact of the grass coverage on runoff and sediment in the Yili river valley. *Acta Agri. U Jiangxi* **2016**, *38*, 995–1101.
45. Zhang, T.Y.; Liu, G.; Duan, X.W.; Wilson, G.V. Spatial distribution and morphologic characteristics of gullies in the Black Soil Region of Northeast China: Hebei watershed. *Phys. Geogr.* **2016**, *37*, 228–250. [CrossRef]
46. Zhao, J.J.; Vanmaercke, M.; Chen, L.Q.; Govers, G. Vegetation cover and topography rather than human disturbance control gully density and sediment production on the Chinese Loess Plateau. *Geomorphology* **2016**, *274*, 92–105. [CrossRef]
47. Desmet, P.J.J.; Poesen, J.; Govers, G.; Vandaele, K. Importance of slope gradient and contributing area for optimal prediction of the initiation and trajectory of ephemeral gullies. *Catena* **1999**, *37*, 377–392. [CrossRef]
48. Torri, D.; Poesen, J.; Borselli, L.; Knapen, A. Channel width-flow discharge relationships for rills and gullies. *Geomorphology* **2006**, *76*, 273–279. [CrossRef]
49. Torri, D.; Poesen, J. A review of topographic threshold conditions for gully head development in different environments. *Earth Sci. Rev.* **2014**, *130*, 73–85. [CrossRef]

50. Zhang, B.J.; Xiong, D.H.; Su, Z.G.; Yang, D.; Dong, Y.F.; Xiao, L.; Zhang, S.; Shi, L.T. Effects of initial step height on the headcut erosion of bank gullies: A case study using a 3D photo-reconstruction method in the dry-hot valley region of southwest China. *Phys. Geogr.* **2016**, *37*, 409–429. [CrossRef]
51. Djuma, H.; Bruggeman, A.; Camera, C.; Eliades, M.; Kostarelos, K. The impact of a check dam on groundwater recharge and sedimentation in an ephemeral stream. *Water* **2017**, *9*, 813. [CrossRef]
52. Calianno, M.; Fallot, J.M.; Ben Fraj, T.; Ben Ouezdou, H.; Reynard, E.; Milano, M.; Abbassi, M.; Ghram Messedi, A.; Adatte, T. Benefits of water-harvesting systems (Jessour) on soil water retention in Southeast Tunisia. *Water* **2020**, *12*, 295. [CrossRef]
53. Samani, A.A.N.; Khalighi, S.; Arabkhedri, M.; Farzadmehr, I. Indigenous knowledge and techniques of runoff harvesting (Bandsar and Khooshab) in arid and semi-arid regions of Iran. *J. Water Resour. Protect.* **2014**, *6*, 784–792. [CrossRef]
54. Yami, T.L. Enhancement of small-scale runoff harvesting ponds in East Africa. *J. Nat. Sci. Res.* **2019**, *9*, 1–19.
55. Zhu, Q.; Gould, J.; Li, Y.; Ma, C.X. *Rainwater Harvesting for Agriculture and Water Supply*; Science Press: Beijing, China, 2015; pp. 3–43.
56. Zhang, J.Z.; Yu, Z.Q.; Yu, T.F.; Si, J.H.; Feng, Q.; Cao, S.X. Transforming flash floods into resources in arid China. *Land Use Policy* **2018**, *76*, 746–753. [CrossRef]
57. Nasri, S.; Albergel, J.; Cudennec, C.; Berndtsson, R. Hydrological processes in macrocatchment water harvesting in the arid region of Tunisia: The traditional system of tabias. *Hydrol. Sci. J.* **2004**, *49*, 272. [CrossRef]
58. Kimiti, D.W.; Riginos, C.; Belnap, J. Low-cost grass restoration using erosion barriers in a degraded African rangeland. *Restor. Ecol.* **2017**, *25*, 376–384. [CrossRef]
59. Nyssen, J.; Veyret-Picot, M.; Poesen, J.; Moeyersons, J.; Haile, M.; Deckers, J.; Govers, G. The effectiveness of loose rock check dams for gully control in Tigray, Northern Ethiopia. *Soil Use Manag.* **2006**, *20*, 55–64. [CrossRef]
60. Sayl, K.N.; Muhammad, N.S.; Yaseen, Z.M.; El-shafie, A. Estimation the physical variables of rainwater harvesting system using integrated GIS-based remote sensing approach. *Water Resour. Manag.* **2016**, *30*, 3299–3313. [CrossRef]

Article

On the Origin of Deep Soil Water Infiltration in the Arid Sandy Region of China

Yiben Cheng [1,*], Wenbin Yang [2,3,*], Hongbin Zhan [4], Qunou Jiang [1], Mingchang Shi [1] and Yunqi Wang [1,*]

1 School of Soil and Water Conservation, Beijing Forestry University, Beijing 100083, China; jiangqo@bjfu.edu.cn (Q.J.); shimc@dtgis.com (M.S.)
2 Institute of Desertification Control, Chinese Academy of Forestry, Beijing 100093, China
3 Inner Mongolia Low Vegetation Coverage Sand Control Technology Development Co., Ltd., Hohho 010000, China
4 Department of Geology and Geophysics, Texas A&M University, College Station, TX 77843-3115, USA; zhan@geos.tamu.edu
* Correspondence: chengyiben@bjfu.edu.cn (Y.C.); nmlkyywb@163.com or yangwb@caf.ac.cn (W.Y.); wangyunqi@bjfu.edu.cn (Y.W.)

Received: 27 July 2020; Accepted: 24 August 2020; Published: 27 August 2020

Abstract: Soil water moisture is one of the most important influencing factors in the fragile ecosystems in arid sandy regions, and it serves as a bridge connecting the rainfall and groundwater, two important water sources in arid sandy regions. The hydrological process of an arid sandy region occurs sporadically and is highly non-uniform temporally, making it difficult to monitor and predict. The deep soil recharge (DSR) at a sufficiently deep soil layer (usually greater than 200 cm below ground surface) is an important indicator for groundwater recharge in the arid sandy region, and thus the quantitative determination of DSR is of great significance to the evaluation of water resources and the study of water balance in the arid sandy region. Due to the large amount of evaporation, small amount of precipitation, and the long term of the frozen-soil period in the winter and spring, the monitoring of infiltration and determination of DSR in the arid sandy region become challenging. This study selects the Ulanbuh desert plots in northern China to monitor DSR, precipitation and seasonal frozen soil thickness change, and reaches the following conclusions: Even though the annual precipitation is only 48.2 mm in the arid sandy region, DSR will still occur and replenish groundwater. The daily threshold of precipitation for generating measurable DSR is lower than 4 mm, where the DSR value is defined as the downward flux over a unit area per day hereinafter. DSR continues during the frozen period of the winter and spring seasons, and it is generated from water vapor transport and condensation in the deep sandy layer. Summer rainstorms do no show an obvious correlation with DSR, which is unexpected. This study reveals the characteristics of the dynamic water resources movement and transformation in the arid sandy area in Ulanbuh Desert and can serve as an important guideline for the quantitative assessment of water resources in arid sandy regions.

Keywords: arid sandy land; infiltration; precipitation; deep soil recharge; freeze–thaw

1. Introduction

The United Nations Environment Program defines regions with a drought index below 0.65 as arid areas, which account for 40% of the global land area. The arid zone is not a restricted zone of life [1]. The net primary productivity of the arid zone is close to 40% of the world total [2,3]. It provides human beings with food, energy and forest products, and has an important regulatory effect on global change and ecosystems' evolution [4,5]. Among the world's total arid region, 10–20% belongs to arid sandy regions, which are affected by desertification considerably [6,7]. Therefore, research on the ecohydrological process in arid sandy regions becomes increasingly important. On a global scale, the overall vegetation in the arid regions tends to degenerate [8,9]. In recent years, some studies have also shown that some arid regions have become revegetated [10,11]. In general, the spatial and temporal changes of the vegetation in the arid regions and their driving mechanisms have not been well recognized.

In arid sandy regions, due to the relatively dry shallow soil layers and the fragile ecological environment, the terrestrial surface water process significantly impacts the ecological environment, soil characteristics and human activities [12,13]. Therefore, study of the soil water dynamic and its relationship with the overlying surface process (such as precipitation and runoff) and the underlying groundwater flow is an important scientific issue that is related to regional development.

In arid sandy regions, due to the small amount of precipitation and large amount of evaporation, the residence time of terrestrial surface water is short, and the terrestrial surface water process has become the main limiting factor of ecosystem development in this area [14]. Any subtle changes of the terrestrial surface water process may cause drastic changes in the arid sandy region. A single precipitation process in an arid sandy region can cause significant fluctuations in surface energy transport and plant physiological characteristics [15,16]. The precipitation amount in the arid sandy region was relatively small, and the evaporable water resources were limited, and, for these reasons, water vapor plays a significant role in regulating the ecohydrological process in arid sandy regions [16,17]. The water vapor could be absorbed by the plant leaves to replenish the vegetation water. Dew accumulation in some areas exceeds precipitation amount [16]. Due to the relatively permeable nature of sandy soil, the transport mechanism of soil moisture was different from other soil types [18]. Groundwater in an arid sandy region can also be evaporated to replenish shallow soil water through evaporation [19]. Arid sandy regions were generally accompanied by a desert oasis, and the shallow moisture and vapor transport process behaved quite differently in the two landforms (bare sandy soil and oasis) [20]. Due to the lack of long-term observation of the replenishment of DSR in arid sandy regions, there were considerable disagreements on issues such as (1) what is the original source of water in an arid sandy region? And (2) how much water is there? In particular, what are the original sources of lakes and groundwater in arid sandy regions and what are their transport pathways? At the same time, due to the dry soil and scarce vegetation, the terrestrial surface water process in the dry sandy region has very limited capability to regulate the temperature change of the climate system, making the climate system in the region more sensitive to global warming. Compared with other humid areas, ecosystems in arid sandy regions are more vulnerable to global warming.

Since the beginning of the 20th century, a large number of scientists have begun to study the surface water process of the dry land and have conducted continuous monitoring works on the natural precipitation, evaporation, capillary suction, and deep soil recharge (DSR), and have obtained certain understandings about arid land water cycle [21–24]. For instance, it is found that condensation water in an arid sandy region has an important ecological significance, even exceeding the supply of precipitation [17,25,26]. Vegetation distribution is scarce in arid sandy regions, but the range of vegetation root activity will generate a soil moisture enrichment layer [27,28], which is extremely important to the arid land ecosystem [29]. However, as far as the current research progress is concerned, observational study and theoretical analysis of terrestrial surface ecosystems are obviously inadequate, and the understanding of the water cycle process in arid sandy regions is still very limited. There are still lots of doubts, including the arid sand replenishment effect and replenishment intensity of precipitation in deep soil water in the arid sandy land [30,31]. Soil water is the link between surface precipitation and groundwater, and plays an important role in the formation, transformation and consumption of the arid land water resources. Studying the changes and sources of soil moisture in an arid sandy region is of great research value for understanding the characteristics of arid sandy land, maintaining regional water balance, fixing mobile sandy lands, thereby controlling desertification, and curbing natural disasters such as sandstorms.

The objective of this study is to explore the redistribution process of precipitation moisture in shallow soil in an arid sandy region, because the fragile ecosystems in arid sandy regions are extremely sensitive to water deficits. There are a number of questions that we are trying to look for answers to in this study. First, does the deep soil layer receive any detectable DSR during the five-month freeze–thaw period of the shallow soil layer in the arid sandy region of this study? Second, what is the ratio of annual DSR to the annual precipitation? Third, how important is the precipitation in the rainy season and how much does it contribute to the overall annual DSR?

2. Seasonal Characteristics of Shallow Soil in an Arid Sandy Region

The soil moisture in an arid sandy region determines the occurrence and reversal of desertification and is the main environmental regulator of desertification [32]. The water stored in sandy soil layer is the link between atmospheric water, surface water and groundwater, and plays an important role in the formation, conversion and consumption of water resources cycle [33,34]. Precipitation infiltration is the main water source of replenishment of soil moisture in sandy regions [35]. Deep soil moisture can be stored in the soil without evaporation and can recharge groundwater [36]. Moreover, deep soil moisture can alleviate the demand for water resources of sand-fixing vegetation in several dry season, reduce water deficit, and maintain the minimum demand for vegetation in the arid sandy regions [37,38]. At present, there are few studies on the soil moisture dynamics and monitoring precipitation infiltration process in arid sandy regions; however, the deep layer infiltration process in arid sandy regions is often overlooked. In particular, there is very few studies on the long-term monitoring of deep soil moisture infiltration in arid sandy regions.

In the middle and high latitudes regions, winter cold weather conditions can form a certain depth of seasonal frozen soil. The seasonal frozen soil can affect the infiltration process of precipitation and snowmelt. When the frozen soil moisture content is relatively low, a great deal of pores have not been filled by the frozen soil moisture and thus can retain a relatively high permeability for infiltration to occur. When the frozen soil moisture content is relatively high, a great deal of pores have been occupied by the frozen soil moistures; thus, the frozen soil may become much less permeable or even completely impermeable to infiltration, resulting in surface runoff, which is usually not seen during the non-frozen period in arid sandy regions. Although a large number of experimental and modelling studies have shown that the seasonal frozen soil has a significant effect on infiltration, these studies largely rely on the use of an estimated infiltration rate, which can significantly affect the distribution of soil moisture during the frozen process [39]. Up to present, there are almost no studies concerning the energy balance and soil moisture redistribution at the lower boundary of the frozen soil. If a certain thickness of seasonal frozen soil has formed, snowfall becomes the main factor for moisture input at the upper interface (and evaporation is the main factor of soil moisture output at the upper interface), and DSR becomes the only exit for soil moisture output at the lower boundary.

Snowfall accumulates on the upper boundary of the frozen shallow soil. If the solar radiation is strong enough, the snowmelt will infiltrate into the frozen soil layer, and continue to fill the soil voids and increase the moisture content of the seasonal frozen soil. Because of the temperature difference between the deep soil and the frozen soil layer, energy exchange will occur at the lower boundary of the frozen soil layer, causing the frozen soil layer to melt. The melted water becomes DSR to recharge the underneath groundwater. To verify this hypothesis, this study measures the depth change of the lower boundary of the frozen soil layer and the change of DSR in an arid sandy region that experiences the seasonal freeze–thaw process.

3. Material and Methods

3.1. Overview of the Experimental Plot

The experiments were carried out in Ulanbuh Desert, Inner Mongolia, China (40°26′41.2″ N 106°44′25.9″ E), as shown in Figure 1. The average altitude of Ulanbuh desert is 1046 m above mean sea level (a.m.s.l.). The landform is a chain-like sandy dune. The height of the dune is generally 1–3 m, and the terrain is relatively flat. The climate type is a temperate desert climate with both continental and seasonal precipitation characteristics. The annual precipitation is concentrated from June to September, and there is significant snowfall in the winter. Multi-year observation records show that the maximum temperature is 39 °C and the minimum temperature is minus 29.6 °C. The annual average temperature is 7.6 °C, and the highest temperature usually occurs in July. The average temperature in July is 23.8 °C. The average temperature in January is minus 10.8 °C. The annual sunshine hours are 3000 h. The annual average wind speed is 3.7 m/s and the maximum wind speed is 15 m/s. The southwest wind is dominant in the region. The annual average precipitation is 102.9 mm, and there is snow accumulation in the winter and the annual average potential evaporation is 2551.9 mm. The depth of groundwater is 6–8 m [40]. The heterogeneity of sand in the soil layer below 20 cm of the test site is small and porosity of the soil layer within 200 cm is quite high, in the range of 44.09–45.63%, the saturated hydraulic conductivity of the in-site sandy layer is 12.85 cm/h [41].

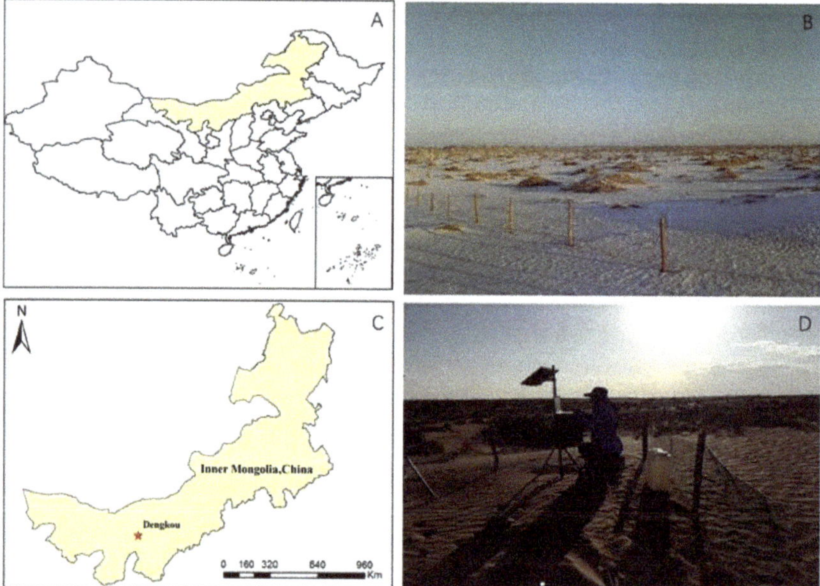

Figure 1. Geographic location of the experimental area. (**A**) refers to the location of the study area in China, (**B**) refers to the area covered by a large amount of snow in winter, (**C**) refers to the location of the study area in the arid and semi-arid area of Inner Mongolia, and (**D**) refers to the landscape of the experimental site.

3.2. *Experiment Design*

3.2.1. Observation of Annual Evapotranspiration

Evaporation is an integral part of the surface water balance, and it is also one of the most directly affected by land use and climate change in the hydrological cycle. At the same time, evaporation is also a critical component in energy balance. Therefore, studying the change of evaporation is indispensable to understand the changes of climate and hydrological cycle characteristics in the region. In this study, measuring the daily evaporation can be used to compare the amount of precipitation and evaporation during this period. The average annual temperature of the experimental site reached minus zero for up to 150 days, which can profoundly affect the evaporation during this period. An American Class A evaporating pan is used to measure the amount of potential evaporation. The daily measurement range is 0–250 mm, and the measurement accuracy is 0.2 mm. The actual evaporation is often smaller than the measured potential evaporation with an empirical reduction coefficient. The results of the Class A evaporating pan experiments by researchers in China and Australia show that the empirical reduction coefficient is 0.68–0.76 for the arid sandy region of this study [42].

3.2.2. Observation on Annual Precipitation and DSR

Precipitation is an important source of water resources in the study site, a precipitation sensor (rain gauge, AV-3665R, AVALON, United States; precision: 0.2 mm) was placed above ground at the site to monitor annual precipitation. The experimental plot is an arid sandy region with large evaporation. The soil type in the experimental area is sandy soil, the infiltration rate is fast, and the shallow dry sand layer can suppress the evaporation. The study shows that, when the thickness of the shallow dry sand layer reaches more than 5 cm, the shallow dry sand layer becomes the most important factor controlling the evaporation [43,44]. As the thickness of the shallow dry sand layer increases, the amount of soil evaporation gradually decreases. For example, a shallow dry sand layer with 5 cm thickness can suppress the evaporation up to 70.6%, and a shallow dry sand layer with 30 cm thickness can suppress the evaporation up to 92.38%. There is clearly identifiable linear relationship between the thickness of the shallow dry sand layer and the amount of evaporation. This indicates that once precipitation moisture enters the deep soil layer, water resources can be stored in the deep soil layer and become part of ecological water resource, because of the suppression of evaporation at the shallow dry soil layer.

Because of the relatively low precipitation and the sporadic and highly variable precipitation strength in the region, the traditional method of using a correlation coefficient to estimate the annual DSR based on the annual precipitation can cause huge errors [11]. In this study, a newly designed lysimeter is used to directly measure DSR at a depth of 200 cm (or Depth A in Figure 2 is 200 cm). The choice of 200 cm here is to ensure that any downward infiltration that can pass Depth A will continue to move downward to recharge the deep groundwater [12]. The newly designed lysimeter has a small size, easy to install, and can measure DSR automatically with high accuracy [36,37]. As shown in Figure 2, the conventional lysimeter uses an impermeable container (constructed all the way from ground surface downward) to wrap the soil column, blocking the horizontal flow of the soil layer; thus, there is inevitable water potential difference existing inside and outside the container of the Lysimeter. The advantage of this newly designed instrument is that it can be directly installed at a depth of 2 m, and there is no need to wrap a soil column like a conventional Lysimeter to block the horizontal flow of soil. Because the layered structure of the native soil is destroyed when filling the soil into the lysimeter, and fissures are easily generated between the filling soil and the wall of lysimeter, precipitated water can easily flow downward along those fissures, resulting in an overestimation of soil infiltration and DSR. The newly designed lysimeter has a water balance part (from Depth A to Depth B in Figure 2), which uses a cylindrical impermeable side wall to wrap the original soil column and a measurement part (below Depth B).

There are several formulas for calculating the water holding height of capillaries (or capillary rise) in sandy soil. This study used the following formula to calculate the water holding height of capillaries based on the soil particle size [45]:

$$H_C = \frac{\sigma n}{\sqrt{2\mu\rho g K}} \cos\alpha + (1-n)h_a \tag{1}$$

where H_c is capillary water rise, σ is the surface tension of water, α is the advancing contact angle, ρ is the density of water, g is the gravity constant, K is the saturated hydraulic conductivity, h_a is air entry head, and n is the porosity of soil. The capillary rises computed for the experimental plot vary from 28.4 to 44.6 cm. Based on this, the length of the balance part (the distance from Depth A to Depth B) used in this study is set to be 50 cm, which is slightly larger than the maximum capillary rise at the study site. The advantage of this new design of lysimeter has been explained in details in Cheng et al. (2017, 2018). The measurement part uses a gauge with an accuracy of 0.2 mm to measure DSR. The groundwater water in the study site is sufficiently deep and will not affect the measurement of DSR. After installing the lysimeter and backfilling the excavated site all the way to ground surface using the in situ soil, one usually has to wait for one year to ensure that soil settlement has approximately reached its pre-installation status.

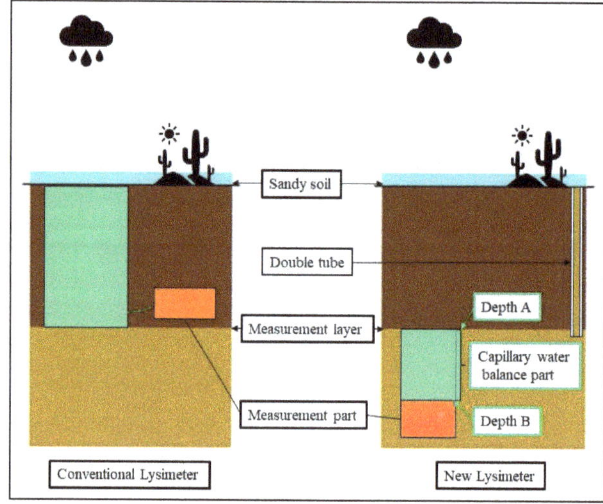

Figure 2. Design of the new lysimeter and experimental plot design.

3.2.3. Observation on Frozen Soil Thickness Change

The surface soil is usually frozen during winter and spring seasons at the study site, making it difficult to monitor the soil moisture change and DSR. To overcome this difficulty, we have designed a double-tube apparatus to measure the depth of the freeze–thaw layer. The double-tube apparatus consists of a hollow barrel (diameter: 20 mm) with a vent hole and a rubber tube (diameter: 18 mm) inside the barrel. First, the hollow barrel with the vent hole is buried vertically in the experimental plot. The upper level of the barrel is at the ground surface and the length of the barrel is 200 cm. The rubber tube filled with water is placed in the cylinder, and it will be extracted at 8 P.M. every day to record the frozen depth, which can be used to interpret the thickness of the frozen soil at that moment at the study site. The freeze–thaw period in the study is relatively long, lasting about 150 days per year, and there is significant snowfall in the winter, and the snowmelt could be an important water resource replenishing the shallow soil moisture. The frozen soil serves as an impermeable layer, which essentially isolates the hydraulic connection above and below this layer. In another word, the DSR measurements recorded during the winter and spring seasons (when the frozen soil layer exists) should reflect the information of soil moisture variation caused by the freezing–thawing process below the frozen soil layer.

4. Results and Analysis

4.1. The Annual Tendency of Change on Evaporation

The main factors influencing the amount of evaporation include sunshine hours, air temperature, and air saturation difference. The annual evaporation in 2013 is shown in Figure 3. The annual cumulative potential evaporation in 2013 reached 2610.3 mm, where the annual cumulative potential evaporation is defined as the upward flux due to potential evaporation per unit area per year. The maximum daily evaporation amount was 26 mm, and the minimum daily evaporation value was 0 mm. The annual evaporation in the experimental plot has shown an upward trend since March.

It began to decline after reaching the maximum value in July and reached the minimum value in December. In summer, the amount of evaporation is large, and the precipitation is quickly evaporated, resulting in a decrease in the amount of infiltration. The accumulation of snowmelt on the surface may increase the amount of infiltration during the freeze–thaw period.

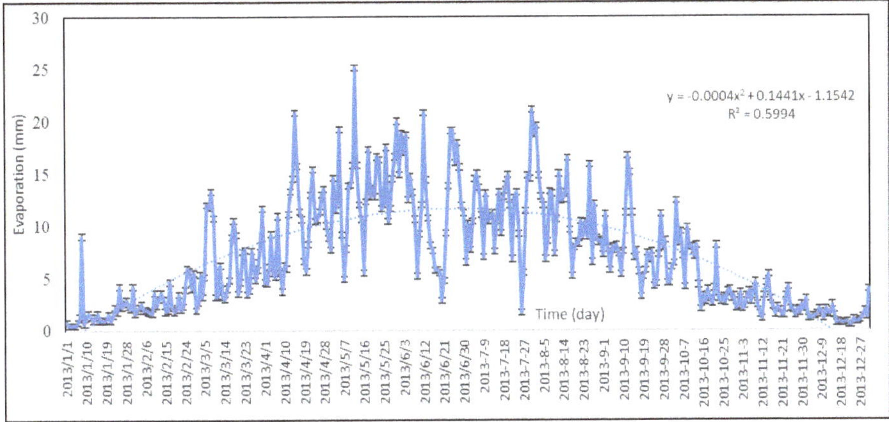

Figure 3. Daily evaporation in the year 2013.

4.2. The Inter-Annual Trends of Precipitation and DSR

The statistical results of annual precipitation and DSR are shown in Table 1. The annual precipitations from 2013 to 2019 are 53, 79, 48.2, 131.2, 89.7, and 66.6 mm, respectively; the DSR values from 2013 to 2019 are 2.6, 3.4, 4.6, 5.8, 3.4, and 5.6 mm, respectively. The average annual precipitation of the experiment site is 102.9 mm, and it can be seen that the annual precipitation fluctuates considerably from year to year. From 2013 to 2016, the ratios of DSR to the annual precipitation were 4.9%, 4.3%, 9.54%, and 4.42%, respectively. In 2015, the annual precipitation was only 48.2 mm, which was the minimum value over the period of 2013–2019, but the amount of DSR reached 4.6 mm, which accounted for 9.54% of the annual precipitation, the largest ratio over the six year period (2013–2019). The precipitation in 2016 was 131.2 mm, which was the largest annual precipitation over the six year period (2013–2019), but the amount of DSR was only 5.8 mm, which accounted for 4.42% of the precipitation. It can be seen from Table 1 that precipitation and DSR are not linearly related at all. More specifically, DSR does not appear to be related to the annual precipitation at all. From this observation, we can infer that it is inaccurate to use a coefficient to determine the amount of annual DSR based on the annual precipitation the arid sandy region of this study. This conclusion is consistent with other recent studies conducted by the authors in arid sandy regions (Cheng et al., 2017; Cheng et al., 2018).

Table 1. Changes in precipitation and deep soil recharge (DSR) from 2013 to 2019.

Time	Precipitation (mm)	DSR (mm)	Infiltration/Precipitation (%)
2013	53	2.6	4.9
2014	79	3.4	4.3
2015	48.2	4.6	9.54
2016	131.2	5.8	4.42
2017	89.7	3.4	3.8
2019	66.6	5.6	8.41

To examine the connection (or disconnection) of the annual precipitation and the amount of DSR in more details, we conducted a year-by-year analysis of precipitation and DSR from 2013 to 2016. After installing the lysimeter in May 2012, and backfilling the excavated site all the way to ground surface using the in situ soil, one usually has to wait for one year to ensure that soil settlement has approximately reached its pre-installation status. As shown in Figure 4, the precipitation in 2013 was concentrated between 11 May and 28 June, with a maximum precipitation event of 10.4 mm/d, a single maximum DSR of 0.4 mm, and three DSR events in May. DSR events are relatively concentrated and there were detected DSR events between September and November in 2013. As the precipitation and DSR data are only available from May to December in 2013, they may not be representative of this study site for other time periods. Nevertheless, one can see that even when the annual precipitation is as low as 53 mm in 2013 in the arid sandy region of the study site, DSR still occurs.

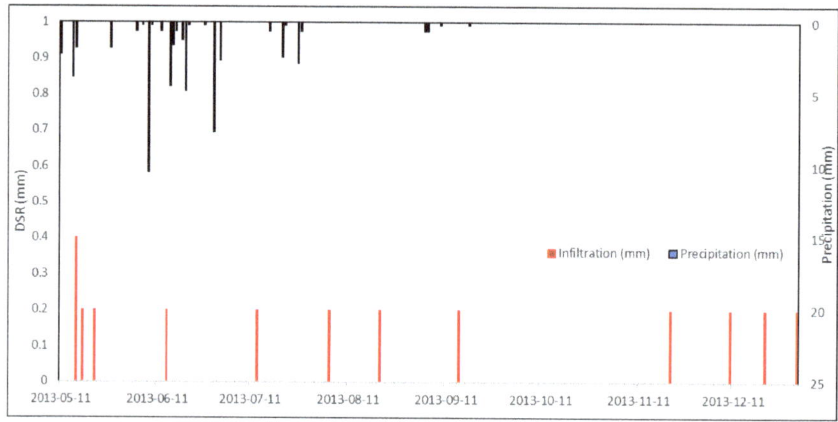

Figure 4. Relationship between precipitation and DSR in 2013.

The distributions of precipitation and DSR in 2014 are shown in Figure 5. The annual precipitation in 2014 was 79 mm, the annual DSR was 3.4 mm, and DSR accounted for 4.3% of the annual precipitation. In 2014, the precipitation was concentrated from 16 April and 30 August. The maximum daily precipitation was 14.2 mm on 30 August, and the minimum daily precipitation was 0.2 mm, which occurred many times throughout the year. In 2014, 17 DSR events were monitored with the same DSR value of 0.2 mm. As shown in Figure 5, DSR events were concentrated from 8 January to 24 March 2014, with a total of 9 DSR events, accounting for more than half of the total DSR events. From April to June and September to November, there were no DSR events at all. This implies that melting in the shallow frozen soil will not replenish the deep soil moisture. One can conclude that DSR generated during the spring and winter seasons (when a frozen soil layer exists) is greater than that of during the summer and fall seasons (when a frozen soil layer does not exist). The snow and ice accumulated in the winter and spring has no obvious replenishment effect on DSR during the melting seasons. From this observation, we can infer that there are other forms of water redistribution and movement below the frozen soil layer, such as deep soil layer water vapor condensation and infiltration.

The distributions of the annual precipitation and DSR in 2015 are shown in Figure 6. The total annual precipitation in 2015 was 48.2 mm, the annual DSR was 4.6 mm, and annual DSR accounted for 9.54% of the annual precipitation in this year. The precipitation in 2015 was distributed between 6 March and 16 November. The maximum daily precipitation of 6.2 mm occurred on 6 April, and the minimum daily precipitation was 0.2 mm, which occurred many times throughout the year. In 2015, we have recorded 23 DSR events with the same DSR of 0.2 mm. Relative to the average annual precipitation of 102.9 mm over 2013–2019, the annual precipitation of 48.2 mm in 2015 signifies a very dry year. Despite this, DSR can still occur in this year when the maximum daily precipitation event of 6.2 mm occurs. This implies that the threshold of daily infiltration for generating measurable DSR is probably around 6.2 mm. Comparing the precipitation-DSR relation from 2014 to 2015, it can be found that the precipitation in 2015 is evenly distributed. The experiment shows that the evaporation during the freezing period is quite limited, indicating that snowfall may accumulate without evaporation on the surface and become a potentially important source to recharge the deep soil during this time the frozen soil layer disappears. From April to June and from October to December of 2015, there were no DSR events, indicating that the accumulated snowfall on the surface layer actually did not replenish the deep soil water during the melting season. Based on this observation, one can conclude that the threshold of daily precipitation for generating detectable DSR is 6.2 mm, and the snowfalls in the winter and spring have no replenishment effect on the deep soil moisture.

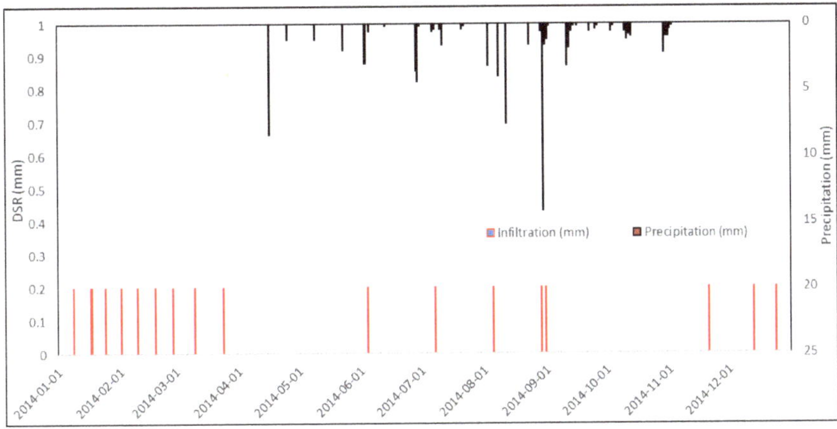

Figure 5. Relationship between precipitation and DSR in 2014.

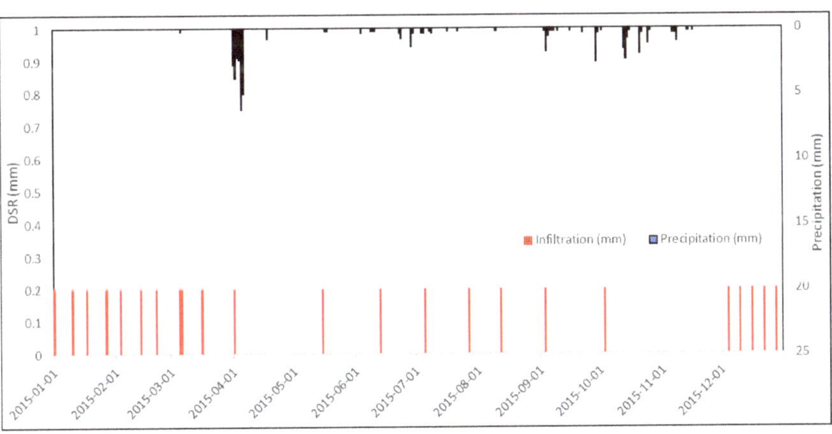

Figure 6. Relationship between precipitation and DSR in 2015.

The distribution of precipitation-DSR in 2016 is shown in Figure 7. The annual precipitation in 2016 was 131.2 mm, the annual DSR was 5.8 mm, and the annual DSR accounted for 4.42% of the annual precipitation of this year. The precipitation was distributed from 21 May to 31 October 2016, with a maximum daily precipitation of 46.4 mm occurring on 13 June. Unfortunately, the heavy precipitation on 17 August did not infiltrate into the deep soil layer, and a large amount of surface runoff was observed at the experimental plot on that day. The minimum daily precipitation is 0.2 mm, which occurred multiple times throughout 2016. We have detected 23 DSR events with the same daily DSR value of 0.2 mm in 2016. From April to June, October to December of 2016, there were no DSR events.

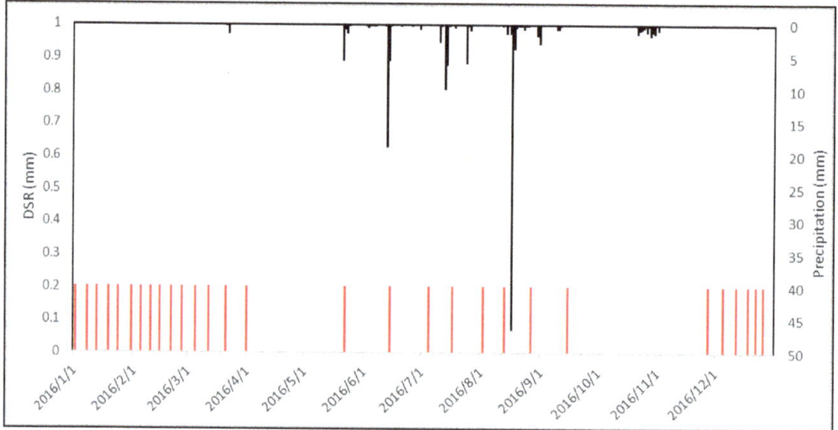

Figure 7. Relationship between precipitation and DSR in 2016.

It is interesting to see that heavy precipitation over a short period of time in arid sandy soil does not always lead to considerable DSR, as expected. This is because of the following reasons. The sandy soil in the arid region of this study has a relatively large value of porosity, and a great portion of the pore spaces are occupied by air, before the start of a heavy precipitation event. When a heavy precipitation event occurs, the shallow soil layer was filled with water so quick that the air underneath the shallow saturated soil layer was trapped there [46]. When the trapped air was suppressed, its pressure increased, and such trapped air then becomes an impermeable layer for further downward water infiltration [47,48]. When the downward infiltration becomes difficult or completely ceases, the precipitated water will start to pond on the ground surface, leading to surface runoff or evaporated [49].

The relationship between the annual precipitation-DSR in 2017 is shown in Figure 8. The annual precipitation in 2017 was 89.7 mm, the annual DSR was 3.5 mm, and the annual DSR accounted for 3.8% of the annual precipitation in this year. The precipitation in 2017 was distributed differently from the previous three years. In 2017, the distribution of precipitation in each month is more uniform than the previous three years. The maximum daily precipitation of 19.1 mm occurred on June 5. Surface runoff associated with such a heavy precipitation event of 19.1 mm was observed, but the amount of DSR did not increase significantly as a result of this precipitation event. This observation once again showed that heavy precipitation in an arid sandy region did not always have a significant replenishment effect on the DSR. The reason for this is associated with the trapped air in the sandy soil, as explained above.

The surface data logger instrument was destroyed in 2018, probably due to strong wind or for other unclear reasons, so there was no data recorded in 2018. The relationship between precipitation and DSR in 2019 is shown in Figure 9. The total precipitation in 2019 was 66.6 mm, the annual DSR was 5.6 mm, and the annual DSR accounted for 8.41% of the annual precipitation in this year. The daily precipitation intensity is generally small in 2019. There were 109 precipitation events throughout the year of 2019, of which the daily maximum precipitation was 4 mm, occurred on August 9. We have recorded 28 DSR events with the same DSR of 0.2 mm. From April to June and from September to December of 2019, there were no DSR events. Observation results again verify that the freeze–thaw process did not make a noticeable contribution to the deep soil moisture. In general, the annual precipitation in this year was not directly related to the amount of DSR.

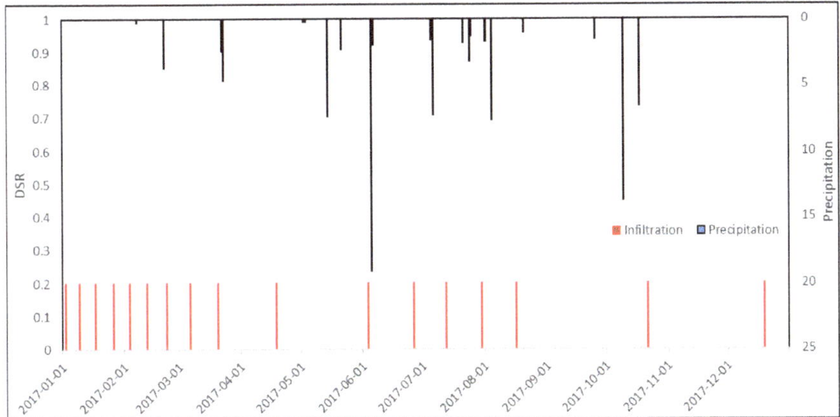

Figure 8. Relationship between precipitation and DSR in 2017.

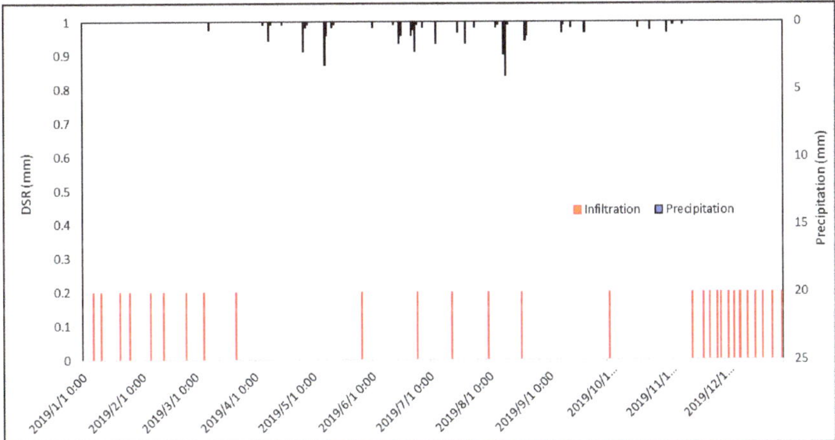

Figure 9. Relationship between precipitation and DSR in 2019.

Figure 10 is a summary of monthly precipitation and DSR from 2013 to 2019. A few general observations can be made. First, the precipitation events were concentrated in the non-frozen period, from April to October, while the DSR events were concentrated in the frozen period, from December to March. Second, precipitation did not show a direct correlation with the DSR in this region. Third, the amount of DSR generated in the freeze–thaw period was greater than that in the rainy season. This result implies that the vapor moisture condensation below the frozen soil layer in the freeze–thaw period season was the main source of DSR, and the replenishment of winter infiltration on DSR exceeds the replenishment of summer precipitation on DSR.

According to the above observations, the exchange of soil moisture between the frozen interface and the deep soil during the frozen period in winter is the main cause of the DSR in winter. Figure 11A,B shows the change in frozen soil layer thickness in the winter and spring of 2015–2016. The intermittent winter frozen period in 2015 was divided into two parts, from 1 January to 14 March and from 12 November to 31 December. One can find that from 12 November to 31 December, the thickness of the frozen soil gradually increased, and DSR was recorded. After the surface soil freezes, it forms an impervious layer that continues to thicken. The water vapor coming from the deep soil continues to convert into condensed water and infiltrated downward. From 1 January to 14 March is the freeze–thaw period. As the time approaches 14 March, the soil ice is melting, but there was no obvious DSR in the deep soil during the late freeze–thaw period. We speculated that the radiation from the surface makes the frozen layer melt, and the melted water evaporated upward into the atmosphere.

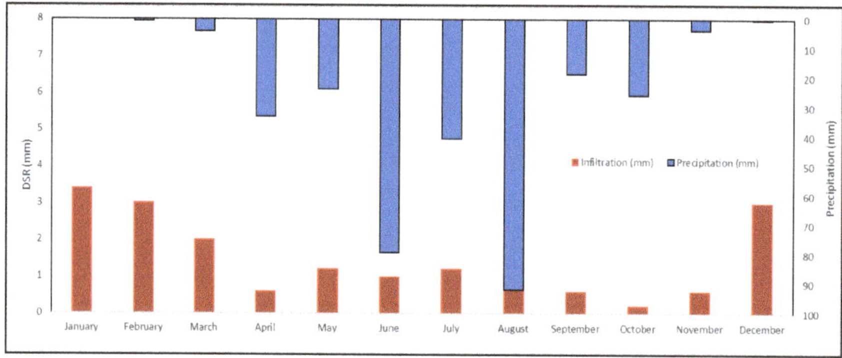

Figure 10. Every month add together.

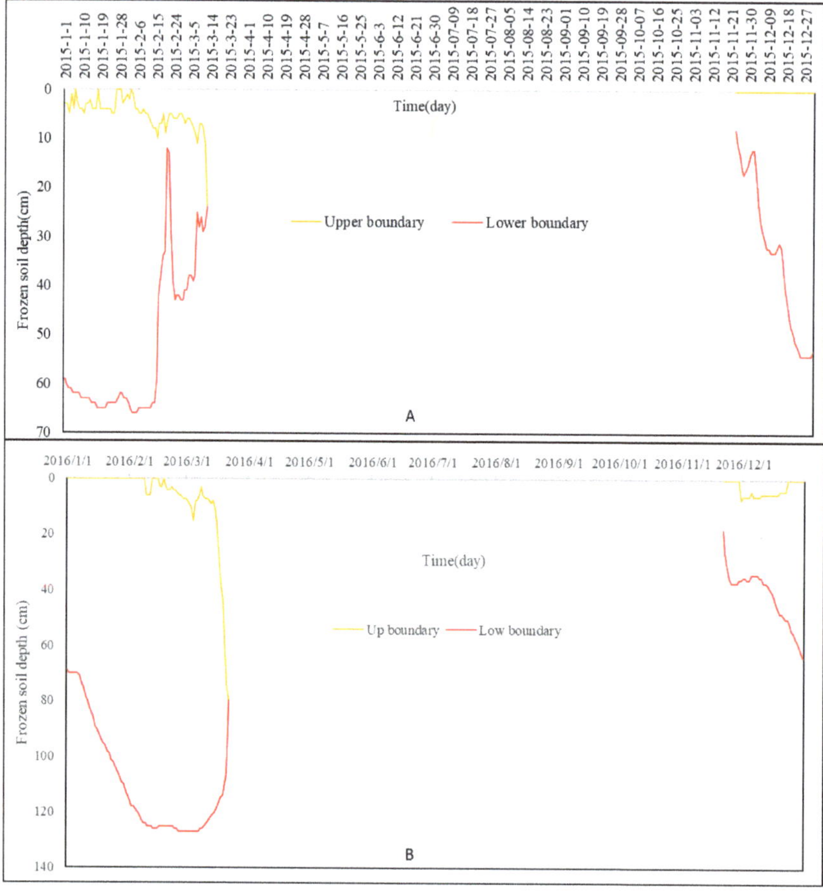

Figure 11. Frozen soil boundary of the field in 2015. (**A**) frozen soil boundary changes in 2015; (**B**) frozen soil boundary changes in 2016.

5. Discussion

There were many studies on the deep soil moisture in arid sandy regions, and the characteristics of precipitation infiltration and water redistribution under different vegetation cover conditions [50,51], but there were few studies on the effect of DSR replenishment. This study used a newly designed Lysimeter to measure the DSR at a depth of two meters in the sandy land and track the changes in the thickness of the frozen soil during the freeze–thaw period. Through the observation of DSR in specific soil layers, the results show that precipitation is not the main source of DSR in arid sandy land, and condensation water is the main source of replenishment for DSR. In dry years (the years when the annual precipitation is lower than the multi-year average precipitation is a dry year), taking 2015 as an example, the DSR in the summer rainfall season is 1.4 mm, and the DSR in the freeze–thaw period (condensate water is the main source of DSR replenishment) is 3.2 mm. In wet years (the years when the annual precipitation is higher than the multi-year average precipitation), taking 2016 as an example, the DSR in the summer rain season is 1.6 mm, and the DSR in the freeze–thaw period is 4.2 mm.

Under the global climate change conditions in perceivable future, extreme weather will occur more frequently. Due to the limited amount of DSR caused by precipitation, extreme high temperatures will further increase evaporation in arid sandy areas, the sandy soil in the arid region may become more dried and groundwater level may decline continuously. Consequently, the ecological degradation becomes inevitable when the groundwater level declines continuously, and it is indeed important to systematically study the recharge process of DSR under dynamically changing precipitation conditions.

Various investigators have conducted controlled precipitation experiments in arid sandy regions to study the infiltration process of precipitation, the sources of vegetation water used, and have divided precipitation to effective and ineffective precipitation for the arid sandy regions [51,52], where the ineffective precipitation refers to precipitation that cannot generate sufficiently strong infiltration that can penetrate a soil layer deeper than 40 cm. The results of this study show that, under natural conditions, instantaneous rainstorms cannot infiltrate into the deep soil layer. Most of the instantaneous rainstorms become shallow infiltration and runoff, which eventually returns to the atmosphere via evapotranspiration. This is because the instantaneous rainstorms fully saturate the shallow sandy soils quickly and form an airtight saturated layer that essentially block the escape routes of air underneath the shallow sandy soils. In return, the blocked air in the deep soil layer becomes an effective barrier for further infiltration to deep soil layers. After the rainstorms, the intense evapotranspiration in the arid sandy regions depletes the surface soil moisture. This was obviously inconsistent with the experimental results of simulated precipitation experiment infiltration. Under the condition of vegetation coverage in the sandy area of the arid region, almost no surface runoff will be formed [53,54], studies have shown that precipitation intensity at 46.4 mm/d will not form surface runoff in dry sand areas [55], and almost no precipitation intensity reached 46.4 mm/d in this study area. The surface runoff in the sandy area is affected by the covering, and the crust will increase the possibility of surface runoff formation [56,57].

During the freezing and thawing period, the DSR accounts for a great portion of the annual infiltration. During this period, the surface frozen soil layer prevents the infiltration of the surface snowmelt, suggesting that the DSR in this period is originated from the condensation of water vapor in the deep sand layer, not the infiltration of snowmelt. The results of this study show that in arid sandy areas, the condensate formed by the condensation of water vapor in the soil voids has an important effect on the replenishment of deep soil moisture. The replenishment effect of condensed water on deep soil exceeds the replenishment effect of rainfall moisture on deep soil moisture.

There is a notable limitation of this study that should be improved in the future. When measuring DSR, a gauge with a measurement accuracy of 0.2 mm was used to automatically record the amount of DSR. The measuring mechanism of this gauge was that when the accumulated amount of DSR reached a certain amount (0.2 mm), which is the downward volumetric flux over a unit area over a certain time lapse, then a data point will be recorded. With this measuring mechanism, it is impossible to know precisely the DSR variation over that time lapse. In the future, we need a more sensitive measuring apparatus that can precisely record the DSR variation with time in a higher precision.

6. Conclusions

This study carried out a 6 year (2013–2019) deep soil layer infiltration observation in an arid sandy soil region, which provided a reliable basis for accurately assessing the effect of precipitation on the replenishment of DSR. The replenishment process of precipitation to DSR in the arid sandy region was very complicated. Environmental factors and the heterogeneous soil make this research particularly difficult. The following conclusions can be obtained from this research.

1. Even with an annual precipitation of 48.2 mm in a very dry year in the arid sandy region, DSR is still recorded at a depth of 200 cm, signifying that groundwater recharge is possible during such a dry year. The threshold of daily precipitation that can induce DSR is around 4 mm in the arid sandy region, and DSR can be recorded under frozen surface layer conditions in winter and spring seasons.
2. There is no direct correlation between the amount of DSR and the annual precipitation in the arid sandy region. Most recorded DSR events are concentrated in the freeze–thaw period in winter and spring seasons. The instantaneous rainstorm in the arid sandy region has no obvious replenishment effect on DSR.
3. The ice and snow accumulated in the frozen season in winter and spring will not replenish the deep soil moisture in the freeze–thaw season. Instead, the condensed water in the deep soil is the main source of DSR in the freeze–thaw season. The amount of DSR replenishment during the rainy season in the arid sandy region is less than that during the winter and spring freeze–thaw season.

Author Contributions: Data curation, Y.C., Y.W., W.Y.; Formal analysis, Y.C., M.S.; Funding acquisition, W.Y. and Q.J.; Investigation, H.Z.; Writing—review & editing, Y.C. All authors have read and agreed to the published version of the manuscript.

Funding: This research received funding from the Demonstration of Near-Natural Restoration Technology for Sandy Ecosystem (Inner Mongolia~2019ZD003), the National Nonprofit Institute Research Grant of Chinese Academy of Forestry (grant number CAFYBB2018ZA004), and the National Science and Technology Major Project and Ministry of Science and Technology of China (No. 2017ZX07101004, No. 2018YFC0507100, NO. 2015ZCQ-SB-01, NO. 2019ZD003).

Acknowledgments: This research was supported by research grants from the Fundamental Research Funds for the Central Universities (BLX201814). We gratefully acknowledge the Beijing Municipal Education Commission for their financial support through Innovative Transdisciplinary Program "Ecological Restoration Engineering". Thanks to the experimental site provided by Inner Mongolia Dengkou Desert Ecosystem National Observation Research Station, the Experimental Center of Desert Forestry, Chinese Academy Forestry.

Conflicts of Interest: The authors declare no conflict of interest.

References

1. Zarch, M.A.A.; Malekinezhad, H.; Mobin, M.H.; Dastorani, M.T.; Kousari, M.R. Drought monitoring by reconnaissance drought index (RDI) in Iran. *Water Res. Manag.* **2011**, *25*, 3485. [CrossRef]
2. Ren, W.; Tian, H.; Chen, G.; Liu, M.; Zhang, C.; Chappelka, A.H.; Pan, S. Influence of ozone pollution and climate variability on net primary productivity and carbon storage in China's grassland ecosystems from 1961 to 2000. *Environ. Pollut.* **2007**, *149*, 327–335. [CrossRef] [PubMed]
3. Cramer, W.; Kicklighter, D.W.; Bondeau, A.; Iii, B.M.; Churkina, G.; Nemry, B.; Ruimy, A.; Schloss, A.L. Intercomparison, Comparing global models of terrestrial net primary productivity (NPP): Overview and key results. *Glob. Change Biol.* **1999**, *5* (Suppl. S1), 1–15. [CrossRef]
4. LeBauer, D.S.; Treseder, K.K. Nitrogen limitation of net primary productivity in terrestrial ecosystems is globally distributed. *Ecology* **2008**, *89*, 371–379. [CrossRef]
5. Haberl, H.; Erb, K.-H.; Krausmann, F. Human appropriation of net primary production: Patterns, trends, and planetary boundaries. *Ann. Rev. Environ. Res.* **2014**, *39*, 363–391. [CrossRef]
6. Seifan, M. Long-term effects of anthropogenic activities on semi-arid sand dunes. *J. Arid Environ.* **2009**, *73*, 332–337. [CrossRef]

7. Ma, Z.; Xie, Y.; Jiao, J.; Wang, X. The construction and application of an Aledo-NDVI based desertification monitoring model. *Proc. Environ. Sci.* **2011**, *10*, 2029–2035. [CrossRef]
8. Modarres, R.V.; da Silva, P.R. Rainfall trends in arid and semi-arid regions of Iran. *J. Arid Environ.* **2007**, *70*, 344–355. [CrossRef]
9. Cao, S.; Chen, L.; Shankman, D.; Wang, C.; Wang, X.; Zhang, H. Excessive reliance on afforestation in China's arid and semi-arid regions: Lessons in ecological restoration. *Earth-Sci. Rev.* **2011**, *104*, 240–245. [CrossRef]
10. Chen, C.; Park, T.; Wang, X.; Piao, S.; Xu, B.; Chaturvedi, R.K.; Fuchs, R.; Brovkin, V.; Ciais, P.; Fensholt, R. China and India lead in greening of the world through land-use management. *Nat. Sustain.* **2019**, *2*, 122–129. [CrossRef]
11. Cao, S.; Chen, L.; Yu, X. Impact of China's grain for green project on the landscape of vulnerable arid and semi-arid agricultural regions: A case study in northern Shaanxi Province. *J. Appl. Ecol.* **2009**, *46*, 536–543. [CrossRef]
12. Jiang, Z.; Lian, Y.; Qin, X. Rocky desertification in Southwest China: Impacts, causes, and restoration. *Earth-Sci. Rev.* **2014**, *132*, 1–12. [CrossRef]
13. Wang, Y.; Shao, M.A.; Zhu, Y.; Liu, Z. Impacts of land use and plant characteristics on dried soil layers in different climatic regions on the Loess Plateau of China. *Agric. For. Meteorol.* **2011**, *151*, 437–448. [CrossRef]
14. Kleidon, A.; Fraedrich, K.; Heimann, M. A green planet versus a desert world: Estimating the maximum effect of vegetation on the land surface climate. *Clim. Chang.* **2000**, *44*, 471–493. [CrossRef]
15. Huxman, T.E.; Snyder, K.A.; Tissue, D.; Leffler, A.J.; Ogle, K.; Pockman, W.T.; Sandquist, D.R.; Potts, D.L.; Schwinning, S. Precipitation pulses and carbon fluxes in semiarid and arid ecosystems. *Oecologia* **2004**, *141*, 254–268. [CrossRef] [PubMed]
16. Agam, N.; Berliner, P.R. Dew formation and water vapor adsorption in semi-arid environments—a review. *J. Arid Environ.* **2006**, *65*, 572–590. [CrossRef]
17. Pan, Y.-X.; Wang, X.-P.; Zhang, Y.-F. Dew formation characteristics in a revegetation-stabilized desert ecosystem in Shapotou area, Northern China. *J. Hydrol.* **2010**, *387*, 265–272. [CrossRef]
18. Kidron, G.J. Analysis of dew precipitation in three habitats within a small arid drainage basin, Negev Highlands, Israel. *Atmos. Res.* **2000**, *55*, 257–270. [CrossRef]
19. Tweed, S.; Leblanc, M.; Cartwright, I.; Favreau, G.; Leduc, C. Arid zone groundwater recharge and salinisation processes; an example from the Lake Eyre Basin, Australia. *J. Hydrol.* **2011**, *408*, 257–275. [CrossRef]
20. Gran, M.; Carrera, J.; Massana, J.; Saaltink, M.W.; Olivella, S.; Ayora, C.; Lloret, A. Dynamics of water vapor flux and water separation processes during evaporation from a salty dry soil. *J. Hydrol.* **2011**, *396*, 215–220. [CrossRef]
21. Jolly, I.D.; McEwan, K.L.; Holland, K.L. A review of groundwater–surface water interactions in arid/semi-arid wetlands and the consequences of salinity for wetland ecology. *Ecohydrol. Ecosyst. Land Water Process Interact. Ecohydrogeomorphol.* **2008**, *1*, 43–58. [CrossRef]
22. Wang, G.-X.; Cheng, G.-D. The characteristics of water resources and the changes of the hydrological process and environment in the arid zone of northwest China. *Environ. Geol.* **2000**, *39*, 783–790. [CrossRef]
23. Sternberg, T. Environmental challenges in Mongolia's dryland pastoral landscape. *J. Arid Environ.* **2008**, *72*, 1294–1304. [CrossRef]
24. Wheater, H. *Hydrological Processes, Groundwater Recharge and Surface-Water/Groundwater Interactions in Arid and Semi-Arid Areas. Groundwater Modeling in Arid and Semi-Arid Areas*, 1st ed.; Cambridge University Press: Cambridge, UK, 2010; pp. 5–37.
25. Zhuang, Y.; Zhao, W. Advances in the condensation water of arid regions. *Adv. Earth Sci.* **2008**, *23*, 31–38.
26. Cui, Y.; Shao, J. The role of ground water in arid/semiarid ecosystems, Northwest China. *Groundwater* **2005**, *43*, 471–477. [CrossRef]
27. Brodersen, C.; Pohl, S.; Lindenlaub, M.; Leibundgut, C.; Wilpert, K.V. Influence of vegetation structure on isotope content of throughfall and soil water. *Hydrol. Process.* **2000**, *14*, 1439–1448. [CrossRef]
28. Bostic, E.; White, J. Soil phosphorus and vegetation influence on wetland phosphorus release after simulated drought. *Soil Sci. Soc. Am. J.* **2007**, *71*, 238–244. [CrossRef]
29. Rodriguez-Iturbe, I.; D'odorico, P.; Porporato, A.; Ridolfi, L. On the spatial and temporal links between vegetation, climate, and soil moisture. *Water Res. Res.* **1999**, *35*, 3709–3722. [CrossRef]

30. Gerten, D.; Schaphoff, S.; Haberlandt, U.; Lucht, W.; Sitch, S. Terrestrial vegetation and water balance—hydrological evaluation of a dynamic global vegetation model. *J. Hydrol.* **2004**, *286*, 249–270. [CrossRef]
31. Musa, A.; Ya, L.; Anzhi, W.; Cunyang, N. Characteristics of soil freeze–thaw cycles and their effects on water enrichment in the rhizosphere. *Geoderma* **2016**, *264*, 132–139. [CrossRef]
32. Allington, G.; Valone, T. Reversal of desertification: The role of physical and chemical soil properties. *J. Arid Environ.* **2010**, *74*, 973–977. [CrossRef]
33. Feng, Q.; Cheng, G.; Endo, K. Water content variations and respective ecosystems of sandy land in China. *Environ. Geol.* **2001**, *40*, 1075–1083.
34. Qi, F.; Kunihiko, E.; Guodong, C. Soil water and chemical characteristics of sandy soils and their significance to land reclamation. *J. Arid Environ.* **2002**, *51*, 35–54. [CrossRef]
35. Wang, X.P.; Li, X.R.; Xiao, H.L.; Berndtsson, R.; Pan, Y.X. Effects of surface characteristics on infiltration patterns in an arid shrub desert. *Hydrol. Process. Int. J.* **2007**, *21*, 72–79. [CrossRef]
36. Cheng, Y.; Zhan, H.; Yang, W.; Dang, H.; Li, W. Is annual recharge coefficient a valid concept in arid and semi-arid regions? *Hydrol. Earth Syst. Sci.* **2017**, *21*, 5031. [CrossRef]
37. Cheng, Y.; Li, Y.; Zhan, H.; Liang, H.; Yang, W.; Zhao, Y.; Li, T. New comparative experiments of different soil types for farmland water conservation in arid regions. *Water* **2018**, *10*, 298. [CrossRef]
38. Li, S.; Xiao, H.; Cheng, Y.; Wang, F. Water use measurement by non-irrigated Tamarix ramosissima in arid regions of Northwest China. *Sci. Cold Arid Reg.* **2015**, *7*, 146–156.
39. Zhao, L.; Gray, D. Estimating snowmelt infiltration into frozen soils. *Hydrol. Process.* **1999**, *13*, 1827–1842. [CrossRef]
40. Chun, X.; Chen, F.-H.; Fan, Y.-X.; Xia, D.-S.; Zhao, H. Formation of Ulan Buh Desert and its environmental evolution. *J. Desert Res.* **2007**, *6*, 193–199.
41. Tian, Y.; He, Y.; Guo, L. Soil water carrying capacity of vegetation in the northeast of Ulan Buh Desert, China. *Front. For. China* **2009**, *4*, 309–316. [CrossRef]
42. McMahon, T.; Peel, M.; Lowe, L.; Srikanthan, R.; McVicar, T. Estimating actual, potential, reference crop and pan evaporation using standard meteorological data: A pragmatic synthesis. *Hydrol. Earth Syst. Sci.* **2013**, *17*, 1331–1363. [CrossRef]
43. Zuo, X.; Zhao, X.; Zhao, H.; Zhang, T.; Guo, Y.; Li, Y.; Huang, Y. Spatial heterogeneity of soil properties and vegetation–soil relationships following vegetation restoration of mobile dunes in Horqin Sandy Land, Northern China. *Plant Soil* **2009**, *318*, 153–167. [CrossRef]
44. Zhang, J.; Zhao, H.; Zhang, T.; Zhao, X.; Drake, S. Community succession along a chronosequence of vegetation restoration on sand dunes in Horqin Sandy Land. *J. Arid Environ.* **2005**, *62*, 555–566. [CrossRef]
45. Liu, Q.; Yasufuku, N.; Miao, J.; Ren, J. An approach for quick estimation of maximum height of capillary rise. *Soils Found.* **2014**, *54*, 1241–1245. [CrossRef]
46. Puigdefábregas, J.; Sole, A.; Gutierrez, L.; del Barrio, G.; Boer, M.J.E.-S.R. Scales and processes of water and sediment redistribution in drylands: Results from the Rambla Honda field site in Southeast Spain. *Earth Sci. Rev.* **1999**, *48*, 39–70. [CrossRef]
47. Massmann, G.; Sültenfuß, J. Identification of processes affecting excess air formation during natural bank filtration and managed aquifer recharge. *J. Hidrol.* **2008**, *359*, 235–246. [CrossRef]
48. Ferguson, B.K. *Stormwater Infiltration*; CRC Press: Boca Raton, FL, USA, 1994.
49. Chen, X.; Hu, Q.J.J. Groundwater influences on soil moisture and surface evaporation. *J. Hidrol.* **2004**, *297*, 285–300. [CrossRef]
50. Cerdà, A. Seasonal variability of infiltration rates under contrasting slope conditions in southeast Spain. *Geoderma* **1996**, *69*, 217–232. [CrossRef]
51. Huang, J.; Wu, P.; Zhao, X. Effects of rainfall intensity, underlying surface and slope gradient on soil infiltration under simulated rainfall experiments. *Catena* **2013**, *104*, 93–102. [CrossRef]
52. Casenave, A.; Valentin, C. A runoff capability classification system based on surface features criteria in semi-arid areas of West Africa. *J. Hydrol.* **1992**, *130*, 231–249. [CrossRef]
53. Yair, A.; Lavee, J.S.P. Runoff generative process and runoff yield from arid talus mantled slopes. *Earth Surf. Process.* **1976**, *1*, 235–247. [CrossRef]
54. Jordán, A.; Zavala, L.M.; Gil, J.J.C. Effects of mulching on soil physical properties and runoff under semi-arid conditions in southern Spain. *Catena* **2010**, *81*, 77–85. [CrossRef]

55. Kidron, G.J.; Yair, A. Rainfall–runoff relationship over encrusted dune surfaces, Nizzana, Western Negev, Israel. *Earth Surf. Process. Landf. J. Br. Geomorphol. Group* **1997**, *22*, 1169–1184. [CrossRef]
56. Yair, A.; Almog, R.; Veste, M. Differential hydrological response of biological topsoil crusts along a rainfall gradient in a sandy arid area: Northern Negev desert, Israel. *Catena* **2011**, *87*, 326–333. [CrossRef]
57. Lange, J.; Leibundgut, C.; Greenbaum, N.; Schick, A.W.R.R. A noncalibrated rainfall-runoff model for large, arid catchments. *Water Resour. Res.* **1999**, *35*, 2161–2172. [CrossRef]

Article

Does the Process of Passive Forest Restoration Affect the Hydrophysical Attributes of the Soil Superficial Horizon?

Nayana Alves Pereira [1], Simone Di Prima [2,3], Renata Cristina Bovi [1], Laura Fernanda Simões da Silva [1], Gustavo de Godoy [1], Rafaela Pereira Naves [1] and Miguel Cooper [1,*]

1 Luiz de Queiroz School of Agriculture, University of São Paulo, Av. Pádua Dias 11, Piracicaba 13418-900, Brazil; nayanaalves.esalq@usp.br (N.A.P.); renata.bovi@usp.br (R.C.B.); laurafsimoes@usp.br (L.F.S.d.S.); gustavodegodoy@usp.br (G.d.G.); rafaelafloresta@usp.br (R.P.N.)

2 Agricultural Department, University of Sassari, Viale Italia, 39, 07100 Sassari, Italy; sdiprima@uniss.it

3 Univ Lyon, Université Claude Bernard Lyon 1, CNRS, ENTPE, UMR 5023 LEHNA, F-69518 Vaulx-en-Velin, France

* Correspondence: mcooper@usp.br; Tel.: +55-19-3429 4171 or +55-19-3417-2102

Received: 17 April 2020; Accepted: 10 June 2020; Published: 12 June 2020

Abstract: There has been an increase in the area of secondary tropical forests in recent years due to forest restoration in degraded areas. Recent analyses suggest that the success of passive forest restoration is highly uncertain and needs to be better understood. This study aimed to investigate the behavior of saturated hydraulic conductivity (*Ks*) and some hydrophysical soil attributes between agricultural land uses, restored forests, and a degraded forest fragment. The areas evaluated are located in the municipality of Rio Claro, São Paulo, Brazil, under different types of land use: (i) two areas in the process of passive forest restoration: one of 18 and another of 42 years (NR18 and NR42); (ii) a degraded forest fragment (FFD); (iii) pasture (P), and (iv) sugarcane (SC). The hydraulic soil conductivity characterization was performed using the Beerkan method. Dry soil bulk density (*BD*), total porosity (*Pt*), macroporosity (*Mac*), microporosity (*Mic*), penetration resistance (*PR*), mean aggregate diameter (*MWD*), and soil organic carbon (*OC*) were also determined. The comparative analysis of the hydrophysical attributes of the soil superficial horizon in agricultural land uses (P and SC), restored forests (NR18 and NR42), and a degraded forest (DFF) confirms that the recovery of soil hydrological functioning in ongoing forest restoration processes can be a relatively slow process. In addition, the intensity of previous land use leaves footprints that can affect passive restoration areas for decades after agriculture abandonment, increasing the time for the recovery of *Ks* and soil hydrophysical attributes.

Keywords: soil hydraulic conductivity; aggregate stability; soil porosity; soil penetration resistance

---

1. Introduction

A substantial increase in the area of secondary tropical forests has been occurring in the last 20 years [1,2]. There is promising evidence that these areas have been increasing as a result of the ambitious goals implemented worldwide for the recovery of degraded and deforested land, such as the Bonn Challenge (2011) and the New York Declaration on Forest (2014). Many degraded areas, especially those that have not been subjected to many years of monoculture, are recovered using passive restoration techniques. One of these techniques is natural forest regeneration, which takes advantage of the resilience of degraded areas by stimulating the germination of local seed banks, seed dispersal processes, regrowth of trunks, or seedlings that resisted the disturbances. This restoration approach is economically advantageous because it reduces implementation costs, such as purchase of seedlings, inputs, application of silvicultural techniques, maintenance of plantations, and labor [3].

The potential of passive restoration to recover some groups of biodiversity in tropical regions (such as plants, birds, and invertebrates) and structural parameters of vegetation (density, biomass, and height) is well reported in the literature [4–6]. Furthermore, large-scale forest cover, either natural or restored, has impacts on climate, soil, and hydrology [7]. Soil is the substrate for forest restoration, and the knowledge of its functioning is essential for understanding and managing restoration processes. Thus, the success of forest restoration processes depends on the presence of a healthy soil environment in which seedlings and trees can develop successfully. The recovery of soil hydrophysical attributes is fundamental to maximize the success of restoration efforts [8]; however, it is still not fully understood in forest restoration because soil is still under-investigated in these environments. Monitoring soil hydrophysical attributes in forest restoration environments, such as soil bulk density, total porosity, macro and microporosity, and aggregate stability, can supply data on soil-related functions, such as root growth, water infiltration and drainage, gas exchange, biological activity, water retention, and carbon stock supply [9].

The hydraulic conductivity of saturated soil (*Ks*) is also a key parameter that describes the movement of water in the soil and exerts a dominating influence on the partitioning of rainfall into vertical and lateral flow paths. Therefore, estimates of *Ks* are essential for describing and modeling hydrological processes [10]. Moreover, *Ks* is dependent on the soil structure, an attribute deeply affected by soil management [11,12]. In forest ecosystems, it is expected that there will be greater input and decomposition of the litter, which is the main supply of organic matter to the soil and regulates the cycling of nutrients. Higher concentrations of organic matter in the soil also favor microbial and soil fauna activities and provide more suitable conditions for soil protection [13]. The formation and evolution of soil aggregates are driven by biological and microbial activities, and the supply of organic matter under forests helps to maintain the stability of soil aggregates [14]. The literature reports some studies that evaluated areas in different stages of passive restoration in tropical forests [4,7,15–19]. In some areas, improvements were observed in the contribution of soil biomass and biodiversity [4,17], lower soil bulk density, and greater total soil porosity [17,19], which probably influenced the formation of more stable aggregates in the soil and higher *Ks*. In general, the results reinforce the importance of expanding knowledge about the hydrophysical attributes in forest restoration areas, since the forest's age alone is not sufficient to restore soil functioning. Besides, different stages of forest restoration may show distinct signs of soil recovery.

Among Brazilian biomes, the Atlantic Forest suffered the most significant degradation, due to the conversion of primary forests to sugarcane and coffee plantations [20]. Currently, it has a high number of degraded fragments and only 12% of the original coverage [21]. In this biome, the current forest cover is a heterogeneous forest mosaic of different ages, located in different landscape conditions and, therefore, with varying levels of disturbance [22]. Understanding the degree of tropical forest disturbances and their recovery by restoration processes is challenging. In addition, the time that natural regeneration takes to recover the hydrophysical soil attributes disturbed by intense previous land use after deforestation is very variable [19]. These processes need to be better understood [18,23–26].

Considering that the soil structure and functioning are essential for the improvement of *Ks* [8,17,18] and that the impact of passive restoration on the soil hydrophysical functioning is still little known, this study aimed to answer the following question: Is passive restoration capable of improving soil *Ks*, as well as the physical attributes of the soil superficial horizon? To answer this question, we investigated the behavior of *Ks* and some soil physical attributes of the soil superficial horizon in forests of different ages undergoing passive restoration by natural regeneration, a degraded forest fragment, a pasture, and a sugarcane field.

2. Material and Methods

2.1. Study Area

The study was carried out in the county of Rio Claro (22°20′ S, 47°34′ W), São Paulo State, Southeast Brazil (Figure 1). The area is inserted in the Corumbataí River hydrographic basin, and has more than 1700 km^2 and more than 200 years of land-use change [22], with an altitude between 562 and 643 m a.s.l. The climate is classified as Cwa, according to Köppen classification, characterized by dry winters and rainy and hot summers [27]. The wettest period is from December to February, while the driest, from June to August. The average annual precipitation is 1344 mm, with an average annual temperature of 20.5 °C [27]. According to the geomorphological map of São Paulo state, the area is located in the Peripheral Depression of the São Paulo state [28]. The native forest cover is classified as semi-deciduous rainforest [29], belonging to the Atlantic Forest biome [30].

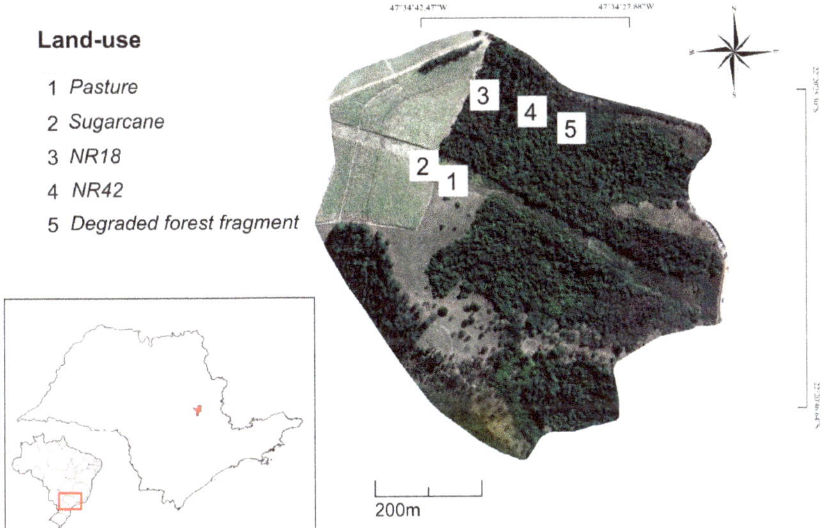

Figure 1. Location of the study areas in the state of São Paulo, Southeast Brazil. NR18: Natural regeneration 18 years; NR42: Natural regeneration 42 years.

The exploitation of the Brazilian Atlantic Forest was accentuated in the 20th century, resulting in severe changes to the ecosystems, primarily by reducing and increasing the pressure on biodiversity [31]. After almost 500 years of land-use changes, the Atlantic Forest, which covered about 150 million hectares in Brazil, currently has less than 12% of its original forest cover (1.2 million km^2) [21]. Nowadays, more than 80% of forest fragments are smaller than 50 ha [21], isolated, unprotected, and severely altered [32]. The few fragments larger than 100 ha are located on steep terrain, where human occupation is difficult [33].

The intense deforestation in the study area resulted from the exploitation and expansion of coffee, cotton, and pastures [32] during the first half of the twentieth century [31]. According to 1962 satellite images [22], the actual restored forest cover in the study area was previously occupied by pastures. Restoration by natural regeneration processes occurred due to two main reasons: (1) the abandonment of areas, which had low agricultural suitability, due to the predominance of steep slopes, and the presence of sandy and rocky soils [22]; and (2) environmental planning efforts to comply with the Brazilian Forest Code [22,30]. Currently, land use in this area is dominated by Urochloa decumbens

Stapf pastures (43.68%), sugar cane (25.57%), fragments of native forest (12.36%), and other types of land uses (14.5%) [34].

For the development of this study, we selected five different land uses: (a) pasture (P); (b) sugarcane (SC); (c) 18-year-old passive forest restoration by natural regeneration (NR18); (d) 42-year-old passive forest restoration by natural regeneration (NR42), and (e) degraded forest fragment (DFF) (Figures 1 and 2). The DFF area is currently isolated, but before this measure, there were intense cattle invasions associated with frequent burning in the contiguous sugar cane fields during the dry season that invaded the remnant fragment. Oxisols and Ultisols represent the soils of the study area, according to Soil Taxonomy, 2014 [35] with sandy loam and loamy, sandy textures (Table 1). The soil texture of the superficial horizon did not show important variations among the study sites being considered homogeneous.

Figure 2. Image of the study areas. P: Pasture; SC: Sugarcane; NR18: Passive restoration by natural regeneration 18 years; NR42: Passive restoration by natural regeneration 42 years; DFF: Degraded forest fragment.

Table 1. Soil classification, soil particle density (ρs), particle size distribution, and textural classes for the soil superficial horizon (0.10 m) of the study areas.

Study Area	Clay (%)	Silt (%)	Sand (%)	ρs g cm^{-3}	Textural Classes	Soil Classification
P	18	4	78	2.63	Sandy loam	Oxisols
SC	19	5	76	2.66	Sandy loam	Oxisols
NR18	11	6	83	2.67	Loamy sand	Oxisols
NR42	14	15	71	2.64	Sandy loam	Ultisols
DFF	11	13	76	2.65	Sandy loam	Oxisols

P: Pasture; SC: Sugarcane; NR18: Natural regeneration 18 years; NR42: Natural regeneration 42 years; DFF: Degraded forest fragment.

2.2. Soil Sampling and Analytical Procedures

The field campaign and soil sampling were performed during July 2018. Within each study area, five sampling points were selected, distant 5.5 m from each other. The soil water infiltration was measured using the Beerkan method [36]. In each study area, ten water infiltration measurements were performed (two infiltration measurements at each sampling point), using a steel cylinder with an internal diameter of 0.16 m. The sampled infiltration surfaces were assumed to be large enough to allow the determination of representative *Ks* values that accounted for soil heterogeneity and the contribution of biopores [37]. The location of the insertion of the steel cylinder was previously prepared by removing the litter and other materials to expose the soil surface. Subsequently, the steel ring was inserted 0.01 m into the soil. For each measurement, a volume of 150 mL of water was added repeatedly into the steel cylinder. The time for complete infiltration of each poured volume was logged. This procedure was repeated until the difference between the infiltration times of two or three consecutive tests underwent minimal variation. In total, 50 infiltration runs (10 runs × 5 areas) of cumulative infiltration (I) as a function of time (t) were performed.

The values of *Ks* were estimated by the simplified Beerkan infiltration method (SSBI—Steady version of the Simplified method based on a Beerkan Infiltration run) [38]. The SSBI method estimates *Ks* as follows:

$$K_s = \frac{i_s}{\frac{\gamma \gamma_w}{r \alpha^*} + 1} \quad (1)$$

where i_s (L T^{-1}) is the slope of the linear regression fitted to the final portion of the cumulative infiltration time series, r (L) is the ring radius, γ_w is a dimensionless constant related to the shape of the wetting front often fixed at 1.818 [39], γ is an infiltration constant, often fixed at 0.75 [40,41], and α^* (L^{-1}) is the sorptive number, that expresses the relative importance of the capillary over gravity forces during water movement in unsaturated soil [42,43]. According to a previous investigation carried out on similar soils in the Atlantic forest [44], we assumed a value of $\alpha^* = 0.0012$ mm^{-1}, also taking into account that it represents the suggested first approximation value for most field soils [45].

At each sampling point, disturbed and undisturbed soil samples were collected at 0–0.10 m depth. The disturbed samples were collected to determine the gravimetric moisture content of the saturated soil (*Ug*, %), the organic carbon (*OC*, g kg^{-1}), and the soil particle density (ρs, g cm^{-3}). Undisturbed samples were collected using Kopeck metal rings with cutting edges measuring 0.05 × 0.05 m (diameter × height). These samples were used to determine: (a) the initial volumetric moisture content (θi, cm^3 cm^{-3}), (b) total porosity (*Tp*, cm^3 cm^{-3}), (c) macroporosity (*Mac*, cm^3 cm^{-3}), (d) microporosity (*Mic*, cm^3 cm^{-3}), (e) dry soil bulk density (*Pt*, g cm^{-3}) and (f) penetration resistance (*PR*, MPa). Bulk density (*BD*) was determined by the ratio of the dry soil mass and the cylinder volume [46]. The volumetric moisture content of the saturated soil (θs, cm^3 cm^{-3}) was obtained using the measure of *BD* ($\theta s = Ug \times BD$). We also computed the depths of wetting fronts for all infiltration experiments at the end of the experiment, Z_{wf} (L). For this purpose, we considered a piston flow width displacement of the volume of water in the soil porosity [47].

In the laboratory, each undisturbed soil sample was gradually saturated with water by capillary action and then weighed. The total porosity (*Tp*) was calculated by the ratio between *BD* and ρs [48]. The soil particle density (ρs) was determined by a helium gas pycnometer (ACCUPYC 1330, Micromeritics Instrument Corporation®, Norcross, GA, USA). *Mic* was estimated using the tension table with the soil matric potential of −6 kPa ($\Psi = -0.006$), and *Mac* was determined by the difference between *Tp* and *Mic*.

PR was determined using an electronic benchtop penetrometer (CT3 Texture Analyzer, Brookfield, Middlebore, MA, USA), equipped with a 25 kg load cell and a metal rod with a 30° semi-angle cone tip, a basal diameter of 3.81 mm, and penetration velocity of 10 mm min^{-1}. *PR* measurements were obtained through an automated data acquisition system. Such measures were collected at the center of each undisturbed soil sample with soil moisture equivalent to a water pressure of 10 kPa. For all samples, the measurements obtained from the upper (1 cm) and lower (1 cm) were discarded, following the procedures of Imhoff [49].

The determination of the aggregate means weight diameter (*MWD*) was performed using the dry methodology [50]. For this, 50 g of soil were collected from the 0–0.10 m depth. These samples were separated into size classes by sieving in a dry medium through a Solotest vibratory mechanical stirrer containing a set of mesh sieves: 4.0, 2.0, 1.0, 0.50, and 0.25 mm. The calculations of (*MWD*) were obtained using the equation of Kemper [51]:

$$MWD = \sum X_{is}.W_{is} \tag{2}$$

where *MWD* = mean weight diameter of the soil aggregates (mm); X_{is} = mean diameter of each class (mm); W_{is} = proportion of aggregates in each sieve class (i) (%).

2.3. Data Analysis

For each studied variable (*Ks, BD, Tp, Mac, Mic, PR, MWD*, and *OC*) analysis of variance (ANOVA) was performed, considering the type of land use as an explanatory variable (P, SC, NR18, NR42, and DFF), after having checked that the assumptions of normality of residues and homogeneity of variances were met, through the Shapiro–Wilk test and the Barllet test, respectively. In the case of significance at 10%, the Tukey test was applied, which compares the means two by two. For the variables *Ks* and *PR*, normality was determined using the natural logarithm of the value obtained initially, due to the high variability of the data [52]. For the *Mic*, the assumptions of ANOVA were not met, even after data transformation. A Kruskal–Wallis non-parametric test was then performed, also at a 10% significance level. To simultaneously compare the hydrophysical attributes of the soil between the different uses, a principal component analysis (PCA) was performed for standardized data. All analyses were performed in the R software (R Core Team, Vienna, Austria, 2018).

3. Results

Soil Hydrophysical Attributes of the Study Areas

The soils in the study area did not present soil particle distribution differences down to the depth of 0.30 m (Table 1). Among the different types of land use studied in this investigation, there were statistically significant differences for the variables *Ks, Mic, PR, MWD*, and *OC* (Table 2 and Figure 3). The *BD, Tp*, and *Mac* variables did not show statistically significant differences among the studied land uses (Table 2).

Table 2. Average values of the hydrophysical attributes of the soils in the study areas. *BD*: bulk density (g cm^{-3}); *Tp*: total porosity (cm^3 cm^{-3}); *Mac:* macroporosity (cm^3 cm^{-3}); *Mic*: microporosity (cm^3 cm^{-3}); *PR*: penetration resistance (Mpa); *MWD*: mean weight diameter of the soil aggregates (mm); *OC*: soil organic carbon content (g kg^{-1}), and *Ks*: saturated soil hydraulic conductivity (mm h^{-1}).

		Land Uses				
Attributes	**Statistic**	**P**	**SC**	**NR18**	**NR42**	**DFF**
BD	Mean	1.51 a	1.53 a	1.47 a	1.48 a	1.39 a
	CV	2.98	4.46	4.40	9.54	5.97
Tp	Mean	0.42 a	0.42 a	0.44 a	0.45 a	0.47 a
	CV	3.51	6.20	5.07	10.35	6.61
Mac	Mean	0.13 a	0.16 a	0.13 a	0.16 a	0.18 a
	CV	19.16	14.39	12.98	17.00	23.28
Mic	Mean	0.29 ab	0.25 b	0.30 a	0.28 ab	0.29 ab
	CV	6.30	6.21	5.26	17.55	5.37
PR	Mean	2.64 a	1.60 a	0.83 b	0.66 b	0.68 b
	CV	23.92	32.37	38.30	33.53	16.95
MWD	Mean	0.99 bc	0.76 c	0.75 c	1.34 ab	1.68 a
	CV	2.98	16.17	12.86	6.84	11.06
OC	Mean	14.00 a	7.33 c	8.66 bc	11.33 ab	9.33 bc
	CV	14.28	15.74	13.32	10.18	16.36
Ks	Mean	28.46 b	39.93 ab	70.30 a	124.22 a	36.28 ab
	CV	104.89	59.60	80.50	115.08	73.28

The letters refer to the comparison test of means two by two in the Tukey test at the 90% confidence level. Averages followed by the same letter do not differ statistically. CV: coefficient of variance (%). Reference values of CV: Low: <10%; medium: between 10 and 20%; high: between 20 and 30%, and very high: >30% [53]. Note: For *BD, Tp, Mac, Mic*, and *PR*, the number of soil samples = 5; for *Ks* = 10; for *MWD* and *OC* = 3.

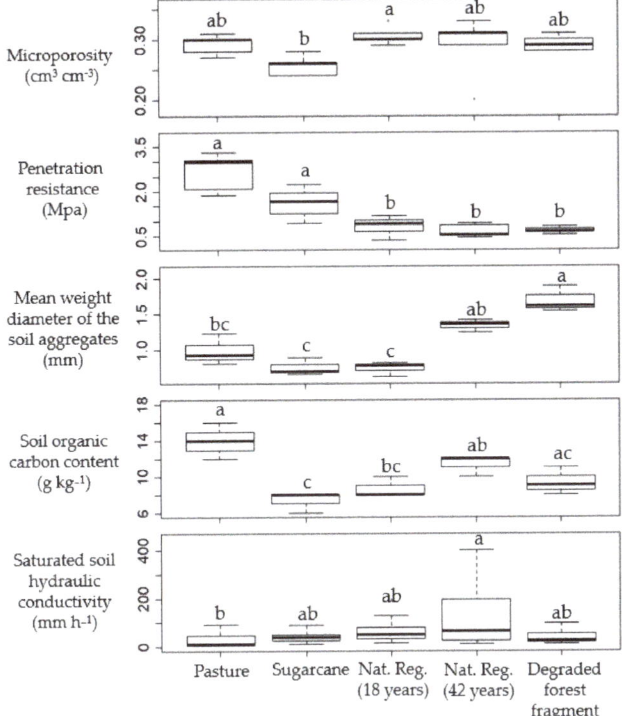

Figure 3. Boxplot of hydrophysical attributes for the different land uses. Line within the box is the median. The bars represent standard error. Note: For this figure, we only considered the soil physical attributes that presented significant differences, according to Table 2. Nat.Reg = natural regeneration.

Considering soil *Mic*, SC only differed from NR18. For this attribute, NR42 presented a medium value of the coefficient of variance (CV) and the rest a low CV. Significant differences were observed for *PR* between the agricultural (P and SC) and the passive restoration areas (NR18 and NR42). *PR* values for P and SC were higher than NR18, NR42, and DFF (Figure 3). NR18 and DFF had *PR* values three times lower than P, and NR42 had *PR* values four times lower than P. When compared to SC, the passive restoration areas and DFF also presented lower *PR* values. High CV values for *PR* were found for SC, NR18, and NR42.

The land use with the highest passive restoration age—NR42—presented similar *MWD* values to DFF and P and differed from all the other land uses. We noted that the youngest passive restoration area—NR18–presented *MWD* values similar to both agricultural land uses (P and SC). For this attribute, CV values were low for P and NR42 and medium for SC, NR18, and DFF (Table 2; Figure 3).

Considering the soil *OC* content, P had the highest values and was similar to NR42, while SC, NR18, and DFF presented lower values. All land uses had medium CV's (Table 2).

The mean values of the wetting front depths ranged from 219 to 319 mm, depending on the site. Therefore, *Ks* data may be considered representative of the studied upper layers (Table 1). In addition, we did not detect any restrictive layer that may have affected *Ks* predictions [54].

Ks was moderately high for P and presented significant differences with both passive restoration areas. NR18 and NR42 had high *Ks* values, approximately double and four times the *Ks* value of P, respectively. SC and DFF had similar values of *Ks*. All land uses presented high CV values for *Ks*, especially NR42, followed by P (Table 2).

Although no significant differences among land uses were observed for *BD*, a tendency of decrease in this attribute could be observed in the forest land uses when compared to the agricultural land uses (Table 2). The same was valid for *Tp*, where the restoration age showed a tendency to improve, presenting similar values to DFF. The behavior of *Ks* also showed a tendency to increase according to restoration age, considering the mean and absolute values of this attribute (Figure 3).

A PCA was performed to investigate and relate the soil hydrophysical attributes between the different land uses (Table 3 and Figure 4).

Table 3. Principal component analysis (PCA) loadings for the first and second components of the principal component analysis.

Variables	Component 1	Component 2
Ks		0.49
BD	0.556	0.109
Tp	−0.578	
Mac	−0.353	0.512
Mic	−0.262	−0.618
PR	0.391	−0.316

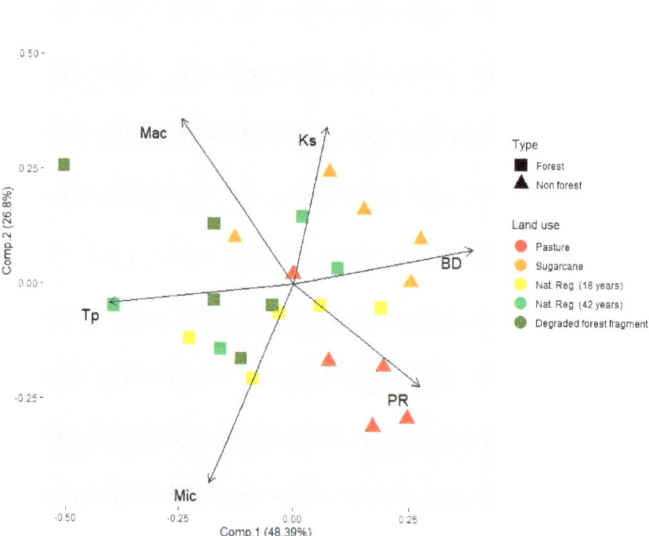

Figure 4. Principal Component Analysis (PCA) biplot based on soil attributes: Log *Ks* (mm h^{-1}), *BD* (g cm^{-3}), *Tp* (cm^3 cm^{-3}), *Mic* (cm^3 cm^{-3}), *Mac* (cm^3 cm^{-3}), *PR* (Mpa). Symbols represent plot sites for each land-cover type: Nat.Reg = natural regeneration.

Component 1 represents 48.39% of the data variation, and the variables most related to this Axis were *BD* (positively) and *Tp* (negatively) (Table 3 and Figure 4). In addition, *BD* was positively correlated to *PR*, while *Tp* was mainly correlated to *Mac* and, to a lesser extent, to *Mic*. *Ks* remained almost perpendicular to component 1; this demonstrated that there was no correlation between this attribute with *BD*, considering all land uses.

The higher values of component 1 indicate higher levels of soil degradation, while the lower values indicate the opposite. On this Axis, the forest areas are positioned to the left, while the non-forest areas are on the right. In this case, the highest values of *BD* and lower values of *Tp* were associated with agricultural land uses (P and SC), while the lower *BD* values were associated with DFF. Therefore, the sequence of the most degraded to the least degraded areas was P, SC, NR18, NR42, and DFF, with a remarkable tendency for less degradation of the two areas in the process of restoration and DFF.

Component 2 explained 26.8% of the data variation. The variables most correlated to this Axis were *Mac* (positively) and *Mic* (negatively), and *Ks* had a low relationship with *Mac*, *Tp*, and *BD*. Considering the land use, it is possible to notice that the forest areas were distributed along the entire Axis, while P was below, and SC was above. Besides that, the highest values of *Mac* were more associated with SC, while *Mic* and *PR* were related to P. As for the restoration areas (NR18 and NR42), we noticed a tendency to decrease *Mic* and increase *Ks*, according to the ages of passive restoration.

4. Discussion

According to the *Ks* classification of the Soil Survey Staff [55], the *Ks* values obtained in the studied areas varied between high (36 < 360 mm h^{-1}) and moderately high (3.6 < 36 mm h^{-1}). At the P site, *Ks* was moderately high due to soil compaction, evidenced by the high *PR* value, high *Mic* (Figures 3 and 4), and low *MWD* (Table 2). *PR* integrates the effects of soil bulk density and moisture, affecting the soil physical conditions for root growth [56]. The *PR* value for P was above 2 Mpa, which is considered a critical limit for root growth in the literature [57,58]. This result reports the degradation of the area, as demonstrated in the PCA (Figure 4). The high content of *OC* is explained by the accumulation of biomass in pastures, promoted by the root system of grasses that are abundant during the year and by the organic matter added by the animals [57,59], but its effect on soil aggregation was not observed due to the high level of compaction in this area. The decrease in *Ks* has been similarly reported in several studies after the establishment of pastures in tropical soils and occurred due to changes in the soil pore distribution caused by animal trampling. A study at the watershed scale in the Atlantic Forest biome showed moderately high values of *Ks* (22 mm h^{-1}) in the surface 0.15 m soil layer in pasture when compared to eucalyptus (40 mm h^{-1}) and a 44-year-old secondary forest (61 mmh^{-1}) [60]. Another study, carried out in the Amazon, also reported a decrease in soil water infiltration after conversion from forest to pasture, from 1258 to 100 mm [61]. The reduction in soil water infiltration and, consequently, lower *Ks*, promote lateral water fluxes [10,62–64], higher runoff and, consequently, higher soil erosion [65].

The absence of soil cover combined with agricultural machinery used in the SC area promoted the lowest values of *MWD* and *OC*, which probably influenced the lower value of *Ks* when compared to NR18 and NR42. Even though *Ks* in SC did not differ statistically from NR18 and NR42, the latter have higher absolute values. This land use was highly associated with soil *Mac* (Figure 4), which was probably associated with soil tillage operations carried out during sugarcane replanting and in the period between ratoon-crop harvests. These soil management practices can have positive effects, in the short term, on the physical quality of the soil, due to the modifications provided to the soil porosity, mainly in the superficial soil horizons [66,67].

The values obtained for *Ks* in the NR18 and NR42 areas were classified as high, according to Soil Survey Staff [55]. Indeed, these *Ks* values indicated a good ability of the soil to infiltrate and percolate plant-available water to the root zone and to drain excess water out of the root zone [64]. These areas showed a tendency to improve the *MWD* and *OC* variables, where the latter is one of the primary agents of formation and stabilization of soil aggregates in the soil superficial horizons, with a direct influence on porosity [68,69]. The *MWD* indicates the largest stable aggregates in the soil in terms of structural organization. Aggregates with *MWD* greater than 0.25 mm are considered stable [68]. The addition of *OC* (root activity, growth, and functioning; death of plant tissues; among others) provides substrate for microbial growth and the production of secondary organic compounds that act as cementing agents. These agents help aggregate formation [70] and stabilization that increase their resistance when submitted to external forces [71]. A tendency for higher soil aggregation was evident in the NR18 and NR42 areas, improving soil structure and, consequently, *Ks* (Figure 3). Indeed, soils with a high degree of aggregate stability usually also have good soil physical quality, with good soil porosity, aeration, water retention and infiltration [72]. In this sense, soils with greater aggregate stability are considered structurally superior to similar soils with weaker aggregation [73].

The high variability of *Ks* obtained in NR42 (Figures 3 and 4) suggested the occurrence of preferential flows in this area, which can occur due to several factors, among them, the presence of continuous biopores created by large roots of decomposed trees or channels created by macrofauna [25,43,74]. This hypothesis corroborates studies developed in forests under tropical climates [18,64,75,76]. However, although preferential flows are considered a common phenomenon in forested soils, future studies are needed to understand better how they work in areas of passive restoration and their effects in space and time, considering that they are the main cause of groundwater pollution and contamination [77].

The set of soil attributes of the soil superficial horizon analyzed in this study demonstrated that the DFF land use was the least degraded (Figure 4). The lower value of *BD* and higher values of *Tp*, *Mac*, and *MWD* (Table 2) show that, although the forest fragment has suffered some degradation in the past, it still maintains its functionality, even though *Ks* presented lower values when compared to NR18 and NR42. The fact that DFF presented lower *Ks* values when compared to the restored areas can be possibly due to other factors that affect this attribute in natural environments, such as plant density and diversity, canopy cover [3,19], geological and topographic variations [78], soil water repellency [44,77], among others.

The joint analysis of the hydrophysical attributes evaluated here, through the PCA, showed that areas in the process of passive restoration had a lower degree of degradation when compared to areas of agricultural use (Figure 4). This lower degree of degradation in restored areas can be interpreted as a general tendency of improvement in the hydrophysical attributes after restoration initiation. High values of *Ks*, aggregate stability, and the improvement tendencies in the other evaluated attributes evidence the soil hydrophysical recovery in the restored forests (Table 2; Figures 3 and 4) [79].

The time scale for the recovery of the hydrophysical attributes in restoration areas in our study was longer than other studies that observed positive changes in these attributes in shorter periods after restoration implementation. In a study in Mexico, areas with passive restoration times of 0, 4, 10, and 20 years after pasture abandonment showed a significant *BD* improvement associated with an increase in water infiltration rates with restoration age [23]. On the other hand, *OC* contents were lower in the 10- and 20-year-old restored forests (3.94 and 3.71%, respectively) when compared to the actual pasturelands (4.79%). Higher water infiltration rates and OC contents were observed with increasing age of forest restoration (15 and 20 years) with values varying between 493 and 2.462 mm h^{-1} when compared to pastures in Costa Rica [75]. Other studies also showed improvements in hydrophysical attributes in shorter periods (8 to 12 years) after restoration initiation [76,80]. The longer time needed to recover the hydrophysical attributes in our study could be explained by the longer period this area was previously occupied by pastures and also by the legacy effect of intense degradation processes that occurred before restoration. This legacy effect phenomenon was observed by Piché and Kelting [14] in soils under 55- to 60-year-old secondary forests. These previous land-use effects could explain the slow recovery of *Mac*, *TP*, and *BD* (Table 2) in NR18 and NR42.

5. Conclusions

The comparative analysis of the soil superficial horizon's hydrophysical attributes in agricultural land uses (P and SC), forests of different ages undergoing passive restoration by natural regeneration (NR18 and NR42), and a degraded forest fragment (DFF) confirms that the evolution of different soil hydrological functions in the ongoing restoration of natural ecosystems can be a relatively slow process.

The observed trend of increasing *Ks* and recovery of soil physical attributes in passive forest restoration of different ages leads us to conclude that the time that the secondary succession takes to recover the soil hydrological properties, disturbed by the previous deforestation and the use of pastures, are not governed just by the restoration time.

The intensity of previous land use leaves footprints that can affect areas submitted to natural regeneration processes for decades after agriculture abandonment and is one of the main critical factors that affect soil physical attributes and *Ks* over time. This is alarming because it is well known that the soil hydrophysical attributes affect fluid flow and transport, and can, consequently, influence runoff generation and water table recharge.

Important ecosystem services, such as water regulation and provision, depend on the recovery of soil hydrological functions in degraded environments. Knowledge of how forest restoration processes, either active or passive, affect soil function, is essential. The data presented and discussed in this study adds to this knowledge and shows that the recovery of soil hydrophysical functioning is a slow process and varies according to the type of soil attribute and the previous land use and its degradation legacy.

Author Contributions: N.A.P. carried out the data collection and wrote the initial draft; M.C. participated in the design of the study and helped to draft and edit the manuscript; L.F.S.d.S. and R.C.B. provided important advice on the concept of the structuring of the manuscript; S.D.P., R.C.B., L.F.S.d.S., G.d.G. and R.P.N. analyzed the data and made great contributions to writing the manuscript. All authors reviewed and approved the final manuscript.

Funding: This research was financed in part by the Coordenação de Aperfeiçoamento de Pessoal de Nível Superior—Brasil—(CAPES), Finance Code 001.

Acknowledgments: Miguel Cooper acknowledges the National Council for Scientific and Technological Development (CNPq) for the fellowship. We would like to thank Aline Aparecida Fransozi, Sergio Esteban Lozano-Baez, Ricardo César Gomes, Silvio F. Barros, and Pedro H.S. Brancalion for support at the beginning of this study and the team "Cooper Trupe" for field support. We would like to thank Vittoria Giannini, who helped with the English revision of the manuscript.

Conflicts of Interest: The authors declare no conflict of interest.

References

1. Chazdon, R. Beyond Deforestation: Restoring Forests and Ecosystem Services on Degraded Lands. *Science* **2008**, *320*, 1458–1460. [CrossRef] [PubMed]
2. Filoso, S.; Bezerra, M.O.; Weiss, K.C.B.; Palmer, M.A. Impacts of forest restoration on water yield: A systematic review. *PLoS ONE* **2017**, *12*, e0183210. [CrossRef] [PubMed]
3. Chazdon, R.L. Regeneração de florestas tropicais. *Boletim Museu Paraense Emílio Goeldi Ciencias Naturais* **2012**, *7*, 195–218.
4. Crouzeilles, R.; Ferreira, M.S.; Chazdon, R.; Lindenmayer, D.B.; Sansevero, J.B.B.; Monteiro, L.; Iribarrem, A.; Latawiec, A.E.; Strassburg, B.B.N. Ecological restoration success is higher for natural regeneration than for active restoration in tropical forests. *Sci. Adv.* **2017**, *3*, e1701345. [CrossRef]
5. Meli, P.; Holl, K.D.; Benayas, J.M.R.; Jones, H.P.; Jones, P.C.; Montoya, D.; Mateos, D.M. A global review of past land use, climate, and active vs. passive restoration effects on forest recovery. *PLoS ONE* **2017**, *12*, e0171368. [CrossRef]
6. Rozendaal, D.M.A.; Bongers, F.; Aide, T.M.; Dávila, E.Á.; Ascarrunz, N.; Balvanera, P.; Becknell, J.M.; Bentos, T.V.; Brancalion, P.H.S.; Cabral, G.A.L.; et al. Biodiversity recovery of Neotropical secondary forests. *Sci. Adv.* **2019**, *5*, eaau3114. [CrossRef]
7. Godsey, S.; Elsenbeer, H. The soil hydrologic response to forest regrowth: A case study from southwestern Amazonia. *Hydrol. Process.* **2002**, *16*, 1519–1522. [CrossRef]
8. Giannini, V.; Bertacchi, A.; Bonari, E.; Silvestri, N. Rewetting in Mediterranean reclaimed peaty soils and its potential for phyto-treatment use. *J. Environ. Manag.* **2018**, *208*, 92–101. [CrossRef]
9. Burger, J.A.; Kelting, D.L. Using soil quality indicators to assess forest stand management. *For. Ecol. Manag.* **1999**, *122*, 155–166. [CrossRef]
10. Zimmermann, A.; Schinn, D.S.; Francke, T.; Elsenbeer, H.; Zimmermann, B. Uncovering patterns of near-surface saturated hydraulic conductivity in an overland flow-controlled landscape. *Geoderma* **2013**, *195*, 1–11. [CrossRef]
11. Mesquita, M.G.B.; Moraes, S.O. The dependence of the saturated hydraulic conductivity on physical soil properties. *Ciência Rural* **2004**, *34*, 963–969. [CrossRef]
12. Ahuja, L.R.; Naney, J.W.; Green, R.E.; Nielsen, D.R. Macroporosity to Characterize Spatial Variability of Hydraulic Conductivity and Effects of Land Management. *Soil Sci. Soc. Am. J.* **1984**, *48*, 699–702. [CrossRef]

13. De Andrade, A.G.; Tavares, S.D.L.; Coutinho, H.D.C. Contribuição da serrapilheira para recuperação de áreas degradadas e para manutenção da sustentabilidade de sistemas agroecológicos. *Embrapa Solos* **2003**, *24*, 55–63.
14. Piché, N.; Kelting, D.L. Recovery of soil productivity with forest succession on abandoned agricultural land. *Restor. Ecol.* **2015**, *23*, 645–654. [CrossRef]
15. Ziegler, A.D.; Giambelluca, T.; Tran, L.T.; Vana, T.T.; Nullet, M.A.; Fox, J.; Vien, T.D.; Pinthong, J.; Maxwell, J.; Evett, S. Hydrological consequences of landscape fragmentation in mountainous northern Vietnam: Evidence of accelerated overland flow generation. *J. Hydrol.* **2004**, *287*, 124–146. [CrossRef]
16. Paul, M.; Catterall, C.P.; Pollard, P.; Kanowski, J. Recovery of soil properties and functions in different rainforest restoration pathways. *For. Ecol. Manag.* **2010**, *259*, 2083–2092. [CrossRef]
17. Nyberg, G.; Bargués-Tobella, A.; Kinyangi, J.; Ilstedt, U. Soil property changes over a 120-yr chronosequence from forest to agriculture in western Kenya. *Hydrol. Earth Syst. Sci.* **2012**, *16*, 2085–2094. [CrossRef]
18. Gageler, R.; Bonner, M.; Kirchhof, G.; Amos, M.; Robinson, N.; Schmidt, S.; Shoo, L.P. Early Response of Soil Properties and Function to Riparian Rainforest Restoration. *PLoS ONE* **2014**, *9*, e104198. [CrossRef]
19. Lozano, S.; Cooper, M.; Meli, P.; Ferraz, S.F.; Rodrigues, R.R.; Sauer, T.J. Land restoration by tree planting in the tropics and subtropics improves soil infiltration, but some critical gaps still hinder conclusive results. *For. Ecol. Manag.* **2019**, *444*, 89–95. [CrossRef]
20. Sparovek, G.; Barretto, A.; Berndes, G.; Martins, S.; Maule, R. Environmental, land-use and economic implications of Brazilian sugarcane expansion 1996–2006. *Mitig. Adapt. Strat. Glob. Chang.* **2008**, *14*, 285–298. [CrossRef]
21. Ribeiro, M.C.; Metzger, J.P.; Martensen, A.C.; Ponzoni, F.J.; Hirota, M.M. The Brazilian Atlantic Forest: How much is left, and how is the remaining forest distributed? Implications for conservation. *Biol. Conserv.* **2009**, *142*, 1141–1153. [CrossRef]
22. Ferraz, S.F.; Ferraz, K.M.; Cassiano, C.C.; Brancalion, P.H.S.; da Luz, D.T.; Azevedo, T.N.; Metzger, J.P. How good are tropical forest patches for ecosystem services provisioning? *Landsc. Ecol.* **2014**, *29*, 187–200. [CrossRef]
23. Peñuela, M.C.; Drew, A.P. A Model to Assess Restoration of Abandoned Pasture in Costa Rica Based on Soil Hydrologic Features and Forest Structure. *Restor. Ecol.* **2004**, *12*, 516–524. [CrossRef]
24. Bruijnzeel, L.A. Hydrological functions of tropical forests: Not seeing the soil for the trees? *Agric. Ecosyst. Environ.* **2004**, *104*, 185–228. [CrossRef]
25. Zwartendijk, B.; Meerveld, H.J.; Ghimire, C.; Bruijnzeel, L.; Ravelona, M.; Jones, J. Rebuilding soil hydrological functioning after swidden agriculture in eastern Madagascar. *Agric. Ecosyst. Environ.* **2017**, *239*, 101–111. [CrossRef]
26. Mendes, M.S.; Latawiec, A.E.; Sansevero, J.B.B.; Crouzeilles, R.; Moraes, L.F.D.; Castro, A.; Alves-Pinto, H.N.; Brancalion, P.H.S.; Rodrigues, R.R.; Chazdon, R.L.; et al. Look down-there is a gap-the need to include soil data in Atlantic Forest restoration. *Restor. Ecol.* **2018**, *27*, 361–370. [CrossRef]
27. Alvares, C.A.; Stape, J.L.; Sentelhas, P.C.; Gonçalves, J.L.M.; Sparovek, G. Köppen's climate classification map for Brazil. *Meteorologische Zeitschrift* **2013**, *22*, 711–728. [CrossRef]
28. De Oliveira, J.B.; Menk, J.R.F.; Barbieri, J.L.; Rotta, C.L.; Tremocoldi, W. *Levantamento Pedologico Semidetalhado de Estado de São Paulo: Quadricula de Araras*; Instituto Agronômico de Campinas: Campinas/São Paulo, Brasil, 1982.
29. Koffler, N.F. Uso das terras da bacia do rio Corumbataí em 1990. *Geografia* **1993**, *18*, 135–150.
30. Rodrigues, R.R. A vegetação de Piracicaba e municípios do entorno. *Circular Técnica IPEF* **1999**, *189*, 1–17.
31. Pinto, L.P.; Bedê, L.; Paese, A.; Fonseca, M.; Paglia, A.; Lamas, I. Mata Atlântica Brasileira: Os desafios para conservação da biodiversidade de um hotspot mundial. In *Biologia da Conservação: Essências*; Rocha, C.F.D., Bergallo, H.G., Sluys, M.V., Alves, M.A.S., Eds.; RiMa Editora: Rio de Janeiro, Brazil, 2006; pp. 91–118.
32. Da Fonseca, G.A. The vanishing Brazilian Atlantic forest. *Biol. Conserv.* **1985**, *34*, 17–34. [CrossRef]
33. Silva, W.G.S.; Metzger, J.P.; Simões, S.; Simonetti, C. Relief influence on the spatial distribution of the Atlantic Forest cover on the Ibiúna Plateau, SP. *Braz. J. Biol.* **2007**, *67*, 403–411. [CrossRef] [PubMed]
34. Valente, R.; Vettorazzi, C. Forest structure assessment in the Corumbataí river basin, SP. *Sci. For.* **2005**, *68*, 45–57.
35. Soil Survey Staff. *Soil Taxonomy: A Basic System of Soil Classification for Making and Interpreting Soil Surveys*; Natural Resources Conservation Service: Whashington, DC, USA, 1999.

36. Lassabatere, L.; Angulo-Jaramillo, R.; Ugalde, J.M.S.; Cuenca, R.; Braud, I.; Haverkamp, R. Beerkan Estimation of Soil Transfer Parameters through Infiltration Experiments-BEST. *Soil Sci. Soc. Am. J.* **2006**, *70*, 521–532. [CrossRef]
37. Angulo-Jaramillo, R.; Bagarello, V.; Lovino, M.; Lassabatère, L. *Infiltration Measurements for Soil Hydraulic Characterization*; Springer International Publishing: Berlim, Germany, 2016. [CrossRef]
38. Bagarello, V.; Di Prima, S.; Iovino, M. Estimating saturated soil hydraulic conductivity by the near steady-state phase of a Beerkan infiltration test. *Geoderma* **2017**, *303*, 70–77. [CrossRef]
39. White, I.; Sully, M.J. Macroscopic and microscopic capillary length and time scales from field infiltration. *Water Resour. Res.* **1987**, *23*, 1514–1522. [CrossRef]
40. Reynolds, W.D.; Elrick, D.E. Pressure infiltrometer. In *Methods of Soil Analysis*; Dane, J.H., Topp, G.C., Eds.; Part 4, Physical Methods, SSSA Book Series, No. 5; Soil Science Society of America: Madison, WI, USA, 2002; Volume 4, pp. 826–836.
41. Haverkamp, R.; Ross, P.J.; Smettem, K.; Parlange, J.Y. Three-Dimensional analysis of infiltration from the disc infiltrometer: 2. Physically based infiltration equation. *Water Resour. Res.* **1994**, *30*, 2931–2935. [CrossRef]
42. Raats, P.A.C. Analytical Solutions of a Simplified Flow Equation. *Trans. ASAE* **1976**, *19*, 683–689. [CrossRef]
43. Di Prima, S.; Stewart, R.D.; Castellini, M.; Bagarello, V.; Abou Najm, M.R.; Pirastru, M.; Giadrossich, F.; Iovino, M.; Angulo-Jaramillo, R.; Lassabatere, L. Estimating the macroscopic capillary length from Beerkan infiltration experiments and its impact on saturated soil hydraulic conductivity predictions. *J. Hydrol.* **2020**. [CrossRef]
44. Lozano, S.; Cooper, M.; Ferraz, S.F.; Rodrigues, R.R.; Lassabatere, L.; Castellini, M.; Di Prima, S. Assessing Water Infiltration and Soil Water Repellency in Brazilian Atlantic Forest Soils. *Appl. Sci.* **2020**, *10*, 1950. [CrossRef]
45. Elrick, D.E.; Reynolds, W.D. Methods for Analyzing Constant-Head Well Permeameter Data. *Soil Sci. Soc. Am. J.* **1992**, *56*, 320–323. [CrossRef]
46. Grossman, R.B.; Reinsch, T. 2.1 Bulk Density and Linear Extensibility. In *Methods of Soil Analysis: Part 4 Physical Methods*; Dane, J.H., Topp, G.C., Eds.; Soil Science Society of America: Madison, WI, USA, 2002; Volume 5, pp. 201–228. [CrossRef]
47. Lassabatere, L.; Di Prima, S.; Angulo-Jaramillo, R.; Keesstra, S.; Salesa, D. Beerkan multi-runs for characterizing water infiltration and spatial variability of soil hydraulic properties across scales. *Hydrol. Sci. J.* **2019**, *64*, 165–178. [CrossRef]
48. Vomocil, J.A. Porosity. In *Methods of Soil Analysis: Part 1 Physical and Mineralogical Properties, Including Statistics of Measurement and Sampling*; Black, C.A., Ed.; Soil Science Society of America: Madison, WI, USA, 1965; Volume 9, pp. 299–314.
49. Imhoff, S.; Da Silva, A.P.; Junior, M.D.S.D.; Tormena, C.A. Quantificação de pressões críticas para o crescimento das plantas. *Revista Brasileira Ciência Solo* **2001**, *25*, 11–18. [CrossRef]
50. Salton, J.C.; Silva, W.M.; Tomazi, M.; Hernani, L.C. *Determinação da Agregação do Solo—Metodologia em Uso na Embrapa Agropecuária Oeste*; Embrapa Agropecuária Oeste. Comunicado Técnico, 184; Embrapa Agropecuária Oeste: Dourados, Brazil, 2012; 8p.
51. Kemper, W.D.; Rosenau, R.C. Aggregate Stability and Size Distribution. In *Methods of Soil Analysis*; Klute, A., Ed.; Soil Science Society of America: Madison, WI, USA, 1986; pp. 425–442.
52. Reynolds, W.; Drury, C.F.; Yang, X.; Tan, C. Optimal soil physical quality inferred through structural regression and parameter interactions. *Geoderma* **2008**, *146*, 466–474. [CrossRef]
53. Gomes, F.P. *Curso de Estatística Experimental*, 12th ed.; Nobel: São Paulo, Brazil, 1990; 467p.
54. Di Prima, S.; Winiarski, T.; Angulo-Jaramillo, R.; Stewart, R.D.; Castellini, M.; Najm, M.R.A.; Ventrella, D.; Pirastru, M.; Giadrossich, F.; Capello, G.; et al. Detecting infiltrated water and preferential flow pathways through time-lapse ground-penetrating radar surveys. *Sci. Total Environ.* **2020**, *726*, 138511. [CrossRef] [PubMed]
55. Soil Science Division Staff. Soil Survey Manual. In *USDA Handbook 18*; Ditzler, C., Scheffe, K., Monger, H.C., Eds.; Government Printing Office: Washington, DC, USA, 2017.
56. Tormena, C.A.; Barbosa, M.C.; Da Costa, A.C.S.; Gonçalves, A.C.A. Densidade, porosidade e resistência à penetração em Latossolo cultivado sob diferentes sistemas de preparo do solo. *Sci. Agric.* **2002**, *59*, 795–801. [CrossRef]

57. Taylor, H.M.; Roberson, G.M.; Parker, J.J. Soil strength-root penetration relations for medium-to coarse-textured soil materials. *Soil Sci.* **1966**, *102*, 18–22. [CrossRef]
58. Taylor, H.M. Effects of soil strength on seedling emergence, root growth and crop yields. In *Compaction of Agricultural Soils*; Barnes, K.K., Ed.; ASAE: St. Joseph, MI, USA, 1971; pp. 292–304.
59. Júnior, M.M.; Melo, W.J. Carbono, carbono da biomassa microbiana e atividade enzimática em um solo sob mata natural, pastagem e cultura do algodoeiro. *Revista Brasileira Ciência Solo* **1999**, *23*, 257–263. [CrossRef]
60. Salemi, L.F.; Groppo, J.D.; Trevisan, R.; De Moraes, J.M.; Ferraz, S.F.; Villani, J.P.; Duarte-Neto, P.J.; Martinelli, L. Land-use change in the Atlantic rainforest region: Consequences for the hydrology of small catchments. *J. Hydrol.* **2013**, *499*, 100–109. [CrossRef]
61. Scheffler, R.; Neill, C.; Krusche, A.; Elsenbeer, H. Soil hydraulic response to land-use change associated with the recent soybean expansion at the Amazon agricultural frontier. *Agric. Ecosyst. Environ.* **2011**, *144*, 281–289. [CrossRef]
62. Biggs, T.; Dunne, T.; Muraoka, T. Transport of water, solutes and nutrients from a pasture hillslope, southwestern Brazilian Amazon. *Hydrol. Process.* **2006**, *20*, 2527–2547. [CrossRef]
63. Chaves, J.; Neill, C.; Germer, S.; Neto, S.G.; Krusche, A.; Elsenbeer, H. Land management impacts on runoff sources in small Amazon watersheds. *Hydrol. Process.* **2008**, *22*, 1766–1775. [CrossRef]
64. Zimmermann, B.; Elsenbeer, H.; De Moraes, J.M. The influence of land-use changes on soil hydraulic properties: Implications for runoff generation. *For. Ecol. Manag.* **2006**, *222*, 29–38. [CrossRef]
65. Germer, S.; Neill, C.; Krusche, A.; Elsenbeer, H. Influence of land-use change on near-surface hydrological processes: Undisturbed forest to pasture. *J. Hydrol.* **2010**, *380*, 473–480. [CrossRef]
66. Satiro, L.S.; Cherubin, M.R.; Safanelli, J.L.; Lisboa, I.P.; Da Rocha Junior, P.R.; Cerri, C.E.P.; Cerri, C.C. Sugarcane straw removal effects on Ultisols and Oxisols in south-central Brazil. *Geoderma Reg.* **2017**, *11*, 86–95. [CrossRef]
67. Topp, G.C.; Reynolds, W.D.; Cook, F.J.; Kirby, J.M.; Carter, M.R. Physical attributes of soil quality. Soil quality for crop production and ecosystem health. In *Development in Soil Science*; Gregorich, E.G., Carter, M.R., Eds.; Elsevier: New York, NY, USA, 1997; Volume 25, pp. 21–58. [CrossRef]
68. Six, J.; Conant, R.T.; Paul, E.A.; Paustian, K. Stabilization mechanisms of soil organic matter: Implications for C-saturation of soils. *Plant Soil* **2002**, *241*, 155–176. [CrossRef]
69. Filho, C.C.; Muzilli, O.; Podanoschi, A.L. Estabilidade dos agregados e sua relação com o teor de carbono orgânico num Latossolo Roxo distrófico, em função de sistemas de plantio, rotações de culturas e métodos de preparo das amostras. *Revista Brasileira Ciência Solo* **1998**, *22*, 527–538. [CrossRef]
70. Bronick, C.; Lal, R. Soil structure and management: A review. *Geoderma* **2005**, *124*, 3–22. [CrossRef]
71. Salton, J.C.; Mielniczuk, J.; Bayer, C.; Boeni, M.; Conceição, P.C.; Fabrício, A.C.; Macedo, M.C.M.; Broch, D.L. Agregação e estabilidade de agregados do solo em sistemas agropecuários em Mato Grosso do Sul. *Revista Brasileira Ciência Solo* **2008**, *32*, 11–21. [CrossRef]
72. An, S.; Mentler, A.; Mayer, H.; Blum, W.E. Soil aggregation, aggregate stability, organic carbon and nitrogen in different soil aggregate fractions under forest and shrub vegetation on the Loess Plateau, China. *Catena* **2010**, *81*, 226–233. [CrossRef]
73. Vezzani, F.M. *Qualidade do Sistema Solo na Produção Agrícola. 184 f. 2001. Tese de Doutorado. Tese (Doutorado em Ciência do Solo)*; Universidade Federal do Rio Grande do Sul: Porto Alegre, Brazil, 2001.
74. Allaire, S.E.; Roulier, S.; Cessna, A.J. Quantifying preferential flow in soils: A review of different techniques. *J. Hydrol.* **2009**, *378*, 179–204. [CrossRef]
75. Deuchars, S.; Townend, J.; Aitkenhead, M.J.; Fitzpatrick, E. Changes in soil structure and hydraulic properties in regenerating rain forest. *Soil Use Manag.* **2006**, *15*, 183–187. [CrossRef]
76. Hassler, S.K.; Zimmermann, B.; Van Breugel, M.; Hall, J.S.; Elsenbeer, H. Recovery of saturated hydraulic conductivity under secondary succession on former pasture in the humid tropics. *For. Ecol. Manag.* **2011**, *261*, 1634–1642. [CrossRef]
77. Doerr, S.H.; Ritsema, C.J.; Dekker, L.W.; Scott, D.F.; Carter, D. Water repellence of soils: New insights and emerging research needs. *Hydrol. Process.* **2007**, *21*, 2223–2228. [CrossRef]
78. Cooper, M.; Medeiros, J.C.; Rosa, J.D.; Soria, J.E.; Toma, R.S. Soil functioning in a toposequence under rainforest in São Paulo, Brazil. *Revista Brasileira Ciência Solo* **2013**, *37*, 392–399. [CrossRef]

79. Hillel, D. Soil structure and aggregation. In *Introduction to Soil Physics*; Academic Press: London, UK, 1980; pp. 40–52, 200–204.
80. Muñoz-Villers, L.E.; Holwerda, F.; Alvarado-Barrientos, M.S.; Geissert, D.; Marín-Castro, B.; Gómez-Tagle, A.; McDonnell, J.J.; Asbjornsen, H.; Dawson, T.; Bruijnzeel, L.A. Efectos hidrológicos de la conversión del bosque de niebla en el centro de Veracruz, México. *Bosque* **2015**, *36*, 395–407. [CrossRef]

Article

Environmental Control on Transpiration: A Case Study of a Desert Ecosystem in Northwest China

Shiqin Xu [1,2] and Zhongbo Yu [1,2,*]

1 State Key Laboratory of Hydrology-Water Resources and Hydraulic Engineering, Hohai University, Nanjing 210098, China; shiqinxu1989@hotmail.com
2 College of Hydrology and Water Resources, Hohai University, Nanjing 210098, China
* Correspondence: zyu@hhu.edu.cn; Tel.: +86-25-8378-6721

Received: 2 March 2020; Accepted: 17 April 2020; Published: 24 April 2020

Abstract: Arid and semi-arid ecosystems represent a crucial but poorly understood component of the global water cycle. Taking a desert ecosystem as a case study, we measured sap flow in three dominant shrub species and concurrent environmental variables over two mean growing seasons. Commercially available gauges (Flow32 meters) based on the constant power stem heat balance (SHB) method were used. Stem-level sap flow rates were scaled up to stand level to estimate stand transpiration using the species-specific frequency distribution of stem diameter. We found that variations in stand transpiration were closely related to changes in solar radiation (R_s), air temperature (T), and vapor pressure deficit (VPD) at the hourly scale. Three factors together explained 84% and 77% variations in hourly stand transpiration in 2014 and 2015, respectively, with R_s being the primary driving force. We observed a threshold control of VPD (~2 kPa) on stand transpiration in two-year study periods, suggesting a strong stomatal regulation of transpiration under high evaporative demand conditions. Clockwise hysteresis loops between diurnal transpiration and T and VPD were observed and exhibited seasonal variations. Both the time lags and refill and release of stem water storage from nocturnal sap flow were possible causes for the hysteresis. These findings improve the understanding of environmental control on water flux of the arid and semi-arid ecosystems and have important implications for diurnal hydrology modelling.

Keywords: sap flow; water-limited ecosystem; transpiration; hysteresis; nocturnal sap flow

1. Introduction

Arid and semi-arid regions covering about 40% of the terrestrial land surface are highly vulnerable to climate change [1]. Evaluating hydrological response to continuous warming and altered precipitation pattern in these regions largely depends on an accurate estimation of evapotranspiration (*ET*), especially plant transpiration which can be a large contributor of *ET*, depending on the ecosystem type [2,3]. Sap flow, an indicator of water transport in the plant xylem, provides species-specific estimates of transpiration rate at the whole tree level and can be scaled up to stand level using the appropriate scaling method [4]. During past four decades, sap flow has been widely used to investigate response of transpiration and canopy conductance to environment variables at both daily and sub-daily temporal scales [5], to characterize spatial heterogeneity of water fluxes within stands [6,7], and to quantify the ecosystem water budget [8,9]. However, prior studies heavily focused on humid forest ecosystems, with few sap flow field studies in arid and semi-arid ecosystems. What is the primary driving force for transpiration in a water-limited ecosystem at sub-daily scale? To what extent is transpiration controlled by environment variables in these ecosystems? Are there species-specific differences in the responses? These questions need to be addressed to improve our understanding of the water, energy and carbon exchange between vegetation and atmosphere, accurate long-term hydrologic modelling, and assess ecosystem adaption in the framework of climate change in arid and semi-arid regions.

Environmental variables that may affect transpiration include solar radiation (R_s), air temperature (T), vapor pressure deficit (VPD), relative humidity (RH), wind speed (u), soil moisture (θ), soil temperature (T_s), and precipitation (P). Huang et al. [10] reported that sap flow of the *Salix psammophila* living in a semi-arid environment in northwestern China was positively related to net radiation (R_n), T, and u, but negatively related to RH. They further analyzed water resources of the *S. psammophila* using a correlation analysis and multiple linear regression and found that Salix bushes can use both soil water and groundwater for transpiration. Zheng et al. [11] found that an increase in reference evapotranspiration (ET_0), PAR, VPD, and T resulted in an exponential increase in transpiration of the *Haloxylon ammodendron* (C. A. Mey.) *(H. ammodendron)*, a dominant desert species. Ma et al. [12] reported that sap flow from four typical species of shelter forest in the desert area exhibited a linear relationship with VPD. Shen et al. [13] reported that the sap flow of the *Populus gansuensis* linearly increased with R_s, and logarithmically increased with VPD and T. All these studies indicate that the response of transpiration to environmental variables is climate- and ecosystem-specific and more research is required to better understand environmental control on transpiration. In addition, diurnal hysteresis between transpiration and environmental variables has been reported in different ecosystems [14–17]. Both transpiration and environmental variables exhibit diurnal changes, reaching peaks at different times and resulting in a hysteresis behavior. Such hysteresis is usually characterized by the looping pattern when measured transpiration time series is plotted against the relevant environmental variables over the course of one day. Plausible explanations include abiotic (such as time lag between radiation and VPD) and biotic factors (such as stem water storage, soil water potential) [15,18]. Investigating the hysteric phenomenon and revealing the underlying causes would help to understand the relations and interactions between vegetation and its surrounding environment [18].

The *H. ammodendron* shrubland is extensively distributed in arid regions of China and plays a crucial role in stabilizing sand dunes and protecting oases from desertification [19]. With the development of irrigated agriculture and rapid population growth in the oasis, over-exploration of water resources has resulted in serious environmental degradations, including a gradually fall of the groundwater table, soil salinization, and desertification [20]. All these environmental problems are threatening the survival of the shrubland. Previous studies have examined the response of the photosynthetic traits and physiological characteristics on enhanced precipitation, nitrogen deposition, and drought stress [21,22], analyzed root distribution characteristics [19], revealed biomass allocation [23], and quantified interactions between canopies and the atmosphere [24]. However, a comprehensive analysis of the relationships between stand transpiration and key environmental variables is still limited.

In the current study, we measured stem sap flow from three dominant species of the *H. ammodendron* shrubland and related environmental variables over the two main growing seasons. We aimed: (i) to clarify the diurnal course of stand transpiration of the selected shrubland; (ii) to explore relationships between transpiration and key environmental variables; and (iii) to reveal hysteresis of transpiration and explore possible causes.

2. Material and Methods

2.1. Site Description

We conducted field experiments within an oasis-desert ecotone in the middle of the Heihe River Basin, Northwestern China (100°08′48″ E, 39°22′07″ N, elevation 1386 m) (Figure 1). The climate in this area is characterized by a typical continental and arid temperate climate with dry and hot summers and cold winters. An analysis of a ten-year meteorological station records (2005–2014) showed that the annual averaged air temperature ranges from −26 °C in January to 39 °C in July, with a mean value of 9 °C; the annual mean precipitation is about 124 mm, with is about 80% of the annual total fall between June and September; the daily averaged sunshine duration is about 8.3 h. The groundwater table in the study area ranged from 4.15 m to 4.29 m during the measuring period.

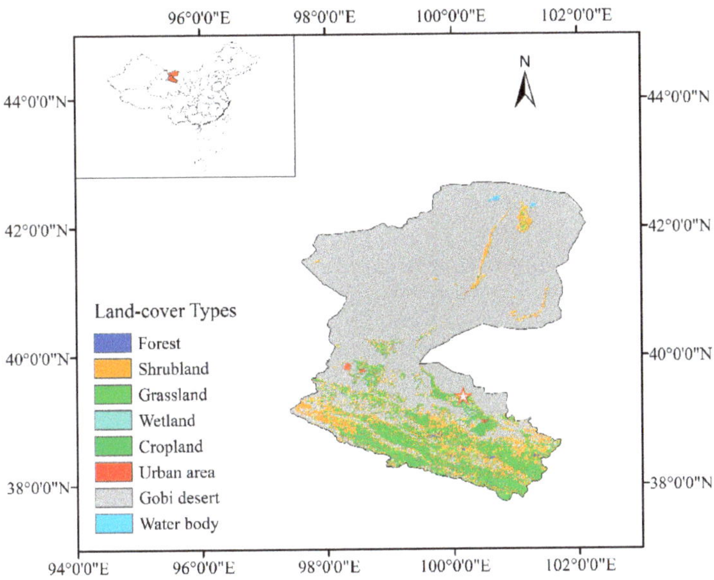

Figure 1. Map of the study area (the land-cover types were obtained from the Moderate Resolution Imaging Spectroradiometer (MODIS) 1 km IGBP land cover product.

2.2. Vegetation Measurement

A representative plot (50 × 50 m) of *H. ammodendron* shrubland was selected to conduct vegetation and sap flow measurements. Vegetation within the study plot is characterized by an open shrub canopy consisting of *H. ammodendron*, *Calligonum mongolicum* (Turcz.) (*C. mongolicum*), *Nitraria tangutorum* (Bobr.) (*N. tangutorum*), and other shrub and subshrub species. During the intensive experimental period in 2014 (DOY (day of year) 196–222), we measured basal diameter (5 cm above the ground), height, crown area, leaf area, and stand density. Details on vegetation survey and stand characteristics can be found in our previous study [24].

2.3. Environmental Measurements

We constructed a standard automatic weather station within the study plot to measure incoming solar radiation (R_s, W/m^2; CNR4, Kipp & Zonen, Delft, Netherlands), photosynthetically active radiation (*PAR*, μmol/m^2/s; LI-190SB, Li-Cor., Lincoln, NE, USA), air temperature (T, °C; HMP155A, Vaisala, Helsinki, Finland), relative humidity (*RH*, %; HMP155A, Vaisala, Helsinki, Finland), wind speed (u, m/s; 1405-PK-052, Gill Instruments Ltd., Lymington, UK), and precipitation (P, mm; TE525, Texas Electronics Inc., Dallas, TX, USA). Soil temperature (T_s, °C) and soil moisture (θ, m^3/m^3) were measured at six depths (10, 20, 40, 60, 80, and 120 cm) using thermistor probes (109-L, Campbell Scientific Inc., Logan, UT, USA) and time domain reflectometers (CS616, Campbell Scientific Inc., Logan, UT, USA), respectively. All the environmental variables were recorded at 30-min intervals using a data logger (CR1000, Campbell Scientific Inc., Logan, UT, USA). Vapor pressure deficit (*VPD*, kPa) was calculated using the following equation [25]:

$$VPD = 0.6108\, exp\left(\frac{17.27T}{T+237.3}\right)\left(1-\frac{RH}{100}\right) \quad (1)$$

2.4. Sap Flow Measurements and Estimation of Stand Transpiration

Sap flow for the three dominant shrub species was measured throughout the growing seasons from DOY 152 to 290 in 2014, and from DOY 141 to 291 in 2015. This ensured that the canopy development of the dominant species at all stages, from leaf emergence to senescence, were sampled. Commercial gauges (Flow32 meters, Dynamax Inc., Houston, TX, USA) based on the constant power stem heat balance (SHB) principle [26] were used to measure sap flow through intact plant stems. We deployed a total of 15 gauges to measure stem sap flow from sample plants of each species. For *H. ammodendron*, sap flow was measured in nine stems using nine gauges (two gauges of each model for 5 (SGA5), 9 (SGA9), and 13-mm (SGA13) and one gauge of each model for 16 (SGB16), 19 (SGB19), and 25-mm (SGB25)); for *N. tangutorum*, sap flow was measured in three stems (three 3-mm gauges (model SGA3)); For *C. mongolicum*, three stems were monitored (one gauge of each model for 5 (SGA5), 9 (SGA9), and 13-mm (SGA13)). We installed gauges on the base of the selected sample stems for each species following the manufacturer's instructions [27]. The data were recorded as 30-min averages by a data logger (CR1000, Cambell Scientific Inc., Logan, UT, USA). We noted that sap flow in the selected stems of *N. tangutorum* in 2015 was recorded for the period of DOY 143–181 due to gauges' malfunction, resulting in a relative underestimation of stand transpiration in this period.

To scale sap flow rates from individual stems of each species to stand level, the sampling strategy should consider stand structure such as stem size, species distribution, and leaf area [28]. Previous studies demonstrated that leaf area is likely a poor scalar in dry ecosystems, as transpiration per leaf area changed considerably during drought [29,30]. We adopted the scaling approach proposed by Allen and Grime [31] in their study of savannah shrubs. They scaled up sap flow rates to whole plot using the frequency distribution of stem diameter and assuming that sap flow rates in each stem were proportional to its cross-sectional area. Their assumption may introduce uncertainties as some studies have observed universal intraspecific variability of sap flow rate among individual trees [28,32]. Future work should conduct individual field measurements including eddy covariance and soil evaporation to comprehensively evaluate the uncertainties. For each species, the stand transpiration (E_c, mm/h) was calculated as follows:

$$E_c = \frac{1}{n}\sum_{i=1}^{n} \frac{0.0001F_i}{A_i \rho_w} A \tag{2}$$

where n is the number of gauged stems of each species, A_i and A are the basal cross-section area (cm^2) of stem i and the basal cross-sectional area per ground (cm^2/m^2), ρ_w is the water density (g cm^{-3}), and F_i is the sap flow rate in stem i (kg/h).

2.5. Characterizing Hysteresis and Modeling Stand Transpiration

The time difference between maximum values of diurnal stand transpiration and environmental variables was defined as the time lags [3]. The hysteretic phenomena are associated with clockwise/anticlockwise loops when measured hourly transpiration rates are plotted against the concurrent environmental variables (such as irradiance, air temperature, and *VPD*) [15,18].

The Principal Component Analysis (PCA) is a classical method for detecting the underlying structure of the multiple co-varying variables and reducing the dimensionality of data. The derived components are independent and retain most of the information in the original variables. Thus, we used PCA to analyze the environmental conditions that drive transpiration (see Appendix A).

The response of stand transpiration to environmental variables was modeled using the multivariate linear regression as follows:

$$E_c = a \times X_1 + b \times X_2 + c \times X_3 + d \tag{3}$$

where X_1, X_2, X_3, and X_4 are environmental variables and a, b, c and d are fitted parameters [14].

A backward stepwise linear regression analysis was used to determine the influence of environmental variables on stand transpiration. The importance of the predictor can be quantified by comparing the determination coefficient (R^2) of the model fit before and after removing a predictor [3].

3. Results and Discussion

3.1. Stand Characteristics and Environmental Conditions

H. ammodendron is the dominant species which contributes about 87% canopy cover; *N. tangutorum* and *C. mongolicum* are subdominant species comprising 6.8% and 3.2% canopy cover, respectively. The frequency distribution of stem diameter for each species in the sample plot indicates that *H. ammodendron* and *C. mongolicum* were dominated by small stems (<2 cm and <1 cm) more than 80% and 73% of each, respectively. The stems of *N. tangutorum* are much smaller, with stem diameters of 0.2–0.4 cm, contributing 79% of total stems.

We performed a correlation analysis to examine correlations between environmental variables. The result show that nearly all the environmental variables were correlated to some extent (Table 1). Soil temperature was highly correlated with T and VPD and moderately correlated with RH and u. Precipitation was moderately correlated with RH. Soil moisture at the surface layer was not correlated with any variable at the 30-min time scale. The first three PCA components explain 73% of the variance in the complete environmental data set (Table 2). The first component explains 46% of the variance in the data and was positively related to R_s, T, VPD, and u and negatively related to RH. The high factor loadings that occurred in the first component occurred on sunny, warm, dry and windy days, representing environmental conditions with high atmospheric evaporative demand (Table 3). Therefore, this component was referred to as an atmospheric evaporative demand index. The second component explained the 14% variance in the data and was positively related to soil temperature and moisture. We refer to this component as a soil index. The third component explains an additional 13% of the variance and was positively related to precipitation. We refer to this component as a precipitation index. R_s, T, and VPD were identified as key environmental variables.

Table 1. Correlations among the 30-min averages of environmental variables during the examined period. Coefficients below absolute value of 0.2 are marked with /. θ_{10cm} represents soil moisture at surface layer (0–10 cm).

Environmental Variables	T	RH	VPD	u	P	T_s	θ_{10cm}
R_s	0.52	−0.44	0.55	0.34	/	/	/
T		−0.59	0.88	0.40	/	0.85	/
RH			−0.81	−0.33	0.21	−0.41	/
VPD				0.41	/	0.71	/
u					x	0.28	/
P						/	/
T_s							/

Table 2. Eigenvalues and the explained variance by the first three PCA components on the completed environmental data.

Axis Number	Eigenvalues	Variance Explained (%)	Cumulative Variance Explained (%)
1	3.707	46	46
2	1.130	14	60
3	1.022	13	73

Table 3. Factor loadings on the first three PCA components.

Environmental Variables	Component Number		
	1	2	3
R_s	0.670	−0.153	−0.084
T	0.874	0.387	0.025
RH	−0.789	0.035	0.288
VPD	0.929	0.217	−0.110
u	0.616	−0.201	0.167
P	−0.066	−0.019	0.956
T_s	0.677	0.517	0.127
θ_{10cm}	−0.057	0.850	−0.043

3.2. Diurnal Courses of Environmental Variables and Transpiration

Figure 2 shows the diurnal variations of meteorological conditions and transpiration on typical sunny and rainy days. It can be seen that on sunny days, R_s increased rapidly in the early morning (between 6:30 and 8:00), peaked in the noon (between 13:00 and 13:30), then declined and dropped near zero at about 20:30. T increased along with R_s and reached the maximum of 31.9 °C between 3:30 and 4:00. VPD varied synchronously with T, which rapidly increased in the morning and reached the maximum of 3.8 kPa between 3:30 and 4:00. Stand transpiration increased rapidly in the early morning, peaked at about 11:00, gradually declined throughout the afternoon, and reached the minimum after midnight. For individual species, transpiration in *H. ammodendron* resembled the temporal pattern of the stand as this species contributed to the largest proportion of the stand transpiration. The maximum transpiration in *N. tangutorum* and *C. mongolicum* occurred later at about 12:00. On rainy days, both stand- and species-specific transpiration considerably decreased due to the decrease in atmospheric evaporative demand represented by lower R_s, T, and VPD. On both sunny and rainy days, hourly stand transpiration was closely related to changes in three environmental variables. We observed a notable midday depression of transpiration during the examined period, which was also reported by other studies in arid and semi-arid habitats [33,34]. The depression is mainly caused by stomata closure in response to high vapor-pressure deficit and irradiance in the midday in water-limited ecosystems [34,35]. During the measuring period, the maximum stand transpiration ranged from 0.009 ± 0.003 to 0.10 ± 0.02 mm/h in 2014 and ranged from 0.006 ± 0.002 to 0.12 ± 0.01 mm/h in 2015.

3.3. Control of Environment Variables on Stand Transpiration

Relationships between hourly stand transpiration and three key environmental variables, namely, hourly R_s, T and VPD, during two-year periods are scatter-plotted in Figure 3. Stand transpiration had a linear correlation with R_s (R^2 = 0.81 in 2014, 0.64 in 2015). Relationships between stand transpiration and T and VPD were well fitted by a two-degree polynomial function (R^2 = 0.41 in 2014, 0.42 in 2015 for T; R^2 = 0.44 in 2014, 0.40 in 2015 for VPD). The general patterns between transpiration and three environmental variables are different from water-unlimited ecosystems [3,36,37]. For example, Wang et al. [3] examined relationships between hourly transpiration and VPD in the Scots pine (*Pinus sylvestris*) and found that an increase in VPD resulted in a linear increase in transpiration. We performed stepwise regression to reveal how the environmental variables interactively control stand transpiration. The determination coefficients (R^2) of the regression analysis and specific equation are given in Table 4 and Table A1, respectively. Three variables together explained 84% of the variance in stand transpiration in 2014, and explained 77% of the variance in 2015. R^2 was significantly decreased when R_s was not considered, suggesting that stand transpiration was primarily controlled by energy availability. The regression analysis for each month showed similar results.

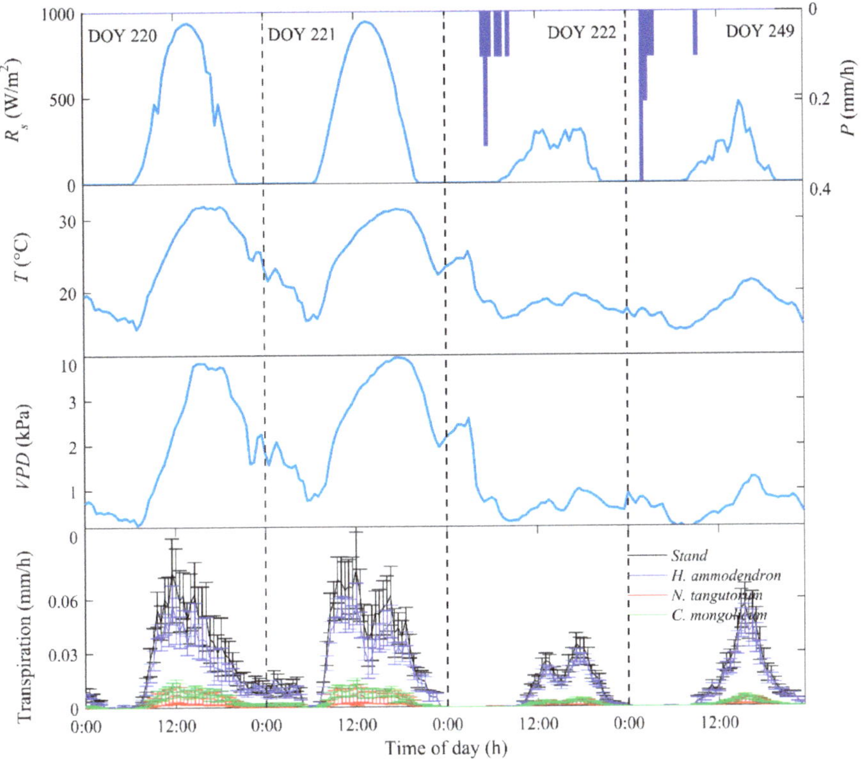

Figure 2. Diurnal courses of half-hourly solar radiation (R_S), air temperature (T), vapor pressure deficit (VPD), precipitation (P), and transpiration for typical sunny days (DOY 220 and 221) and rainy days (DOY 222 and 249) in 2014.

Analyzing the relationship between stand transpiration and VPD, we found that stand transpiration increased rapidly with an increase in VPD while it decreased with $VPD > 2$ kPa. The threshold control of VPD (~2 kPa) on stand transpiration in the two-year examined period suggests a stomata regulation of transpiration [38,39]. The result is generally consistent with the studies by Zheng et al. [11], Bai et al. [14], and Tie et al. [40], although with a somewhat different VPD threshold. However, Shen et al. [13] found that the sap flow density of shelter-belt trees in an arid inland river-basin increased logarithmically with VPD. Du et al. [41] reported that the sap flow density of the three forest species growing in the semiarid Loess Plateau region of China saturated at high VPD. These results suggested that relations between transpiration and VPD are not only climate specific but also ecosystem specific.

Relations between stand transpiration and soil moisture were weak. The correlation coefficient between stand transpiration and surface soil moisture was 0.05 in 2014 and −0.12 in 2015; the corresponding value for the whole layer soil moisture (0–120 cm) was 0.11 in 2014 and 0.19 in 2015. This finding is in line with Du et al. [41]. They analyzed the effects of soil moisture on three native and exotic species living in a semi arid habitat and found that the sap flow density in native species was less sensitive to changes in soil water conditions. However, our result was different from Bovard et al. [5], who observed that hourly sap flow in three forest species declined in dry soil when VPD was higher than 1 kPa and Yin et al. [42] who reported that cumulative sap flow of the *Salix matsudana* living in semi-arid Hailiutu River catchment was significantly and negatively correlated with soil water content. This is mainly because the *H. ammodendron* community is dominated by drought-resistant shrub species which adopted a more conservative water use strategy for preventing excessive water loss during long-period drought [19,41].

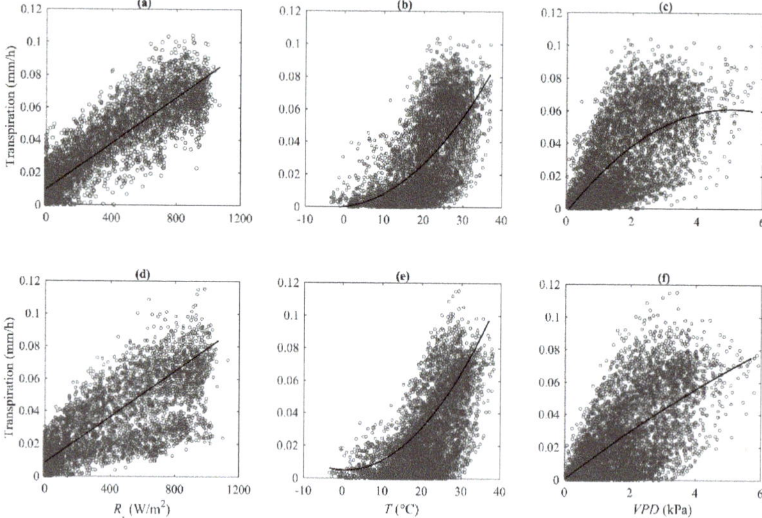

Figure 3. Half hourly stand transpiration against solar radiation (R_S), air temperature (T), and vapor pressure deficit (VPD) during the growing season in (**a–c**) 2014 and (**d–f**) 2015.

Table 4. Determination coefficient (R^2) of the seasonal bivariate and trivariate stepwise regression models in which stand transpiration is the dependent variable and solar radiation (R_S), air temperature (T), and vapor pressure deficit (VPD) are independent variables.

Year	Variable	All-time	Jun	July	August	September	October
2014	R_S, T, VPD	0.84	0.87	0.83	0.84	0.88	0.80
	R_S, T	0.84	0.87	0.83	0.84	0.88	0.79
	R_S, VPD	0.84	0.87	0.83	0.84	0.88	0.79
	T, VPD	0.43	0.49	0.41	0.43	0.43	0.79
2015	R_S, T, VPD	0.77	0.87	0.84	0.84	0.88	0.80
	R_S, T	0.76	0.87	0.83	0.84	0.88	0.79
	R_S, VPD	0.76	0.87	0.83	0.84	0.88	0.79
	T, VPD	0.43	0.49	0.41	0.43	0.43	0.40

3.4. Hysteresis between Stand Transpiration and Environmental Variables

To further elucidate relationships between transpiration of the *H. ammodendron* shrubland and related environmental factors, we examined time lags between diurnal maximum transpiration and three key environmental variables during the two-year periods. During the growing season in 2014, the average time of maximum transpiration, R_S, T, and VPD were 12:00 pm, 13:30 pm, 4:30 pm, and 4:30 pm, respectively. The corresponding times in 2015 were 14:30 pm, 14:00 pm, 4:30 pm, and 4:30 pm, respectively. In 2014, the mean time lags between maximum diurnal transpiration and three variables were 1.5, 4.5, and 4.5 h, respectively, with transpiration occurring before all environmental maxima. Maximum transpiration occurred after maximum R_S for half an hour but before the maximum T and VPD for two hours in 2015. We further examined the seasonality of the time lags and found that it varied among months (Figure 4b). For T and VPD, time lags decreased from the beginning of the growing season until mid-growing season (July in 2014, August in 2015) and then began to increase towards the end of the growing season. The longest time lags between transpiration and VPD (or T) occurred in October, whereas the shortest time lags were in July. Time lags between transpiration and R_s had less seasonal regularity compared to time lags between transpiration and VPD and T in the two study years. Wang et al. [3] and Zheng et al. [43] also observed seasonality of time lags in their study. The driving force, however, is still unclear.

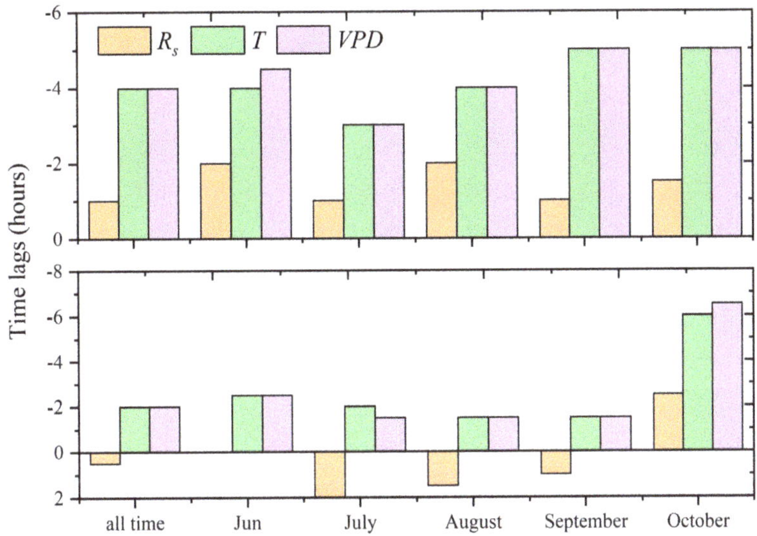

Figure 4. Time lags between maximum stand transpiration and environmental variables including solar radiation (R_S), temperature (T), and vapor pressure deficit (VPD) for different months in (**a**) 2014 and (**b**) 2015. Positive values indicate that maximum stand transpiration occurred later than maximum environmental variable.

Plots of 30-min average stand transpiration against R_s, T, and VPD revealed clockwise hysteresis loops for T and VPD in the two-year examined periods, whereas hysteresis was eliminated or considerably reduced in the R_s plots (Figure 5). These patterns are consistent with all three species. The clockwise hysteresis loops between transpiration and T and VPD were in line with previous studies such as Tie et al., O'Grady et al., and Matheny et al. [40,44,45]. However, the result is contrary to Wang et al.'s [3] finding in Scots pine (*Pinus sylvestris*), where dominant anti-clockwise loops were observed for T and VPD. This is mainly because frequent precipitation during their examined period could reduce limitation of water availability and allow transpiration evaporative demand to be relatively constant. Plants tend to close stomatal aperture in the afternoon under high evaporative demand to avoid excessive water loss and thus inhibit transpiration [14,18,36,44,46].

Another possible reason for the hysteresis may be regulation of the water storage in the stems of the shrub, with the release of storage water as sap flow in the early morning prior to variations of the atmosphere conditions [47,48]. We quantified nocturnal sap flow (defined as PAR equals to zero [39]) to daily total water use and found that the contribution of nocturnal sap flow to daily total water use ranged from 0.01% to 25% with a mean value of 10% in 2014, and ranged from 0.02% to 34% with a mean value of 8% in 2015 (Figure 6). We also observed that nocturnal sap flow was more prominent before midnight, indicating that sap flow before midnight may be used for replenishing the water lost by daytime transpiration (Figure 2). The problem that attributes the nocturnal sap flow to stem water storage is that nighttime sap flow is possibly used for nighttime transpiration [49–51]. We found that VPD and u, the factors that usually driving nighttime transpiration [52–54], together explained 44% and 41% variations of the nighttime sap flow in 2014 and 2015, respectively (Table 5). In summary, both atmospheric drivers and stem water storage are likely causes of the hysteresis.

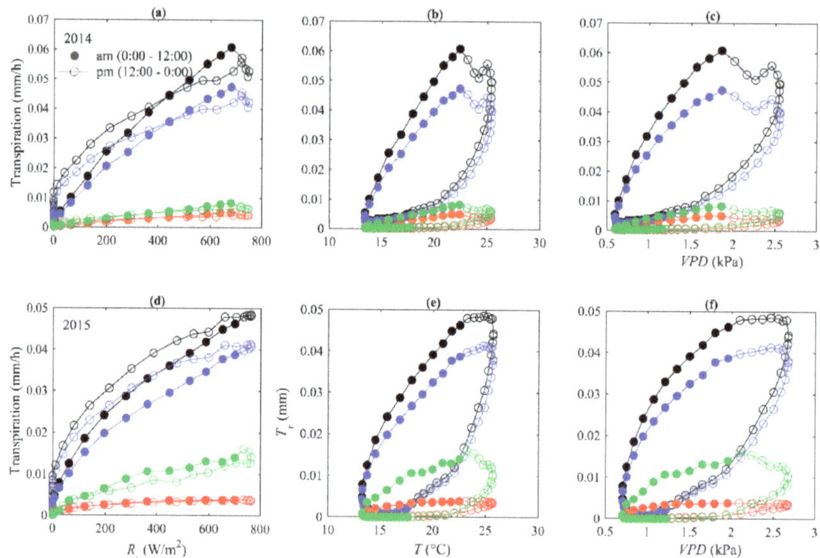

Figure 5. Hysteresis loops between transpiration (T_r) and solar radiation (R_S), air temperature (T), and vapor pressure deficit (VPD) in (**a**–**c**) 2014 and (**d**–**f**) 2015. The black, blue, green, and red cycles represent transpiration in the stand, *H. ammodendron*, *N. tangutorum*, and *C. mongolicum*, respectively.

Table 5. Multivariate linear equation between nocturnal sap flow and solar radiation, air temperature and vapor pressure deficit.

Period	Equation	R^2
2014	$SF = 0.0002 + 0.0137VPD + 0.0040u$	0.44
2015	$SF = -\ 0.0012 + 0.013VPD + 0.0026u$	0.41

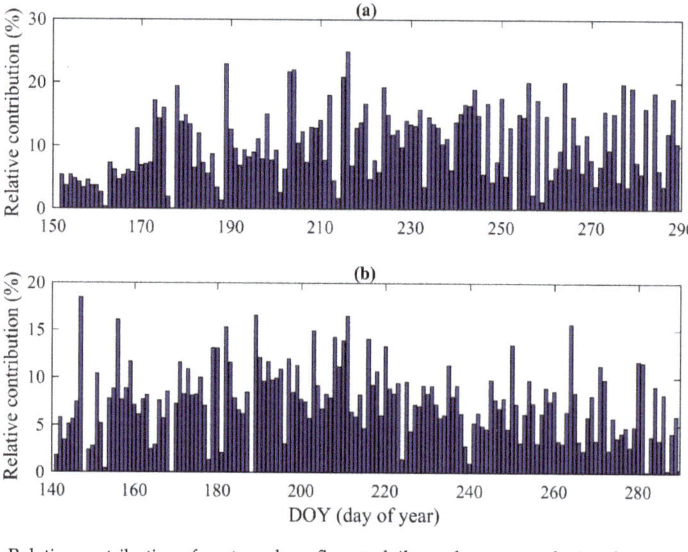

Figure 6. Relative contribution of nocturnal sap flow to daily total water use during the growing season in (**a**) 2014 and (**b**) 2015.

4. Conclusions

In the current study, we comprehensively investigated environmental control on transpiration of a desert ecosystem within an oasis-desert ecotone. We measured sap flow from three dominant shrub species and concurrent environmental variables during the growing season in 2014 and 2015. The PCA analysis showed that atmospheric evaporative demand explained the largest proportion of the variance in the environmental data, followed by soil and precipitation indexes. We observed notable midday depression of transpiration during the examined period. At the hourly scale, variations in stand transpiration were closely related to changes in R_s, T, and VPD. Three factors together explained 84% and 77% variations in stand transpiration in 2014 and 2015, respectively, with R_s being the primary driving force. We observed a threshold control of VPD (~2 kPa) on stand transpiration in the two-year examined periods, which was different from patterns of water-unlimited ecosystem and a part of a water-limited ecosystem. The result suggest a strong stomata regulation of transpiration of the selected desert ecosystem under high evaporative demand conditions. Stand transpiration was not sensitive to changes in soil moisture, as the studied ecosystem was dominated by drought-resistant species which adopt a more conservative water use strategy for preventing excessive water loss during long-period drought. Clockwise hysteresis loops between hourly transpiration and T and VPD were observed during the two growing seasons and exhibited seasonal variations. Both the time lags and stem water storage were possible causes of the hysteresis, while more sophisticated field experiments are required to separate their individual contributions. Our work provides insights into environmental controls on the water flux of arid and semi-arid regions at ecosystem scale and implications for diurnal hydrology modeling, in particular on diurnal transpiration and water stress modeling.

Author Contributions: Conceptualization, Z.Y.; methodology: Z.Y. and S.X.; formal analysis, S.X.; writing—original draft preparation, S.X.; writing—review and editing, Z.Y. All authors have read and agreed to the published version of the manuscript.

Funding: This research was funded by the National Key R&D Program of China (grant number: 2016YFC0402710); the National Natural Science Foundation of China (grant number: 51539003); the Fundamental Research Funds for the Central Universities (grant number: 2018B608X14; 2019B80214); and the Postgraduate Research & Practice Innovation Program of Jiangsu Province (grant number: KYCX18_0580).

Acknowledgments: We thank three anonymous reviewers for their constructive comments and suggestions which improved the manuscript substantially.

Conflicts of Interest: The authors declare no conflict of interest.

Appendix A

Computational procedures of the PCA are as following:

(i) Calculating correlation coefficient matrix of the environmental variables.

$$R = \begin{bmatrix} r_{11}\ r_{12} \cdots r_{1p} \\ r_{21}\ r_{22} \cdots r_{2p} \\ \cdots\cdots\cdots \\ r_{p1}\ r_{p2} \cdots r_{pp} \end{bmatrix} \tag{A1}$$

where r_{ij} $(i, j = 1, 2, \ldots, p)$ is the correlation coefficient between variable x_i and x_j. Noting that all the variables should be normalized before calculating the correlation coefficient matrix.

(ii) Calculating eigenvalues and the corresponding eigenvectors.

Solving the equation $|\lambda I - R| = 0$ to compute eigenvalues, λ, and lay them out in order of increasing size: $\lambda_1 \geq \lambda_2 \geq \lambda_3 \ldots \geq \lambda_p \geq 0$. Then to solve the eigenvector, e_i (i = 1, 2, … , p), of the corresponding λ_i with $\sum_{j=1}^{p} e_{ij}^2 = 1$.

(iii) Calculating explained variance of the component i and cumulative variance.

$$\frac{\lambda_i}{\sum_{k=1}^{p} \lambda_k} (i = 1, 2, \ldots, p) \tag{A2}$$

$$\frac{\sum_{k=1}^{i} \Lambda_k}{\sum_{k=1}^{p} \Lambda_k} (i = 1, 2, \ldots, p) \tag{A3}$$

(iv) Calculating factor loadings of the corresponding component i.

$$\mathrm{l}_{ij} = \mathrm{p}\left(\mathrm{z}_i, \mathrm{x}_j\right) = \sqrt{\Lambda_i} e_{ij} (i, j = 1, 2, \ldots, p) \tag{A4}$$

Appendix B

Table A1. Multivariate linear equation between stand transpiration (E_t) and solar radiation (R_s), air temperature (T), and vapor pressure deficit (VPD).

	Time	R_s, T & VPD	R_s & T	R_s & VPD	T & VPD
2014	all-time	$E_t = 0.001 + 0.0001R_s + 0.0003T + 0.004VPD$	$E_t = -0.003 + 0.0001R_s + 0.0007T$	$E_t = -0.003 + 0.0001R_s + 0.004VPD$	$E_t = -0.004 + 0.0008T + 0.011VPD$
	Jun	$E_t = -0.002 + 0.0001R_s + 0.0004T + 0.005VPD$	$E_t = -0.014 + 0.0001R_s + 0.0014T$	$E_t = -0.014 + 0.0001R_s + 0.005VPD$	$E_t = -0.022 + 0.002T + 0.006VPD$
	July	$E_t = 0.007 + 0.0001R_s - 0.0002T + 0.005VPD$	$E_t = -0.009 + 0.0001R_s + 0.001T$	$E_t = -0.009 + 0.0001R_s + 0.005VPD$	$E_t = -0.033 + 0.0025T + 0.004VPD$
	August	$E_t = -0.003 + 0.0001R_s + 0.0006T + 0.001VPD$	$E_t = -0.007 + 0.0001R_s + 0.001T$	$E_t = -0.007 + 0.0001R_s + 0.001 VPD$	$E_t = -0.041 + 0.0033 T - 0.002 VPD$
	September	$E_t = -0.004 + 0.0001R_s + 0.0008T + 0.001VPD$	$E_t = -0.005 + 0.0001R_s + 0.0009T$	$E_t = -0.005 + 0.001R_s + 0.001 VPD$	$E_t = -0.022 + 0.0001T + 0.003VPD$
	October	$E_t = 0.002 + 0.0001R_s + 0.0005T + 0.003VPD$	$E_t = 0.002 + 0.0001R_s + 0.0002T$	$E_t = 0.002 + 0.0001R_s + 0.001 VPD$	$E_t = 0.001 + 0.001T + 0.005VPD$
2015	all-time	$E_t = -0.009 + 0.0001R_s + 0.0007T + 0.003VPD$	$E_t = -0.012 + 0.0001R_s + 0.001T$	$E_t = -0.012 + 0.0001R_s + 0.003VPD$	$E_t = -0.009 + 0.001T + 0.009VPD$
	Jun	$E_t = -0.002 + 0.0001R_s + 0.0004T + 0.005VPD$	$E_t = -0.014 + 0.0001R_s + 0.0014T$	$E_t = -0.014 + 0.0001R_s + 0.005VPD$	$E_t = -0.022 + 0.0021T + 0.006VPD$
	July	$E_t = 0.007 + 0.0001R_s - 0.0002T + 0.005VPD$	$E_t = -0.009 + 0.0001R_s + 0.0009T$	$E_t = -0.009 + 0.0001R_s + 0.005VPD$	$E_t = -0.033 + 0.0024T + 0.004VPD$
	August	$E_t = -0.003 + 0.0001R_s + 0.0006T + 0.001VPD$	$E_t = -0.007 + 0.0001R_s + 0.0009T$	$E_t = -0.007 + 0.0001R_s + 0.001VPD$	$E_t = -0.041 + 0.0033T - 0.002VPD$
	September	$E_t = -0.004 + 0.0001R_s + 0.0008T + 0.001VPD$	$E_t = -0.005 + 0.0001R_s + 0.0009T$	$E_t = -0.004 + 0.0001R_s + 0.001VPD$	$E_t = -0.022 + 0.0025T - 0.003VPD$
	October	$E_t = 0.002 + 0.0001R_s + 0.0002T + 0.003VPD$	$E_t = 0.002 + 0.0001R_s + 0.0005T$	$E_t = 0.002 + 0.0001R_s + 0.003VPD$	$E_t = 0.001 + 0.001T - 0.005VPD$

References

1. Beuhler, M. Potential impacts of global warming on water resources in southern California. *Water Sci. Technol.* **2003**, *47*, 165–168. [CrossRef] [PubMed]
2. Wang, K.; Dickinson, R.E. A review of global terrestrial evapotranspiration: Observation, modeling, climatology, and climatic variability. *Rev. Geophys.* **2012**, *50*, RG2005. [CrossRef]
3. Wang, H.; Tetzlaff, D.; Soulsby, C. Hysteretic response of sap flow in Scots pine (*Pinus sylvestris*) to meteorological forcing in a humid low-energy headwater catchment. *Ecohydrology* **2019**, *12*, e2125. [CrossRef]
4. Smith, D.; Allen, S. Measurement of sap flow in plant stems. *J. Exp. Bot.* **1996**, *47*, 1833–1844. [CrossRef]
5. Bovard, B.D.; Curtis, P.S.; Vogel, C.S.; Su, H.; Schmid, H.P. Environmental controls on sap flow in a northern hardwood forest. *Tree Physiol.* **2005**, *25*, 31–38. [CrossRef]
6. Dalsgaard, L.; Mikkelsen, T.N.; Bastrup-Birk, A. Sap flow for beech (*Fagus sylvatica L.*) in a natural and a managed forest—effect of spatial heterogeneity. *J. Plant Ecol.* **2011**, *4*, 23–35. [CrossRef]
7. Otieno, D.; Li, Y.; Liu, X.; Zhou, G.; Cheng, J.; Ou, Y.; Liu, S.; Chen, X.; Zhang, Q.; Tang, X. Spatial heterogeneity in stand characteristics alters water use patterns of mountain forests. *Agric. For. Meteorol.* **2017**, *236*, 78–86. [CrossRef]
8. Wilson, K.B.; Hanson, P.J.; Mulholland, P.J.; Baldocchi, D.D.; Wullschleger, S.D. A comparison of methods for determining forest evapotranspiration and its components: Sap-flow, soil water budget, eddy covariance and catchment water balance. *Agric. For. Meteorol.* **2001**, *106*, 153–168. [CrossRef]
9. Zhao, W.; Liu, B.; Chang, X.; Yang, Q.; Yang, Y.; Liu, Z.; Cleverly, J.; Eamus, D. Evapotranspiration partitioning, stomatal conductance, and components of the water balance: A special case of a desert ecosystem in China. *J. Hydrol.* **2016**, *538*, 374–386. [CrossRef]
10. Huang, J.; Zhou, Y.; Yin, L.; Wenninger, J.; Zhang, J.; Hou, G.; Zhang, E.; Uhlenbrook, S. Climatic controls on sap flow dynamics and used water sources of *Salix psammophilain* a semi-arid environment in northwest China. *Environ. Earth Sci.* **2015**, *73*, 289–301. [CrossRef]
11. Zheng, C.; Wang, Q. Seasonal and annual variation in transpiration of a dominant desert species, *Haloxylon ammodendron*, in Central Asia up-scaled from sap flow measurement. *Ecohydrology* **2014**, *8*, 948–960. [CrossRef]
12. Ma, J.; Chen, Y.; Li, W.; Huang, X.; Zhu, C.; Ma, X. Sap flow characteristics of four typical species in desert shelter forest and their responses to environmental factors. *Environ. Earth Sci.* **2012**, *67*, 151–160. [CrossRef]
13. Shen, Q.; Gao, G.; Fu, B.; Lv, Y. Sap flow and water use sources of shelter-belt trees in an arid inland river basin of Northwest China. *Ecohydrology* **2015**, *8*, 1446–1458. [CrossRef]
14. Bai, Y.; Zhu, G.; Su, Y.; Zhang, K.; Han, T.; Ma, J.; Wang, W.; Ma, T.; Feng, L. Hysteresis loops between canopy conductance of grapevines and meteorological variables in an oasis ecosystem. *Agric. For. Meteorol.* **2015**, *214*, 319–327. [CrossRef]
15. Zhang, Q.; Stefano, M.; Gabriel, K.; Amilcare, P.; Dawen, Y. The hysteretic evapotranspiration—Vapor pressure deficit relation. *J. Geophys. Res. Biogeo.* **2014**, *119*, 125–140. [CrossRef]
16. Hong, L.; Guo, J.; Liu, Z.; Wang, Y.; Ma, J.; Wang, X.; Zhang, Z. Time-Lag effect between sap flow and environmental factors of *Larix principis-rupprechtii* Mayr. *Forests* **2019**, *10*, 971. [CrossRef]
17. Bai, Y.; Li, X.; Liu, S.; Wang, P. Modelling diurnal and seasonal hysteresis phenomena of canopy conductance in an oasis forest ecosystem. *Agric. For. Meteorol.* **2017**, *246*, 98–110. [CrossRef]
18. Lin, C.; Gentine, P.; Frankenberg, C.; Zhou, S.; Kennedy, D.; Li, X. Evaluation and mechanism exploration of the diurnal hysteresis of ecosystem fluxes. *Agric. For. Meteorol.* **2019**, *278*, 107642. [CrossRef]
19. Xu, S.; Ji, X.; Jin, B.; Zhang, J. Root distribution of three dominant desert shrubs and their water uptake dynamics. *J. Plant Ecol.* **2017**, *10*, 780–790. [CrossRef]
20. Ji, X.; Kang, E.; Chen, R.; Zhao, W.; Zhang, Z.; Jin, B. The impact of the development of water resources on environment in arid inland river basins of Hexi region, Northwestern China. *Environ. Geol.* **2006**, *50*, 793–801. [CrossRef]
21. Wu, X.; Zheng, X.-J.; Li, Y.; Xu, G. Varying responses of two Haloxylon species to extreme drought and groundwater depth. *Environ. Exp. Bot.* **2019**, *158*, 63–72. [CrossRef]
22. Lü, X.; Gao, H.; Zhang, L.; Wang, Y.; Shao, K. Dynamic responses of Haloxylon ammodendron to various degrees of simulated drought stress. *Plant Physiol. Biochem.* **2019**, *139*, 121–131. [CrossRef] [PubMed]
23. Xu, G.; Yu, D.; Li, Y. Patterns of biomass allocation in *Haloxylon persicum* woodlands and their understory herbaceous layer along a groundwater depth gradient. *For. Ecol. Manag.* **2017**, *395*, 37–47. [CrossRef]

24. Xu, S.; Yu, Z.; Zhang, K.; Ji, X.; Yang, C.; Sudicky, E.A. Simulating canopy conductance of the *Haloxylon ammodendron* shrubland in an arid inland river basin of northwest China. *Agric. For. Meteorol.* **2018**, *249*, 22–34. [CrossRef]
25. Allen, R.G.; Pereira, L.S.; Raes, D.; Smith, M. Crop evapotranspiration-Guidelines for computing crop water requirements-FAO Irrigation and drainage paper 56. *FAO Rome* **1998**, *300*, D05109.
26. Baker, J.M.; Van Bavel, C.H.M. Measurement of mass flow of water in the stems of herbaceous plants. *Plant Cell Environ.* **1987**, *10*, 777–782.
27. Dynamax. Dynagage Sap Flow Sensor User Manual. 2009. Available online: http://dynamax.com/images/uploads/papers/Dynagage_Manual.pdf (accessed on 6 April 2013).
28. Köstner, B.; Granier, A.; Cermák, J. Sapflow measurements in forest stands: Methods and uncertainties. *Ann. For. Sci.* **1998**, *55*, 13–27. [CrossRef]
29. Hatton, T.J.; Wu, H.I. Scaling theory to extrapolate individual tree water use to stand water use. *Hydrol. Process.* **1995**, *9*, 527–540. [CrossRef]
30. Hatton, T.J.; Moore, S.J.; Reece, P.H. Estimating stand transpiration in a *Eucalyptus populnea* woodland with the heat pulse method: Measurement errors and sampling strategies. *Tree Physiol.* **1995**, *15*, 219–227. [CrossRef]
31. Allen, S.; Grime, V. Measurements of transpiration from savannah shrubs using sap flow gauges. *Agric. For. Meteorol.* **1995**, *75*, 23–41. [CrossRef]
32. Ford, C.R.; McGuire, M.A.; Mitchell, R.J.; Teskey, R.O. Assessing variation in the radial profile of sap flux density in Pinus species and its effect on daily water use. *Tree Physiol.* **2004**, *24*, 241–249. [CrossRef] [PubMed]
33. Gartner, K.; Nadezhdina, N.; Englisch, M.; Čermak, J.; Leitgeb, E. Sap flow of birch and Norway spruce during the European heat and drought in summer 2003. *For. Ecol. Manag.* **2009**, *258*, 590–599. [CrossRef]
34. Yu, O.; Goudriaan, J.; Wang, T. Modelling diurnal courses of photosynthesis and transpiration of leaves on the basis of stomatal and non-stomatal responses, including photoinhibition. *Photosynthetica* **2001**, *39*, 43–51. [CrossRef]
35. Rodrigues, T.R.; Vourlitis, G.L.; Lobo, F.d.A.; Santanna, F.B.; de Arruda, P.H.; Nogueira, J.d.S. Modeling canopy conductance under contrasting seasonal conditions for a tropical savanna ecosystem of south central Mato Grosso, Brazil. *Agric. For. Meteorol.* **2016**, *218*, 218–229. [CrossRef]
36. Wullschleger, S.D.; Wilson, K.B.; Hanson, P.J. Environmental control of whole-plant transpiration, canopy conductance and estimates of the decoupling coefficient for large red maple trees. *Agric. For. Meteorol.* **2000**, *104*, 157–168. [CrossRef]
37. Clausnitzer, F.; Köstner, B.; Schwärzel, K.; Bernhofer, C. Relationships between canopy transpiration, atmospheric conditions and soil water availability—Analyses of long-term sap-flow measurements in an old Norway spruce forest at the Ore Mountains/Germany. *Agric. For. Meteorol.* **2011**, *151*, 1023–1034. [CrossRef]
38. Meinzer, F.; Goldstein, G.; Holbrook, N.; Jackson, P.; Cavelier, J. Stomatal and environmental control of transpiration in a lowland tropical forest tree. *Plant Cell Environ.* **1993**, *16*, 429–436. [CrossRef]
39. Motzer, T.; Munz, N.; Küppers, M.; Schmitt, D.; Anhuf, D. Stomatal conductance, transpiration and sap flow of tropical montane rain forest trees in the southern Ecuadorian Andes. *Tree Physiol.* **2005**, *25*, 1283–1293. [CrossRef]
40. Tie, Q.; Hu, H.; Tian, F.; Guan, H.; Lin, H. Environmental and physiological controls on sap flow in a subhumid mountainous catchment in North China. *Agric. For. Meteorol.* **2017**, *240*, 46–57. [CrossRef]
41. Du, S.; Wang, Y.; Kume, T.; Zhang, J.; Otsuki, K.; Yamanaka, N.; Liu, G. Sapflow characteristics and climatic responses in three forest species in the semiarid Loess Plateau region of China. *Agric. For. Meteorol.* **2011**, *151*, 1–10. [CrossRef]
42. Yin, L.; Zhou, Y.; Huang, J.; Wenninger, J.; Uhlenbrook, S. Dynamics of willow tree (*Salix matsudana*) water use and its response to environmental factors in the semi-arid Hailiutu River catchment, Northwest China. *Environ. Earth Sci.* **2014**, *71*, 4997–5006. [CrossRef]
43. Zheng, H.; Wang, Q.; Zhu, X.; Li, Y.; Yu, G. Hysteresis responses of evapotranspiration to meteorological factors at a diel timescale: Patterns and causes. *PLoS ONE* **2014**, *9*, e98857. [CrossRef] [PubMed]
44. O'Grady, A.P.; Worledge, D.; Battaglia, M. Constraints on transpiration of *Eucalyptus globulus* in southern Tasmania, Australia. *Agric. For. Meteorol.* **2008**, *148*, 453–465. [CrossRef]

45. Matheny, A.M.; Bohrer, G.; Vogel, C.S.; Morin, T.H.; He, L.; Frasson, R.P.d.M.; Mirfenderesgi, G.; Schäfer, K.V.; Gough, C.M.; Ivanov, V.Y. Species-specific transpiration responses to intermediate disturbance in a northern hardwood forest. *J. Geophys. Res. Biogeo.* **2014**, *119*, 2292–2311. [CrossRef]
46. Chen, L.; Zhang, Z.; Li, Z.; Tang, J.; Caldwell, P.; Zhang, W. Biophysical control of whole tree transpiration under an urban environment in Northern China. *J. Hydrol.* **2011**, *402*, 388–400. [CrossRef]
47. Meinzer, F.C.; James, S.A.; Goldstein, G. Dynamics of transpiration, sap flow and use of stored water in tropical forest canopy trees. *Tree Physiol.* **2004**, *24*, 901–909. [CrossRef] [PubMed]
48. Goldstein, G.; Andrade, J.; Meinzer, F.; Holbrook, N.; Cavelier, J.; Jackson, P.; Celis, A. Stem water storage and diurnal patterns of water use in tropical forest canopy trees. *Plant Cell Environ.* **1998**, *21*, 397–406. [CrossRef]
49. Dawson, T.E.; Burgess, S.S.; Tu, K.P.; Oliveira, R.S.; Santiago, L.S.; Fisher, J.B.; Simonin, K.A.; Ambrose, A.R. Nighttime transpiration in woody plants from contrasting ecosystems. *Tree Physiol.* **2007**, *27*, 561–575. [CrossRef]
50. Rosado, B.H.; Oliveira, R.S.; Joly, C.A.; Aidar, M.P.; Burgess, S.S. Diversity in nighttime transpiration behavior of woody species of the Atlantic Rain Forest, Brazil. *Agric. For. Meteorol.* **2012**, *158*, 13–20. [CrossRef]
51. Snyder, K.; Richards, J.; Donovan, L. Night-time conductance in C3 and C4 species: Do plants lose water at night? *J. Exp. Bot.* **2003**, *54*, 861–865. [CrossRef]
52. Daley, M.J.; Phillips, N.G. Interspecific variation in nighttime transpiration and stomatal conductance in a mixed New England deciduous forest. *Tree Physiol.* **2006**, *26*, 411–419. [CrossRef] [PubMed]
53. Phillips, N.; Ryan, M.; Bond, B.; McDowell, N.; Hinckley, T.; Čermák, J. Reliance on stored water increases with tree size in three species in the Pacific Northwest. *Tree Physiol.* **2003**, *23*, 237–245. [CrossRef] [PubMed]
54. Phillips, N.G.; Lewis, J.D.; Logan, B.A.; Tissue, D.T. Inter-and intra-specific variation in nocturnal water transport in Eucalyptus. *Tree Physiol.* **2010**, *30*, 586–596. [CrossRef] [PubMed]

MDPI
St. Alban-Anlage 66
4052 Basel
Switzerland
Tel. +41 61 683 77 34
Fax +41 61 302 89 18
www.mdpi.com

Water Editorial Office
E-mail: water@mdpi.com
www.mdpi.com/journal/water

www.ingramcontent.com/pod-product-compliance
Lightning Source LLC
LaVergne TN
LVHW070951170726
843515LV00005B/958

* 9 7 8 3 0 3 6 5 4 7 4 7 3 *